Towards Efficient
Fuzzy Information Processing

Studies in Fuzziness and Soft Computing

Editor-in-chief

Prof. Janusz Kacprzyk
Systems Research Institute
Polish Academy of Sciences
ul. Newelska 6
01-447 Warsaw, Poland
E-mail: kacprzyk@ibspan.waw.pl
http://www.springer.de/cgi-bin/search_book.pl?series=2941

Chongfu Huang · Yong Shi

Towards Efficient Fuzzy Information Processing

Using the Principle of Information Diffusion

With 45 Figures
and 51 Tables

Physica-Verlag

A Springer-Verlag Company

Professor Chongfu Huang
Beijing Normal University
Institute of Resources Science
Beijing 100875
China
hchongfu@bnu.edu.cn

Professor Yong Shi
University of Nebraska at Omaha
College of Information Science and Technology
Omaha, NE 68182
USA
yshi@mail.unomaha.edu

ISSN 1434-9922
ISBN 978-3-7908-2511-4   e-ISBN 978-3-7908-1785-0

Cataloging-in-Publication Data applied for
Die Deutsche Bibliothek – CIP-Einheitsaufnahme
Huang, Chongfu: Towards efficient fuzzy information processing using the principle of information diffusion: with 51 tables / Chongfu Huang; Yong Shi. – Heidelberg; New York: Physica-Verl., 2002
  (Studies in fuzziness and soft computing; Vol. 99)

Physica-Verlag Heidelberg New York
a member of BertelsmannSpringer Science+Business Media GmbH

http://www.springer.de

© Physica-Verlag Heidelberg 2010
Printed in Germany

Hardcover Design: Erich Kirchner, Heidelberg

To our wives and children:

Wei Zhang and Yundong Huang

and

Bailing Gong and Chris Shi

Foreword

When we learn from books or daily experience, we make associations and draw inferences on the basis of information that is insufficient for understanding. One example of insufficient information may be a small sample derived from observing experiments. With this perspective, the need for developing a better understanding of the behavior of a small sample presents a problem that is far beyond purely academic importance.

During the past 15 years considerable progress has been achieved in the study of this issue in China. One distinguished result is the *principle of information diffusion*. According to this principle, it is possible to partly fill gaps caused by incomplete information by changing crisp observations into fuzzy sets so that one can improve the recognition of relationships between input and output. The principle of information diffusion has been proven successful for the estimation of a probability density function. Many successful applications reflect the advantages of this new approach. It also supports an argument that fuzzy set theory can be used not only in "soft" science where some subjective adjustment is necessary, but also in "hard" science where all data are recorded.

The principle of information diffusion is also an important development within the falling shadow theory suggested by myself in 1985. This principle guarantees the possibility of obtaining fuzzy random sets from traditional observations. The falling shadow theory sets up a mathematical link between random sets and fuzzy sets. The principle of information diffusion sets up a visible link between fuzziness and incompleteness. Through the two bridges, many statistical problems could be studied using fuzzy set theory.

This book is the first attempt to present a comprehensive and self-contained theory of information diffusion. One cannot but be greatly impressed by the wide variety of concepts that Chongfu Huang and Yong Shi discuss so well and with so many carefully worked-out examples. Many of the concepts and techniques they describe are original, in particular the concept of information matrix, the soft histogram estimate, and an algorithm of the interior-outer-set model for fuzzy risk analysis.

The information diffusion theory consists of four main parts: information matrix, information distribution, information diffusion, and self-study discrete regression. Applications in earthquake engineering and risk analysis show that this new theory is useful for studying many practical cases.

The framework of information matrix provides a tool for the illustration of a sample. Neurons in the human brain are discrete rather than continuous. As a discrete framework neurons save the memory in the brain. The theory of artificial neural networks also agrees with this assumption. A difference between them is that a trained neural network is a black box while an information matrix is visible.

The approach of information distribution explains the function of fuzzy intervals for constructing a fuzzy-interval matrix. The one-dimensional linear information distribution produces a soft histogram estimate. The two-dimensional linear information distribution produces information matrixes on fuzzy intervals.

The principle of information diffusion guarantees that we can improve pattern recognition through the fuzzification of observations. In Chapter 5, the authors draw an interesting analogy between the electoral votes of the states in presidential elections in the USA and the observations of a sample. The electoral votes and the observations are only representatives. They play a role that is similar to that of spokesmen for groups with vague boundaries. The principle of information diffusion is not only supported by the analogy, but also proven by a series of mathematical proofs.

The model of self-study discrete regression establishes a link between information matrixes and fuzzy approximate inference. The estimation of epicentral intensity by using the normal diffusion self-study shows that the new regression model can give better results than the linear regression model when we only have a small sample.

In Chapter 9, the estimation of isoseismal area demonstrates that using the normal diffusion method, we can change contradictory patterns into more compatible ones that can smoothly and quickly train a backpropagation neural network to get the relationship we want.

In this book, the authors pay much attention to risk issues such as the essence of risk, fuzzy risk definition, and calculation. Three applications on the improvement of a probability-risk estimate, system analysis of a natural disaster, and calculation of a possibility–probability distribution show that the principle of information diffusion can play an important role in fuzzy risk analysis.

In the years ahead, in China, the information diffusion theory is likely to become an important tool of fuzzy system analysis. A research group in surveying and mapping has developed issues related to the theory, tools, and robustness of the parameter estimation. Another research group in biomedical engineering used it for decomposing an electromyography signal and obtained satisfactory results. A method for analyzing meteorological data based on information distribution has been given. Experimental results show that the objective law of meteorological data can be easily obtained by the method. The normal diffusion method has been introduced to resolve problems of non-parametric input in a decision support system. It can gradually improve the system precision and be very significant for applications. The method is also applied to form nonlinear mapping relations between geological conditions and fully mechanized coal mining methods for the identification of geological conditions and the determination of reasonable fully mechanized coal mining methods.

The authors of the book deserve to be recognized for producing a seminal work that makes an important contribution toward the realization of this new branch of fuzzy set theory.

November 2001   *Pei-Zhuang Wang*

Preface

The goal of this book is to propose how to use fuzzy set theory to study small-sample problems in pattern recognition. We focus on the change of a a traditional sample point into a fuzzy set for partly filling the gaps caused by incompleteness of data, so that we can improve the recognition of relationships between input and output. The approach is called the *principle of information diffusion*. The principle is demonstrated by the estimation of a probability density function.

From our perspective, current methods for fuzzy information processing apply only to single-body fuzzy information, whose fuzziness is caused by a fuzzy scale for measuring the objects. The methods of information diffusion dealt with in this book process another kind of fuzzy information, called mass-bodies fuzzy information, whose fuzziness is caused by incompleteness.

This book is a product of 15 years of intensive work and research. For applied scientists who have to deal with real data, it will be of great value in studying small-sample problems using fuzzy set theory. Applied mathematicians will also find the many numerical examples given in the book interesting.

From our initial work on the principle of information diffusion, some 15 years ago, to our recent efforts to complete this book, there have been four important periods.

When fuzzy set theory was first introduced to China, it bothered most scientists and engineers in earthquake engineering that, using the same fuzzy model, different researchers usually obtained different results based on the same observations because their analyses depended on personal engineering experience rather than on the observations. To solve that problem, in 1985 Chongfu Huang and Zhenrong Liu proposed the concept of information distribution. It provided a fuzzy technique in which personal experience could not affect the results. In 1993, under the guidance of Peizhuang Wang, Chongfu Huang developed information distribution into information diffusion and completed his Ph.D. dissertation. In 1995, with the help of Yuncheng Fen, Chongfu Huang introduced a computer simulation technique to study the benefits of information diffusion. During 1999 and 2000, Chongfu Huang and Yong Shi worked together and completed this book.

The book consist of two parts: theory and applications. Part I is devoted to showing where the principle of information diffusion comes from and providing mathematical concepts and proofs of issues related to the principle. Some diffusion models are also shown in this part.

Chapter 1 reviews basic concepts of information, fuzzy sets, and fuzzy systems. A definition of fuzzy information is given and single-body fuzzy information is distinguished from mass-bodies fuzzy information. Chapter 2 introduces a novel approach, called information matrix, to illustrate the information structure of a given small sample. It is clear that, with the fuzzy method proposed in this chapter, we can construct a more intelligent architecture to illustrate all information carried by a given sample. Chapter 3 reviews some concepts from probability theory and statistics which are needed in the following chapters. In this chapter we also review Monte Carlo methods and give some computer programs to generate samples for computer simulation experiments. Chapter 4 presents the method of information distribution that develops the information matrix with respect to fuzzy intervals into the more generalized fuzzy sets. Then, fuzzy sets can be employed to improve probability estimation and some fuzzy relationships can be directly constructed with a given sample without any linear or nonlinear assumptions. Chapter 5 describes the principle of information diffusion. Some mathematical proofs are given to demonstrate this principle. Chapter 6 discusses the property of quadratic information diffusion and suggests a model to choose a diffusion coefficient. In this chapter we also use computer simulation to compare the quadratic information distribution with some other methods. Chapter 7 introduces a simple and practical model called a normal information diffusion that results from the analysis of similarities between molecules and information diffusion. Some formulas to calculate the coefficient in the normal diffusion model are also derived.

Part II focus on exploring application models based on information distribution and information diffusion. Some of them are employed to analyze practical cases in earthquake engineering and risk assessment of floods.

Chapter 8 gives the self-study discrete regression model for constructing a statistical relationship from a given sample. To make it easily understood, a detailed discussion on the estimation of epicentral intensity by magnitude is presented. Chapter 9 proposes a hybrid model integrating information-diffusion approximate reasoning and a backpropagation neural network to solve recognition problems with strongly incompatible patterns. The model is employed to estimate an isoseismal area with respect to earthquake magnitude. Chapter 10 reviews some concepts of risk assessment and gives a definition of a fuzzy risk. The chapter shows a normalized normal-diffusion estimate model which is employed to assess flood risk with a small sample. Chapter 11 gives a fuzzy system analytic model to assess the risk of natural disasters. An information distribution technique is used to calculate basic fuzzy relationships showing historical experience. The model is employed to study the fuzzy risk of an earthquake in a city. Chapter 12 gives an interior-outer-set model and an algorithm to calculate a possibility–probability distribution (PPD). It also gives an application of the PPD in risk management.

This book is the first monograph on the principle of information diffusion. It can also serve as a definitive reference for a variety of courses exemplified by:

- A first course in fuzzy information processing at the junior, senior, or graduate level in mathematics, engineering, management, or computer science.
- A second course in statistics for graduate students for dealing with small-sample problems.
- A second course in risk analysis for graduate students for dealing with fuzzy environments.
- A second course in earthquake engineering for graduate students in pattern recognition with observations.

Prerequisites for understanding the material in this book are a basic knowledge of fuzzy sets, probability, and statistics (though we give a review of these topics in Chapter 3), and some computer literacy.

Beijing Normal University, China   *Chongfu Huang*
University of Nebraska at Omaha, USA   *Yong Shi*

Contents

Part II: Applications

Part I
Principle of Information Diffusion

1. Introduction

This chapter reviews basic concepts of information, fuzzy sets and fuzzy systems. It addresses Shannon's information theory which is useful for problems of information transfer only; however, it cannot imply structure about information that would be very important for recognizing relationships among factors. The essential natural of fuzzy information is structural, but not transferable. The fuzzy set theory as an algebra is introduced to describe the fuzziness in fuzzy information. Statements produced from incomplete data are true or false to only some degree in a continuous or vague logic. Fuzzy systems are relationships among factors that map inputs to outputs.

1.1 Information Sciences

Information in a technically-defined sense was first introduced in statistics by Fisher[6] in 1925 in his work on the theory of estimation. Shannon[24] and Wiener[29], independently, published in 1948 works describing logarithmic measures of information for use in communication theory. In his fundamental paper, Shannon proposed a mathematical scheme in which the concepts of the production and the transmission of information could be defined quantitatively. He then formulated and proved a number of very general results which showed the importance and usefulness of these definitions. This kind of classical information theory as an ensemble notion is generally called the Shannon information theory[24]. Since the notion of information in this theory does not capture the full richness of information in human communication, it is appropriate to refer to it as uncertainty-based information[11]. The theory was produced due to the fact that people know that there are uncertainties in the world. Because of the discovering of randomness, Shannon's theory can use randomness to measure the information in terms of the reduction uncertainty. This is analogous to measuring the reduction of darkness by light.

Although the theory has become more precise and more complete in the past 50 years, no fundamental concepts of major importance have been added to nor significantly altered from those originally proposed by Shannon. The original structure of the theory contains three major parts: information source, information content, and rate of transmission of information. They are almost independent so far as analytical techniques are concerned. However, they must be integrated to solve the communication problem.

An information source is a basic probability space $(\Omega, \mathcal{A}, P)$, that is, a space of elementary events Ω and sigma-field of events $\mathcal{A}$ for which a probability measure P has been defined.

$\forall x \in \Omega$ is called a message or an observation. If a message x has a probability $P(x)$, its information content is defined to be

$$I(x) = -\log P(x). \tag{1.1}$$

Although any base might be used for the logarithm in (1.1), it is common to specify that one unit of information should be associated with a probability of one-half. On this basis, the logarithmic base becomes two and the unit of information is the binary digit or bit.

A discrete information source is one that produces a sequence of symbols from a finite set of symbols, $x_1, x_2, \cdots, x_n$. If these symbols are produced independently, with probabilities of $P(x_i), i = 1, 2, \cdots, n$, then the average information per symbol is

$$H = E[I(x_i)] = -\sum_{i=1}^{n} P(x_i) \log P(x_i), \text{ bits/symbol}, \tag{1.2}$$

where $E[\cdot]$ implies the mathematical expectation. The similarity between this result and certain formulations in statistical thermodynamics led Shannon to use the term *entropy* for the average information, H. It is easy to show that H possesses the following properties:

a. $H = 0$, if and only if $P(x_i) = 1$ for one i and $P(x_j) = 0$ for every $j \neq i$.

b. H is a maximum when $P(x_i) = 1/n, i = 1, 2, \cdots, n$, and has a maximum value of

$$H_{\max} = \log n, \text{ bits/symbol}. \tag{1.3}$$

A continuous information source is one that can assume a continuous range of values. If the random variable is denoted as x and has a probability density function of $p(x)$, the corresponding information is defined as

$$H = -\int_{-\infty}^{\infty} p(x) \log p(x) dx, \text{ bits}. \tag{1.4}$$

Intuitively, the notion of "transmission of information" is relatively simple. We can, for example, envisage an information source and a "transmitter" which makes known to a "receiver," more or less accurately, the output of the information source. In mathematical terms, we define an information source as a finite space (X, x) with a probability distribution $p(x)$ defined over it. To complete the picture, we need a set Y of "received signals" y, and a conditional probability distribution $p(\cdot|x)$, defined over Y for each x. We can now define

$$p(x,y) = p(x)p(y|x), \ p(y) = \sum_X p(x,y), \text{ and } p(x|y) = \frac{p(x,y)}{p(y)}. \qquad (1.5)$$

Suppose now that a given $y = y_0$ has been received. Our total "knowledge" as to which x was transmitted is embodied in the conditional distribution $p(\cdot|y_0)$. The amount of information which is required, on the average over all transmission in which y_0 happens to be received, to determine the transmitted x is therefore given by

$$H(X|y_0) = - \sum_X p(x|y_0) \log p(x|y_0). \qquad (1.6)$$

Averaging this over all y_0 which can be received, we obtain

$$H(X|Y) = - \sum_Y p(y)[\sum_X p(x|y) \log p(x|y)] = - \sum_Y \sum_X p(x|y) \log p(x|y),$$
$$(1.7)$$

which we call the equivocation of the channel with respect to the source (or input) probability $p(x)$. Since the x transmitted at any instant may be said to have been completely known at the channel source before being transmitted, we can consider $H(X|Y)$ to be the average of information lost in the channel during transmission. Similarly, $H(X) - H(X|Y)$ is the average amount of information, per transmission, received through the channel, for which represents the amount of information required, on the average, to specify an x before transmission, minus the average amount required after transmission. Therefore, the term "rate of transmission of information" is taken to mean the quantity $R = H(X) - H(X|Y)$.

In the information theory, the term *information* is not related to the meaning, usefulness, or correctness of a message or an observation, but rather to the uncertainty or randomness of that message or observation. When we recall that information is "information power," however, the problem of measuring the value of the output of a data system which is a potential source of information becomes more difficult. What is taking the place of uncertainty? In what sense is it true?

Let us study a simple case. Suppose we send a telegram that consists of n series of binary digits. If it does not possess redundancy, that is, H is a maximum, the telegram carries n bits of information. Now, suppose some new situation arises and the telegrapher adds one binary digit into the telegram after we sent it, where a binary digit means:

0 — the telegram just sent is all false;
1 — the telegram is all right and available.

According to information theory, it implies that only 1 bit information has been added. Obviously, this explanation contradicts the reality.

Case 1: the original telegram is right, then:
"0" denies all information in the telegram, and it provides negative information;

"1" makes the telegram possess redundancy, and it does not provide any new information.

Case 2: the original telegram is indeed false, then:
"0" has an important meaning;
"1" leads to a wrong way and is negative information.

The information theory cannot represent the meanings of "1" and "0" due to the fact that it considers that information always is a positive amount, and there is no negative information.

Furthermore, take a computer as another example. In some cases, a computer keeps all information contained in input data. In other cases, a computer may lose part of the information due to an approximate calculation. In a point of view of the information theory, a computer cannot add any new information but only, by other language, repeats the information if that exists or even lose information. However, from a practical utility point of view, computers enhance information usage and make the information more useful to users.

There have been many attempts to extend the Shannon concept of information from a pure syntactic level to a semantic and a pragmatic level[25]. Most of these attempts tried to introduce an additional term, for instance weight factors, or to find a more general expression. However, we are far from having a unified concept of information. One is used, for example, to name computer information processing devices whereas one has no idea how they process what we call information. In biology the genetic processes are conceived as an information transfer, but the knowledge of amount of information (counted in bits or bytes) stored in a DNA sequence tells us nothing about the mechanisms of an adequate reaction of the cell on the "received" information. Different disciplines have different concepts - and they are making different use of information within their respective theories.

In spite of that, information still can be regarded as the third new description of the world. With matter and energy the information is used to describe the world. Matter is simply sensitive perceptible and has physical-chemical properties. Energy works on matter within the system, therefore it is potential. Information consists of a material-energetial carrier and what is being carried[26]. Here a new general notion of code will be used. The code is only assignment, fixing, arrangement, settlement, classification, and standardization between a signal (information carrier) and its signification/content (carried) for several sender/receivers. Information generally appears in some kind of code such as letters, data, figures and so on.

In a broad sense, information science is a discipline that deals with the processes of storing and transferring information. It attempts to bring together concepts and methods from various disciplines such as library science, computer science, linguistics, cybernetics, and other technologies in order to develop techniques and devices to aid in handling—that is, in the storage, control, and retrieval—information.

In this book, on one hand, information is philosophically defined to be **the reflection of motion state and existential fashion of objective reality.** For example, a plane in the sky is an object. By using radar, we can monitor its motion and shape. A series of radar images is information. On another hand, information is technically defined as the object coding that can provide knowledge to the receiver. For example, speed and geometry of the monitored plane are the object codings.

1.2 Fuzzy Information

The history of research of fuzzy information is almost as long as that of fuzzy set theory. It is most convenient to introduce some concepts of fuzzy set theory before we study fuzzy information.

1.2.1 Some basic notions of fuzzy set theory

A set is an aggregate, totality, collection of any objects, called elements, which have a common characteristic of property. A set A is said to be a subset of a set B provided every element of the set A is also an element of the set B. In this case we write $A \subset B$ or $B \supset A$ and we say that A is included in B. The relation $\subset$ is called the inclusion relation. For every set X there exists a family of sets $\mathcal{P}$ which consists exactly of all the subsets of the X: $\forall A \in \mathcal{P} \Rightarrow A \subset X$. It is easy to prove that set $\mathcal{P}$ is uniquely determined by X. This set $\mathcal{P}$ is called the *power set* of X and denoted by $\mathcal{P}(X)$.

Let X be a classical set of objects, called the *universe,* whose generic elements are denoted x. Membership in a classical subset A of X is often viewed as a characteristic function μ_A from X to $\{0,1\}$ such that

$$\mu_A(x) = \left\{ \begin{array}{ll} 1, & \text{if } x \in A, \\ 0, & \text{if } x \notin A. \end{array} \right. \tag{1.8}$$

$\{0,1\}$ is called a *valuation set.*

If the valuation set is allowed to be the real interval $[0,1]$, A is called a *fuzzy set*[31]. The function $\mu_A(x)$ is the grade of membership of x in A. The closer the value of $\mu_A(x)$ is to 1, the more x belongs to A. Clearly, A is a subset of X that has no sharp boundary.

A is completely characterized by the set of pairs

$$A = \{(x, \mu_A(x)) | x \in X\}. \tag{1.9}$$

In *fuzzy set theory* classical sets are called *crisp sets* in order to distinguish them from *fuzzy sets.*

The concept of a fuzzy set can be also defined by employing mathematics mapping. A mapping is a law according to which every element of a given set X has been assigned a completely defined element of another given set Y

(X may coincide with Y). Such a relation between the elements $x \in X$ and $y \in Y$ is denoted in form $y = f(x)$. One also writes $f : X \to Y$ and says that the mapping f *operates* from X into Y. The set X is called the *domain* (of definition) of the mapping, while the set $\{f(x)|x \in X\} \subset Y$ is called the *range* (of values) of the mapping. The mapping $f : X \to Y$ is called a *mapping* of the set X into the set Y. Logically, the concept of a "mapping" coincides with that of a **function**, an **operator** or a **transformation**.

Definition 1.1. Let X be a classical set of objects whose generic elements are denoted x. A mapping μ_A from X into the unit interval $[0,1]$:

$$\mu_A : \quad X \to [0,1]$$
$$x \mapsto \mu_A(x), \forall x \in X$$

is called a *fuzzy set* on X and denoted by A.

$\mu_A(x)$ is called the *membership function* of fuzzy set A. A fuzzy set is also called a fuzzy subset sometimes.

A more convenient notion was proposed by Zadeh[33]. When X is a finite set $\{x_1, x_2, \cdots, x_n\}$, a fuzzy set on X is expressed as

$$A = \mu_A(x_1)/x_1 + \cdots + \mu_A(x_n)/x_n = \sum_{i=1}^{n} \mu_A(x_i)/x_i, \quad (1.10)$$

or shortly,

$$A = \sum \mu_A(x_i)/x_i.$$

When X is not finite, we write

$$A = \int_X \mu_A(x)/x, \quad (1.11)$$

or shortly,

$$A = \int \mu_A(x)/x.$$

Two fuzzy sets A and B are said to be equal (denoted $A = B$) if and only if

$$\forall x \in X, \ \mu_A(x) = \mu_B(x).$$

The *support* of a fuzzy set A is the ordinary subset of X:

$$\text{supp } A = \{x|x \in X, \mu_A(x) > 0\}. \quad (1.12)$$

The elements of x in supp A are called *supporting points*. The elements of x such that $\mu_A(x) = \frac{1}{2}$ are the *crossover points* of A. The *height* of A is $\text{hgt}(A) = \sup_{x \in X} \mu_A(x)$, i.e., the least upper bound of $\mu_A(x)$. A is said to be normalized if and only if $\exists x \in X$ such that $\mu_A(x) = 1$. The *empty set* $\emptyset$ is defined as $\forall x \in X, \ \mu_\emptyset(x) = 0$; of course, $\forall x \in X, \ \mu_X(x) = 1$.

Let $\mathcal{F}(X)$ be the set of all fuzzy sets of X. $\mathcal{F}(X)$ is called the *fuzzy power set* of X.

The classical union ($\cup$) and intersection ($\cap$) of ordinary subsets of X can be extended by the following formulas

$$\forall x \in X, \; \mu_{A \cup B}(x) = \max(\mu_A(x), \mu_B(x)), \tag{1.13}$$

$$\forall x \in X, \; \mu_{A \cap B}(x) = \min(\mu_A(x), \mu_B(x)), \tag{1.14}$$

where $\mu_{A \cup B}$ and $\mu_{A \cap B}$ are respectively the membership functions of $A \cup B$ and $A \cap B$.

The *complement* A^c of A is defined by the membership function

$$\forall x \in X, \; \mu_{A^c}(x) = 1 - \mu_A(x). \tag{1.15}$$

When we want to exhibit an element $x \in X$ that typically belongs to a fuzzy set A, we may demand its membership value to be greater than some threshold $\alpha \in]0, 1]$. The ordinary set of such elements is the *α-cut* A_α of A,

$$A_\alpha = \{x | x \in X, \mu_A(x) \geq \alpha\}. \tag{1.16}$$

1.2.2 Fuzzy information defined by fuzzy entropy

The earliest concept of fuzzy information is associated with fuzzy entropy. In other words, fuzzy entropy in terms of depending on probabilities and without concerning the probabilities is regarded as fuzzy information. In fact, it is a simple extension of the Shannon information.

So-called fuzzy entropy in terms of the probabilities is defined based on the probability of fuzzy events[32].

We shall assume for simplicity that X is the Euclidian n-space R^n. Let $\mathcal{A}$ be a Borel field in R^n and P a probability measure on $\mathcal{A}$. A fuzzy event in R^n is a fuzzy set A on R^n whose membership function $\mu_A(x)$ is measurable. The probability of a fuzzy event A is defined by Lebesgue-Stieltjes integral

$$P(A) = \int_{R^n} \mu_A(x) dP. \tag{1.17}$$

Zadeh's fuzzy entropy $H^P(A)$ is defined as follows: given a fuzzy set, A, of the finite set $\{x_1, x_2, \cdots, x_n\}$ with respect to a probability distribution $P = \{p_1, p_2, \cdots, p_n\}$, then

$$H^P(A) = -\sum_{i=1}^{n} \mu_A(x_i) p_i \log p_i. \tag{1.18}$$

When A is a classical subset, $H^P(A)$ is Shannon entropy.

There are some definitions of the fuzzy entropy that do not depend on the probabilities but on a membership function.

Kauffiman[9] defined the entropy of a discrete fuzzy set A having n supporting points as

$$H_K(A) = -\frac{1}{\ln n} \sum_{i=1}^{n} \phi_A(x_i) \log \phi_A(x_i),$$

where

$$\phi_A(x_i) = \mu_A(x_i) / \sum_{i=1}^{n} \mu_A(x_i).$$

De Luca and Termini[4] defined the entropy of a fuzzy subset A as

$$H_{DT}(A) = \frac{1}{n \ln 2} \sum_{i=1}^{n} S_n(\mu_A(x_i)),$$

where

$$S_n(\mu_A(x_i)) = -\sum_{i=1}^{n} \mu_A(x_i) \ln \mu_A(x_i) - (1 - \mu_A(x_i)) \ln(1 - \mu_A(x_i)).$$

It is clear that, if we consider only uncertainty of information, either fuzzy entropy definitions can be used to study fuzzy information. Unfortunately, because the information sense has gone beyond uncertainty defined by Shannon's entropy, any fuzzy entropy definition cannot express the meaning of fuzzy information in this book.

In this book, naturally, the fuzzy information defined by fuzzy entropy is called *narrow sense fuzzy information*.

1.2.3 Traditional fuzzy information without reference to entropy

Recently, many researchers have considered fuzzy information in its general sense rather than the narrow one restricted by entropy.

A popular approach employed to study fuzzy information without reference to entropy is to introduce fuzzy set theory into an information system[15] designed to model, store, and retrieve effectively large amounts of information. In these systems, information is equal to documents or data. For example, the objects of an information retrieval system acquiring, indexing, representing, storing, and retrieving are documents. The notion of document has been extended to images, multimedia, hypermedia, and Internet or World Wide Web. Here fuzzy set theory allows the extension of the classical Boolean model to incorporate weights and partial matches, and adding the idea of document ranking. The index weights can be seen as fuzzy membership functions, mapping a document into the fuzzy subset of documents "about" the concept(s) represented by each term.

It is worthy to note is that fuzzy information may be associated directly with a proposition without reference to entropy. For example[34], the proposition "Mary is young," conveys information about Mary's age by constraining the values that the variable Age (Mary) can take. Because "young" is a fuzzy concept, the proposition conveys fuzzy information about Mary's age.

In general, traditional fuzzy information without reference to entropy relates to propositions including natural language whose representative elements are: "several", "most", "much", "not many", "very many", "not very many", "few", "quite a few", "large number", "small number", "close to five", "approximately ten", and "frequently". In this approach, the proposition "John is a young man" is treated as a piece of fuzzy information. Furthermore, the rule "If a tomato is red, then it is ripe," conveying knowledge about a tomato by a fuzzy relation between color and ripeness, is also a piece of fuzzy information. Sometimes a rule can be regarded as a composition proposition. Similarly, the fuzzy rule in earthquake engineering "If this building is struck by a strong earthquake, then it is almost collapsed" provides a piece of fuzzy information about the building to earthquakes.

This kind of fuzzy information represented by propositions or their derivative is called *single-body fuzzy information*. Its feature is that every component of information carries fuzziness.

For example, the component "young" of information "Mary is young" carries fuzziness. Thus, this proposition is a single-body fuzzy information.

1.2.4 Fuzzy information due to an incomplete data set

From several points of view the concept of data is a fundamental one. Sometimes science is described as the discovery of relations between objects from observing, experiments and data. The set of all values under consideration – that is, all pertinent data – is customarily referred to as a *population*. In general, any set of quantifiable data can be referred to as a population if that set of data constitutes all values of interest. Any population possesses some specific properties such as the central location of all values or relationship between two variables. Since it is impossible or impractical to determine the exact value of the properties of a population, the characteristics of a given population are normally judged by collecting or observing only a limited or restricted amount of the set of all possible values. Data collected or observed for this purpose are called a *sample*. The individual values contained in a sample are often referred to as *"observations"*, and the population from which they come is sometimes called *"parent"* population.

The major task of fuzzy engineering[14] is to deal with data for producing fuzzy rules which represent the fuzzy relation between the input and the output in a fuzzy system.

We restrict ourselves here to the definition that data are verifiable facts about the real world. Taking a step further, we restrict data as observations. Let $X = \{x_1, x_2, \cdots, x_n\}$ be a set of n observations which is employed to analyse some relations between objects of interest. If the observations cannot provide sufficient evidence for discovering the relations precisely, X is called an *incomplete-data set*. For example, let the training set be

$$\{(0,0), (1/4, 1/16), (1/2, 1/4), (3/4, 9/16), (1,1)\}.$$

Training a back-propagation network with topology $1 - K - 1$ on this training set, we could get the result that the trained network is not close to real function $y = x^2$ for many values of x in $[0,1]$. This training set is an incomplete-data set.

The information conveyed by an incomplete-data set is fuzzy information because the discovered relation depends on the data that must be an approximate relation which can be represented by a fuzzy relation. Without confusion, an incomplete-data set X is also called fuzzy information.

For example, tossing a coin n times, we obtain:

$$X = \{x_1, x_2, \cdots, x_n\}$$

where

$$x_i = \left\{ \begin{array}{ll} 1, & \text{if the head of the coin appears} \\ 0, & \text{if the tail of the coin appears.} \end{array} \right. \quad i = 1, 2, \cdots, n$$

If we employ X to estimate the relation between events (head and tail) and probabilities of events occurrence of the coin, for any $n < \infty$, the estimation values are not true ones in statistics terms. Therefore, this X is fuzzy information.

Another example is that, collecting all earthquake records in earthquake magnitude, M, and epicentral intensity, I_o, we may obtain:

$$\begin{aligned} X &= \{x_1, x_2, \cdots, x_n\} \\ &= \{(M_1, I_{o1}), (M_2, I_{o2}), \cdots, (M_n, I_{on})\}. \end{aligned}$$

If we employ X to discover the relationship between magnitude and epicentral intensity, for any $n < \infty$, the relation obtained from X must suffer from some drawbacks. And hence, this X is also fuzzy information.

There are many reasons to cause data to be incomplete, such as mistaking, losing, lacking, and roughing. In principle, mistaking and losing can be overcome if we carefully collect the data, and roughing might be studied by the standard fuzzy methods as linguistic variables. When the incompleteness is caused by a scarcity of data, there exist information gaps for producing the relations. There have been many attempts to fill in these gaps. The most popular one is to employ expert's experience with empirical Bayes methods[3]. Many fuzzy methods[8][23][27], in some degree, were developed to smooth down the gaps.

In this book, a new fuzzy approach is established to unearth fuzzy information which is buried in an incomplete-data set with respect to scarcity. This kind of fuzzy information is called *mass-bodies fuzzy information*. Its feature is that each of the data may be crisp but its aggregation or collective, as a whole, has uncertainty different from randomness and carries fuzziness when we employ them to recognize a relation.

For example, suppose that 100 American women are measured and their ages are recorded. These data in height and age are crisp. However, as a whole, these data has uncertainty when we employ them to recognize the relation between American women age and height. Therefore, the aggregation of these data is a mass-bodies fuzzy information.

When we study the mass-bodies fuzzy information, unless stated otherwise, it is assumed that we are given a set of observations $X = \{x_1, x_2, \cdots, x_n\}$ on a given universe of discourse $U = \{u\}$, which is a domain of value of x_i. X is called a *sample*, x_i $(i = 1, 2, \cdots, n)$ is called an *observation*, and n is called the *sample size of X*.

For a 2-dimension observation (x_i, y_i) results from an experiment, has a number of outcomes with input and output, we call x_i input-component and y_i output-component.

1.2.5 Fuzzy information and its properties

Clearly, fuzzy information is just an extension of traditional information, both in the Shannon sense and a broad sense, allowing for have a capacity of fuzziness. It may be argued that, **fuzzy information is defined as the approximate reflection of motion state and existential fashion of objective reality.** For example, to briefly report a monitored plane, a radar operator says "A big plane are coming in high speed." His report is fuzzy information. In this piece of information, "plane" is an object, "big" and "high speed" are the approximate reflection of existential fashion and motion state, respectively. This reflection is revealed in the form of material or energy, and is perceived by human sense organs directly or indirectly. X denotes the perceived reflection. X is called an information source.

When a radar transmits energy, part of it may be intercepted by an opponent receiver. To protect ourselves against the exposure to enemy, we have to minimize observation time. In this situation, we can receive only a few of radar pulses. The information from these radar pulses is fuzzy.

Because the range of the meaning of fuzzy information is much wider than that of the traditional information, we have to define a boundary for fuzzy information. Fuzzy information in this book is limited just to information that can be accumulated to become experience and knowledge in a certain form. Our main interest is not to study the measurement of fuzzy information, but to analyze the structure of information from which we can know, and by which we may discover as some useful natural laws.

Therefore, X is fuzzy information if and only if it satisfies the following properties:

Property 1. X can be recorded in the form of document or data, i.e., X is recordable;
(e.g. A news film is recorded in the form of document, a credit number is recorded in the form of data.)

Property 2. X can be employed to study something in which we have interest, i.e., X is meaningful;
(e.g. The historical data of Nasdaq Canada Index during 9:30-3:30 on December 1,1999 can be employed to predict Nasdaq Canada Index of December 2, 1999. The proposition "Mary is young" is meaningful that conveys information about Mary's age.)

Property 3. Some or all of the elements of X have been or can be represented in fuzzy concepts or fuzzy sets, i.e., X is fuzzy.
(e.g. The proposition "Mary is young" is a piece of fuzzy information because "young" is a fuzzy concept. There are gaps among a series of non-continuous radar images. Then, it is possible to partly fill the gaps by fuzzy sets representing these images.)

As a general example, a letter written in natural language is fuzzy information. In fact, a letter is a document, it satisfies property 1. From the letter the receiver can know something he/she wants to know, hence the letter satisfies property 2. When the letter is written in natural language there must be some fuzzy concepts such as the *nose* of a human head which is not sharply defined, so the letter satisfies property 3.

Another example is that, a set of data that recorded from the historical earthquakes in a region can be viewed as fuzzy information. Obviously, it satisfies property 1. Because the set of data can tell us something about earthquakes in the region, it satisfies property 2. When the records associate with a fuzzy scale such as earthquake intensity, the set of data satisfies property 3. Even though every element in the set is crisp, for filling the gaps with respect to a scarcity, fuzzy sets can be employed to represent data. In this case, the set of data also satisfies property 3.

1.2.6 Fuzzy information processing

Information processing means to treat information by a particular process into a finished product. In computer science, information processing is also called data processing, in which the relevant problems are how to organize and store data, how to input, output and transmit data, how to efficiently retrieve, classify and look up, and how to maintain data. As information science evolves, information processing becomes a general sense, such as calculation, management, and applications.

Traditionally, fuzzy information processing involves classifying fuzzy objects for abstracting or representing concepts, judging for recognition, inferring consequence, predicting characters, making decisions, and controlling results. Current methods for fuzzy information processing apply only to the single-body fuzzy information whose fuzziness is caused by a fuzzy scale to measure objects. Today, the most common ones are cluster analysis[2], multiple criteria decision-making[5][30], and fuzzy logic[20].

The main task of processing the mass-bodies fuzzy information, for getting a better result, is to unearth (or mine) fuzzy information which is buried in an incomplete-data set. It is different from processing the fuzzy information that is ready to directly use.

Today, many people believe that fuzzy systems must be studied by employing fuzzy methods based on fuzzy set theory. In a sense, the concept of a fuzzy boundary makes us retain some information which embodies the fuzzy boundary, and then we would hold more information than a crisp boundary of information. The extra information due to the fuzzy bounds is called *fuzzy-transition information*.

For example, suppose that John is 25 years old and Henry is 40 years old. In the viewpoint of daily language, we say that John is a young man and Henry is a middle-age man. The bound between "young" and "middle-age" is fuzzy. In some degree, John is also a middle-age man. The possibility that John is also a middle-age man is a piece of extra information. The information showing " John who is 25 years old, in some degree, is also a middle-age man" is fuzzy-transition information.

In many cases, extra information is useful for us to effectively study a fuzzy system. If we cut a fuzzy object to be crisp, the extra information will disappear. It is worthy to mention that the fuzzy-transition information, in terms of a single-body fuzzy information as a proposition with a fuzzy concept can be used as a basis in studying a fuzzy systems. It is easier to process the single-body fuzzy information than to do the mass-bodies fuzzy information, because the extra information of mass-bodies fuzzy information needs to be analyzed and recovered. There may exist many methods to do it. Different fuzzy systems may need different approaches.

In this book, we have proven, in theory and practice, that the *principle of information diffusion* holds in some cases, which lead us to use an observation x_i into a fuzzy set $\mu_{x_i}(u)$. Therefore, a sample X can be changed into *a set-valued sample* $\mathcal{X} = \{\mu_{x_i}(u)|i = 1, 2, \cdots, n\}$. These set-valued observations can, in some degree, fill up the gaps between the observations and then we can get a better result when we employ $\mathcal{X}$ to recognize a relation.

1.3 Fuzzy Function Approximation

From an epistemological point of view, a system makes energy from material. Similarly, there is no sense in discussing information without a system.

Although the word "system" is commonly used in conversation and writing, it is often used indiscriminately. It is apparent that the concept of a system can be hazy and lack in precision and definiteness. A general, but precise, definition[22] of a system for our purpose is as follows: A system is any collection, grouping, arrangement or set of elements, objects or entities that may be material or immaterial, tangible or intangible, real or abstract to which a measurable relationship of cause and effect exists or can be rationally assigned.

The concept of cause is defined as any change, disturbance, perturbation, input, or other external stimulus that exists or is produced within the system's external environment and is or can be applied to or imposed on the system. The concept of effect is defined as any change, disturbance, perturbation, output, or response that is produced by or a consequence of the system's action on its inputs and is perceived and measurable in the external environment of the system.

Causes and effects are also called inputs and outputs, respectively.

For example, written and oral communications are the most important means for humans to of communicate with each other. It is the major means for the transfer of information, observations, feelings, and knowledge itself. The human mind can be considered as a system, with the input "words" in either written or oral form and the output from the mind and an assessment of the meaning and impact of this series data form.

A primary task of system analysis is the determination and evaluation of the relationship between the inputs and outputs. A knowledge and understanding of this relationship is valuable for several reasons. First, an accurate knowledge of the relationship between inputs and outputs for a given system leads to some understanding of the behavior and inner operations or internal mechanics of that system, and is a vital step in full understanding the operation and nature of the system. Second, where the input and output relationship adequately is known, it is possible to explain past performance of the system and to predict the output or response of the system for any permissible future input. Third, with the output response of the system known for any input, it is then possible to develop means for controlling or influencing the system in some desirable or optimal manner.

It is important to distinguish between "relation" and "relationship". The former is employed to represent any connection among objects, and the latter refers to a cause-effect connection with respect to input and output. For example, any fuzzy set on $U \times V$ can be called a fuzzy relation from U to V. It is a relationship if and only if it shows some cause-effect connection for two variables x and y, and U, V are the domain and the range, respectively. Certainly, a relationship is a relation.

Mathematically, a relationship between inputs and outputs can be described abstractly by employing a mapping from the space of the input to that of the output. A mapping is also called a function.

If a complete and accurate function is known for a given system, then the system may be considered to be completely understood and fully predictable from a system's view. However, in general, the full function is never completely or exactly known. It is common, therefore, at least for real systems, to make simplifying assumptions and approximations about the system and its interaction with its environment.

However, a lot of systems are too complicated to be simplified for the desired purpose. If we choose to do so, we may face the risk caused by a mistaken simplicity. For the system that cannot be simplified, the analytical methods of fuzzy system are useful.

The concepts of fuzzy system appeared very early in the literature of fuzzy sets; it was originated by Zadeh in 1965 [31]. Fuzzy systems can take many forms. Minimal systems act as lone rules or associations. They map fuzzy subsets or concepts of one space to fuzzy subsets of a second space as when degrees of high pressures map to degrees of high temperatures. Most fuzzy systems are more complicated.

Research on fuzzy systems seems to have developed in three main directions. The first is rather formal and considers fuzzy systems as a generalization of nondeterministic systems. These have been studied within the same conceptual framework as classical systems. This approach has given birth to a body of abstract results in such fields as minimal realization theory[18] and formal automata theory[28].

The second direction of research is the linguistic approach to fuzzy systems, in which a fuzzy model is viewed as a linguistic description by means of fuzzy logical propositions. A first extensive outline of the linguistic approach was given by Zadeh [33]. Since then it has been applied to the synthesis of linguistic controllers by Mamdani and Assilian[16] followed by many others.

The third is more constructive and regards a fuzzy system as a set of if-then fuzzy rules that maps inputs to outputs[13]. Each fuzzy rule defines a fuzzy patch in the input-output state space of the function. The fuzzy system $F : X \to Y$ approximates a function $f : X \to Y$ by covering its graph with rule patches and averaging patches that overlap. In this point of view, all fuzzy systems give some type of patch covering. The standard additive model[10] is a type example that averages the patches that overlap by adding them, and then a centroid or other operation converts the patch covering to the function $F : R^n \to R^p$.

It may be argued that, **a fuzzy system is defined as a set of fuzzy relationships between measurable inputs and outputs.**

For example, a Fuzzy Controller is a fuzzy system.

Again, note that we employ the term "relationship" to distinguish a relation which has some physical sense from a general relation in pure mathematical sense. In a fuzzy system, a state can be a fuzzy set due to both the nature, such as the human language, and the epistemology concerning to simplifying complexity with some fuzzy descriptions. In general, a fuzzy system is an approximate representation of a complicated process that is not itself necessarily fuzzy. We think that, the human ability to perceive complicated phenomena stems from the use of a certain type of fuzzy language to summarize information. The notion of probabilistic system corresponds to a different point of view: all the available information at any time is modeled by probability distributions, built from repeated experiments.

For simplicity, we shall restrict our attention to the universes of fuzzy sets when we are talking about whether a fuzzy system is discrete or continuous rather than the time-process. Namely, a fuzzy system is called a discrete fuzzy system if and only if the universes of fuzzy sets in the fuzzy system are discrete. Meanwhile, a fuzzy system is called a continuous fuzzy system if and only if the universes of fuzzy sets in the fuzzy system are continuous.

In a broad sense, to a system, the meaning of the term "relationship" is the same as one of "function." However, if we restrict ourselves only to employ a fuzzy matrix to represent a relationship, there is some difference between the two terms. For example, in the standard additive model[10], no fuzzy matrix as a fuzzy relation can be employed to represent the function as a mapping from inputs to outputs. Therefore, in case of there may be some confusion, we prefer "function" to "relationship."

Undoubtedly, it is one of the main tasks in the study of fuzzy systems to search for a fuzzy function that approximately represents the mapping between the inputs and output. This is called a *fuzzy function approximation.*

The history of the fuzzy function approximation formally started in the early 1990s and tracked back the history of neural function approximation that started in the late 1980s. These fuzzy systems are considered to have a left-to-right feedforward structure that acts much as a feedforward multilayer neural network acts. Nodes combine inputs and pass those signals to the next layer of nodes. The same learning schemes can use the same sample data to tune the neural system $N : R^n \to R^p$ or the fuzzy system $F : R^n \to R^p$. We often call such adaptive fuzzy systems "neural fuzzy systems" or "fuzzy neural systems." This wrongly suggests that all use of known concepts: unsupervised clustering or supervised gradient descent depends on neural networks. Feedforward neural and fuzzy systems have similar architectures and even have the same architecture in the case of most radial basis function network[17]. So it is no surprise that results in fuzzy function approximation have quickly tracked results in neural function approximation.

In 1989 Hornik and White [7] first used the Stone-Weierstrass theorem of functional analysis[21] to show uniform convergence of such neural networks. The Stone-Weierstrass theorem states that $A = C(X)$ if $C(X)$ is the sup-norm space of continuous functions on a compact and Hausdorf space X, and the set of functions $A \subset C(X)$ is closed algebra where A is self-adjoint and separates points and contains the constant functions.

However, these neural results do not show how to build or learn real neural systems. The first fuzzy approximation theorem[12] appeared in 1991. It showed that additive fuzzy systems could define simple functions and so can uniformly approximate bounded measurable functions. Then even bivalent expert systems can act as universal approximates if they use enough binary rules. The fuzzy approximation theorem[13] for additive fuzzy systems appeared in 1992. It states that an additive fuzzy system $F : X \rightarrow Y$ uniformly approximates $f : X \rightarrow Y$ if X is compact and f is continuous.

So far scientists and engineers of fuzzy system are good at dealing with the single-body fuzzy information to construct the function and have met with great success in many fields such as fuzzy control[19], pattern recognition[1], filtering impulsive noise[10] and so on.

This book presents a principle for illustrating a fuzzy function based on the mass-bodies fuzzy information as a new tool in the field of fuzzy sets and systems.

1.4 Summary

1. In Shannon's information theory, the term information is not related to the meaning, usefulness, or correctness of a message or an observation, but rather to the uncertainty or randomness of that message or observation. In this book, information is defined to be the reflection of motion state and existential fashion of objective reality.

2. A fuzzy set is a mathematical mapping for a classical set of objects, called universe, into the unit interval [0,1].

3. Fuzzy information is defined as the approximate reflection of motion state and existential fashion of objective reality.

4. The fuzzy information that is associated with propositions including the fuzzy concepts is called *single-body fuzzy information*. For example, the proposition "John is a young man" is treated as a piece of single-body fuzzy information.

5. The fuzzy information that is associated with an incomplete-data set with respect to scarcity is called *mass-bodies fuzzy information*. For example, a sample obtained from tossing a coin n times is treated as a mass-bodies fuzzy information if we want to employ the sample to estimate the relation between events (head and tail) and probabilities of events occurrence of the coin.

6. When we study mass-bodies fuzzy information, symbol "X" is employed to denote a set of data, i.e., $X = \{x_1, x_2, \cdots, x_n\}$, where x_i is called an observation result from an experiment, and X is called a sample. The domain of definition of X is called the universe of discourse of X, denoted by $U = \{u\}$.

7. X is called fuzzy information if and only if it is recordable, meaningful and fuzzy.

8. A fuzzy system is defined as a set of fuzzy relationships between measurable inputs and outputs.

References

1. Bezdek, J.C. (1981), Pattern Recognition with Fuzzy Objective Function Algorithms. Plenum, New York
2. Bezdek, J.C. (1993), A review of probabilistic, fuzzy, and neural models for pattern recognition. Journal of Intelligent and Fuzzy Systems 1 (1), pp.1-25
3. Carlin, B.P. and Louis, T.A. (1996), Bayes and Empirical Bayes Methods for Data Analysis. Chapman & Hall, London
4. De Luca A. and Termini S. (1972), A definition of non-probabilistic entropy in the setting of fuzzy sets theory. Information and Control **20**, pp.301-312
5. Dubois, D. and Prade, H. (1980), Fuzzy Sets and Systems: Theory and Applications. Academic Press, New York
6. Fisher, R.A. (1925), Theory of statistical estimation. Proc. Camb. Phil. Soc. **22**, pp.700-725
7. Hornik, K., Stinchcombe, M., and White, H. (1989), multilayer feedforward networks are universal approximators. Neural Networks **2**, pp.35-366
8. Inoue, H. (1995), Randomly weighted sums for exchangeable fuzzy random variables. Fuzzy Sets and Systems **69**, pp.347-354
9. Kauffiman, A. (1975), Introduction to the theory of fuzzy subsets. Academic Press, New York, Vol.1
10. Kim, H.M. and Kosko, B. (1996), Fuzzy prediction and filtering in impulsive noise. Fuzzy Sets and Systems **77**, pp.15-33
11. Klir, G.J. (1995), Principles of uncertainty: What are they? Why do we need them?. Fuzzy Sets and Systems **74**(1), pp.15-31
12. Kosko, B. (1991), Neural Networks and Fuzzy Systems: A Dynamical System Approach to Machine Intelligence. Prentic Hall, Englewood Cliffs, New Jersey
13. Kosko, B. (1994), Fuzzy systems as universal approximators. IEEE Transactions on Computers **43** (11), pp.1329-1333
14. Kosko, B. (1996), Fuzzy Engineering. Prentic Hall, Englewood Cliffs, New Jersey
15. Kraft, D.H. and Petry, F.E. (1997), Fuzzy information systems: managing uncertainty in databases and information retrieval systems. Fuzzy Sets and Systems **90**, pp.183-191
16. Mamdani, E.H. and Assilian, S. (1975), An experiment in linguistic synthesis with a fuzzy logic controller. Int. J. Man-mach. Stud. **7**, pp.1-13
17. Moody, C. and Darken, C. (1989), Fast learning in networks of locally tuned processing units. Neural Computation **1**, pp.281-294
18. Negoita, C.V. and Ralescu, D.A. (1975), Application of fuzzy sets to systems analysis. ISR. 11, Birkhaeuser, Basel
19. Palm, R., Driankov, D. and Hellendoorn H. (1997), Model Based Fuzzy Control. Springer-Verlag, Berlin
20. Ruan, D. (1996), Fuzzy Logic Foundations and Industrial Applications. Kluwer Academic Publishers, Boston
21. Rudin, W. (1973), Functional Analysis. McGraw-Hill, New York
22. Sandquist, G.M. (1985), Introduction to System Science. Prentice-Hall, Inc., Englewood Cliffs, New Jersey
23. Schmucker, K.J. (1984), Fuzzy Sets, Natural Language Computations, and Risk Analysis. Computer Science Press, Rockvill, Maryland
24. Shannon, C.E. (1948), A mathematical theory of communication. Bell System Technical Journal **27**, pp.379-423, pp.623-656

25. Thiele, H. (1972), Einige Bemerkungen zur Weiterentwicklung der Information-stheorie. Scharf, J.H. (ed.): Informationtik. Vorträge anläßlich der Jahresver-sammlung vom 14-17, Oktober 1971 zu Halle. *Nova Acta Leopoldina*-Neue Folge Nr.206, Band 37/1, Barth, Leipzig, pp.473-502

26. Völz, H. (1996), Meaning outline of the term information. Kornmaches, K. and Jacoby, K. (eds): Information: new questions to a multidisciplinary concept. Akademie Verlag Gmbh, Berlin, pp.19-39

27. Wang, P.Z. (1985), Fuzzy Sets and Falling Shadows of Random Sets. Beijing Normal University Press, Beijing

28. Wee, W.G. and Fu, K.S. (1969), A formulation of fuzzy automata and its application as a model of learning systems. IEEE trans. Syst. Sci. Cybern. **5**, pp.215-223

29. Wiener, N. (1948), Cybernetics. John Wiley & Sons, New York

30. Yu, P.L.(1985), Multiple criteria Decision Making: Concepts, Techniques and Extensions. Plenum, New York

31. Zadeh, L.A. (1965), Fuzzy sets. Information Control **8**, pp.338-353

32. Zadeh, L.A. (1968), Probability measures of fuzzy events. Journal of Mathe-matics Analysis and Applications **23** (1), pp.421-427

33. Zadeh, L.A. (1972), A fuzzy set theoretic interpretation of linguistic hedges. Journal of Cybernetics **2**, pp.4-34

34. Zadeh, L.A. (1997), Toward a theory of fuzzy information granulation and its centrality in human reasoning and fuzzy logic. Fuzzy sets and Systems **90**, pp.111-127

2. Information Matrix

This chapter introduces a novel approach, called information matrix, to illustrate a given small-sample for its information structure. There are three kinds of information matrixes, respectively, on discrete universes of discourse, crisp intervals and fuzzy intervals. This chapter is organized as follows: in section 2.1, we briefly discuss small-sample problems. In section 2.2, we describe the concept of the information matrix. In this section, we also describe the information matrix on discrete universes of discourse. In section 2.3 and 2.4, we discuss the information matrixes, respectively, on crisp intervals and fuzzy intervals. In section 2.5, we analyze three kinds of information matrixes regarding them to grab the observation's information. In section 2.6, we review other four approaches to describe or construct relationships in systems, and then, in section 2.7, we compare the approach of the information matrix with them.

2.1 Small-Sample Problem

It is well known that statistical methods are usually employed to deal with engineering problems with sample observations. For example, employing the linear-regression method, we can obtain a relation between earthquake intensity, I, and earthquake magnitude, M, which is linear:

$$I = aM + b$$

where a and b are constants which are calculated by using earthquake records observed in a seismic zone. If the linear relation can show the inherent relationship between I and M in the seismic zone, it is very useful in earthquake engineering.

Whether a statistical result, as $I = aM + b$, is available or not, in general, is determined by two conditions: the assumption formula and the size of given sample. If the assumption formula is correct and the size is large, then the corresponding statistical result is available, else it is unavailable.

Although there are some powerful tools to test an assumption, it is difficult to find a reasonable assumption formula when we study a complicated nonlinear system. For example, several studies[13] have demonstrated that it is impossible to find an assumption formula to represent the relationship between isoseismal area, S (in km^2), with intensity, I, and magnitude, M.

In principle, if the size of given sample is large (more than 30) and the assumption is correct, a researcher can get a statistical result. The larger the size, the more precise the statistical result. However, in many cases, it is very difficult to get a correct assumption and a large sample.

For example, in earthquake engineering, besides the nonlinear problem of seismotectonic structures that causes difficulty to assume formulae, we know that destructive earthquakes are infrequent events with very small probability of occurrence. Therefore, the size of a sample observed in a seismic zone must be small unless if we collect all earthquake records or extend the limits in geography. It is well known that small magnitude earthquakes are frequent in a seismic zone, which cause that the main part of given sample consists of small magnitude earthquakes. When we using the so-called large sample to support a statistical model, we cannot see any laws controlled by destructive earthquakes. Obviously, a sample that consists of earthquake records collected from a wide zone will miss the intricate effects of seismotectonic structures on earthquakes. The statistical result so obtained is of little use.

Let X be a sample which would be employed to support a mathematical model to find a relationship among factors. If the size of X is small, the produced relation is unavailable. It called a *small-sample problem.*

In parametric statistical theory, when the size of a random sample is small, the error between the estimated parameter and the population parameter is large. It is also called the small-sample problem. The interval estimation is suggested to quantify the problem. The aim of the method is to select a region in the parameter space and specify the probability that the estimated values of a set of parameters will lie within the selected region.

Let random variable x be with mean λ and variance σ^2. It is well known that for small sample size n,

$$t = \frac{\bar{x} - \lambda}{\sigma/\sqrt{n}}$$

is the Student's t-distribution, with $f = n - 1$ degrees of freedom.

Hence,

$$[\bar{x} - t_{\alpha/2}\frac{\sigma}{\sqrt{n}}, \bar{x} + t_{\alpha/2}\frac{\sigma}{\sqrt{n}}]$$

is a $1 - \alpha$ confidence interval estimator for λ. $\forall \alpha \in [0,1]$, t_α is the t-value corresponding to the probability, α, of committing the Type I error.

For example, a sample of size 4 is taken from a normal distribution $N(\mu, 0.16)$. The observations are 7.6, 8.4, 7.9, and 8.1. To determine the 95% confidence interval for μ, we note that

$$1 - \alpha = 0.95, \quad \alpha/2 = 0.025.$$

From the t-distribution table for the normal distribution (i.e., degrees of freedom is ∞), we have $t_{\alpha/2} = 1.96$. Because

$$\bar{x} = 32.0/4 = 8.0, \quad \sigma/\sqrt{4} = 0.4/2 = 0.2,$$

the 98% confidence interval is given by

$$8.0 - 1.96(0.2) \text{ to } 8.0 + 1.96(0.2),$$

that is $[7.6, 8.4]$.

In this fashion, a statistician is able to label his estimators with a measure of confidence.

However, in many cases, engineers are interested in a more accurate estimator instead of a label to the estimator.

Empirical Bayes and related techniques come into play when data are generated by repeated execution of the same type of random experiment.

For example, in the preparation of an insulation material, measurements by suitable choice of units are made of the conductivity using an instrument of known standard deviation which we can suppose to be one. Prior knowledge of the production process suggests that most likely the conductivity will lie between 15 and 17. Therefore it seems reasonable to suppose a prior distribution of conductivity that is $N(16, 1/4)$; that is, $\mu_0 = 16$, $\sigma_0 = 1/2$. If ten readings are given: 16.11, 17.37, 16.35, 15.16, 18.82, 18.12, 15.82, 16.34, 16.64, 15.01, with a mean of 16.57, then, $n = 10$, $\sigma = 1$, $\bar{x} = 16.57$. From Bayes corollary for the normal distribution:

$$\mu_n = \frac{n\bar{x}/\sigma^2 + \mu_0/\sigma_0^2}{n/\sigma^2 + 1/\sigma_0^2}, \quad \frac{1}{\sigma_n^2} = \frac{n}{\sigma^2} + \frac{1}{\sigma_0^2},$$

we have

$$\mu_{10} = \frac{10 \times 16.57 + 4 \times 16}{10 + 4} = 16.41,$$

and

$$\frac{1}{\sigma_{10}^2} = 10 + 4 = 14, \quad \sigma_{10} = 1/\sqrt{14} = 0.27.$$

Hence the posterior distribution is $N(16.41, 0.27^2)$. In the basis of this, it can be said that the mean conductivity of the material likely lies between 15.87 and 16.95, while the most probable value is 16.41. Notice that the posterior mean at 16.41 occupies a position between the prior mean and the sample mean, but nearer the latter than the former because the sample mean has precision (n/σ^2) of 10 and the prior precision $(1/\sigma_0^2)$ is only 4.

Empirical Bayes methods provide a way by which such historical data can be used in the assessment of current results. A fundamental problem in the pure Bayesian approach is the specification of the prior distribution. In many cases, however, historical data are generally insufficient or unavailable for a reliable specification of the prior distribution.

Now, we develop another approach to process the small-sample problem as follows.

2.2 Information Matrix

Let X be the set of data for a given sample with input value x and output value y,

$$X = \{(x_1, y_1), (x_2, y_2), \cdots, (x_n, y_n)\}. \tag{2.1}$$

In this section, we introduce the concept of an information matrix to illustrate its information structure.

Let U and V be input and output universe of discourse respectively, then whose Cartesian space is

$$U \times V = \{(u, v) | u \in U, v \in V\}. \tag{2.2}$$

Denote

$$q_{uv}(x_i, y_i) = \begin{cases} 1, & \text{if } x_i = u \text{ and } y_i = v, \\ 0, & \text{otherwise.} \end{cases} \tag{2.3}$$

$q_{uv}(x_i, y_i)$ is called *fallen information* from observation (x_i, y_i) on Cartesian space point (u, v). In short, q is called fallen information.

We call (2.3) the *falling formula*.

Let

$$Q_{uv} = \sum_{i=1}^{n} q_{uv}(x_i, y_i), \tag{2.4}$$

where n is the size of the sample X. We say that X gives information gain at Q_{uv} to Cartesian space point (u, v).

When U and V are discrete, such as

$$\begin{aligned} U &= \{u_1, u_2, \cdots, u_t\}, \\ V &= \{v_1, v_2, \cdots, v_l\}, \end{aligned} \tag{2.5}$$

in short,

$$q_{u_j v_k}(x_i, y_i) \text{ is written as } q_{jk}(x_i, y_i),$$

and

$$Q_{u_j v_k} \text{ is written as } Q_{jk}.$$

In this case, we can use a matrix to illustrate all information gains from X on $U \times V$. The matrix is called the information matrix of X on $U \times V$, which is written as Q.

Definition 2.1 Given a sample $X = \{(x_i, y_i) | i = 1, 2, \cdots, n\}$. Let U be input space and V be output spaces. Let $u_j, j = 1, 2, \cdots, t$, and $v_k, k = 1, 2, \cdots, l$ be discrete points in U and V, respectively. If sample X gives information gain at Q_{jk}, by formulas (2.3) and (2.4), to Cartesian space point (u_j, v_k), then matrix

$$Q = \begin{array}{c} \\ u_1 \\ u_2 \\ \cdots \\ u_t \end{array} \begin{array}{cccc} v_1 & v_2 & \cdots & v_l \\ \left(\begin{array}{cccc} Q_{11} & Q_{12} & \cdots & Q_{1l} \\ Q_{21} & Q_{22} & \cdots & Q_{2l} \\ \cdots & \cdots & \cdots & \cdots \\ Q_{t1} & Q_{t2} & \cdots & Q_{tl} \end{array} \right) \end{array} \qquad (2.6)$$

is called the *simple information matrix* of X on $U \times V$.

We say that information matrix Q illustrates the given sample X on the Cartesian space $U \times V$.

For example, Table 2.1 showing 1996 popular vote summary of the U.S. presidential election is an information matrix, where $U = \{AL, AK, \cdots, WY\}$, and $V = \{$Clinton, Dole, Perot$\}$.

Table 2.1[1] 1996 popular vote summary of the U.S. presidential election

	Clinton	Dole	Perot
AL	662,165	769,044	92,149
AK	80,380	122,746	26,333
...	...	...	...
WY	77,934	105,388	25,928

Example 2.1 There are 6 students in a group. Measuring their height, h in meters, and weight, w in kilograms, we obtain a sample

$$\begin{aligned} G &= \{(h_1, w_1), (h_2, w_2), \cdots, (h_6, w_6)\} \\ &= \{(1.6, 50), (1.7, 65), (1.65, 55), (1.7, 60), (1.7, 65), (1.6, 55)\}. \end{aligned}$$

Taking

$$H = \{H_1, H_2, H_3\} = \{1.6, 1.65, 1.7\},$$

and

$$W = \{W_1, W_2, W_3, W_4\} = \{50, 55, 60, 65\}$$

to be input h and output w universe of discourse respectively, we can calculate the information matrix of G on $H \times W$.

Using the falling formula in (2.3), we can obtain all fallen information q. For example, for Cartesian space point $(H_1, W_1) = (1.6, 50)$, from observation $(h_1, w_1) = (1.6, 50)$, we obtain fallen information $q_{11}(x_1, y_1) = 1$ because $h_1 = H_1 = 1.6$ and $w_1 = W_1 = 50$. Obviously, $q_{11}(x_i, y_i) = 0, i = 2, 3, 4, 5, 6$. Therefore, the information gain from X on Cartesian space point (H_1, W_1) is

$$Q_{11} = \sum_{i=1}^{6} q_{11}(x_i, y_i) = 1 + 0 + 0 + 0 + 0 + 0 = 1.$$

Illustrating all information gains from G on $H \times W$, we obtain the information matrix of G on $H \times W$ as

---

[1] From http://www.fec.gov/pubrec/elecpop.htm

$$Q = \begin{array}{c} \\ H_1 \\ H_2 \\ H_3 \end{array} \begin{array}{cccc} W_1 & W_2 & W_3 & W_4 \\ \left(\begin{array}{cccc} 1 & 1 & 0 & 0 \\ 0 & 1 & 0 & 0 \\ 0 & 0 & 1 & 2 \end{array} \right). \end{array}$$

Definition 2.2 An information matrix is called complete if and only if $\forall(x_i, y_i) \in X$, $\exists(u_j, v_k) \in U \times V$ such that $x_i = u_j$ and $y_i = v_k$.

For example, all of Alabama votes in Table 2.1 is

$$662,165 + 769,044 + 92,149 = 1,523,358.$$

However, in this state, 1996 total popular vote of the U.S. presidential election is 1,534,349. That is, some polls cannot be shown by this table. Therefore, the information matrix given by Table 2.1 is not complete.

Theorem 2.1 The sum of all information gains of a complete information matrix is equal to the size of given sample.

Proof Let

$$Q = \{Q_{jk} | j = 1, 2, \cdots, t; k = 1, 2, \cdots, l\}$$

is a complete information matrix of given sample

$$X = \{(x_1, y_1), (x_2, y_2), \cdots, (x_n, y_n)\}$$

on Cartesian space $U \times V$, where $U = \{u_1, u_2, \cdots, u_t\}$, $V = \{v_1, v_2, \cdots, v_l\}$.

Because Q is complete, $\forall(x_i, y_i) \in X$, $\exists(u_j, v_k) \in U \times V$ such that $x_i = u_j$ and $y_i = v_k$. Hence, according to the falling formula in (2.3), $\forall(x_i, y_i) \in X$, $\exists(u_j, v_k) \in U \times V$ such that $q_{jk}(x_i, y_i) = 1$.

On the other hand, if $j' \neq j$ or $k' \neq k$, according to the property of the Cartesian space, we know that $x_i \neq u_{j'}$ or $y_i \neq v_{k'}$ must be true. Therefore, if $q_{jk}(x_i, y_i) = 1$ and, $j' \neq j$ or $k' \neq k$, then $q_{j'k'}(x_i, y_i) = 0$. That is, $\forall(x_i, y_i) \in X$, we obtain

$$\sum_{j=1}^{t} \sum_{k=1}^{l} q_{jk}(x_i, y_i) = 1$$

Hence,

$$\sum_{j=1}^{t} \sum_{k=1}^{l} Q_{jk} = \sum_{j=1}^{t} \sum_{k=1}^{l} \sum_{i=1}^{n} q_{jk}(x_i, y_i) = \sum_{i=1}^{n} \sum_{j=1}^{t} \sum_{k=1}^{l} q_{jk}(x_i, y_i) = \sum_{i=1}^{n} 1 = n$$

$\square$

In many cases, for given sample and Cartesian space, we cannot obtain a complete information matrix. In other words, $\exists(x_i, y_i) \in X$ such that $q_{jk}(x_i, y_i) = 0$, $\forall(u_j, v_k) \in U \times V$. For example, if we add a new student to the group in Example 2.1, whose height is 1.66 metre and weight is 62 kilogram, then we cannot illustrate all observations on the given Cartesian space because $1.66 \notin H$ and $62 \notin W$.

2.3 Information Matrix on Crisp Intervals

One way to produce a complete information matrix is to construct a new Cartesian space that is made of intervals whose bounds are crisp.

Let $U = \{u\}$ and $V = \{v\}$ be input and output universe of discourse respectively. There is no loss in generality when we suppose that $U, V \subseteq \mathbb{R}$ (set of real numbers). In this case, given an origin u_0 ($\in U$) and a bin width (step length) h_x, we can construct the intervals $U_j = [u_0 + (j-1)h_x, u_0 + jh_x[$, $j = 1, 2, \cdots, t$; also, for V, we can obtain $V_k = [v_0 + (k-1)h_y, v_0 + kh_y[$, $k = 1, 2, \cdots, l$. In the case that $U \nsubseteq \mathbb{R}$ or $V \nsubseteq \mathbb{R}$, we also can construct similar subsets to classify the universes. Then a new Cartesian space can be defined as

$$\mathbf{U} \times \mathbf{V} = \{(U_j, V_k)|j = 1, 2, \cdots, t; k = 1, 2, \cdots, l\}. \tag{2.7}$$

When $U, V \subseteq \mathbb{R}$, we know that (U_j, V_k) is a box in R^2. In general, we regard (U_j, V_k) as a point in Cartesian space $\mathbf{U} \times \mathbf{V}$.

Corresponding to the falling formula in (2.3), we construct a so-called characteristic falling formula.

Given an observation $(x_i, y_i) \in X$ and Cartesian space point $(U_j, V_k) \in \mathbf{U} \times \mathbf{V}$, we can calculate fallen information of (x_i, y_i) on (U_j, V_k) by

$$\chi_{U_j V_k}(x_i, y_i) = \begin{cases} 1, & \text{if } x_i \in U_j \text{ and } y_i \in V_k, \\ 0, & \text{otherwise.} \end{cases} \tag{2.8}$$

In short, $\chi_{U_j V_k}(x_i, y_i)$ is written as $\chi_{jk}(x_i, y_i)$. We call (2.8) *characteristic falling formula*.

Let

$$E_{jk} = \sum_{i=1}^{n} \chi_{jk}(x_i, y_i), \tag{2.9}$$

we say that X gives a *bump*, in height E_{jk}, to Cartesian space point (U_j, V_k).

Definition 2.3 Given a sample $X = \{(x_i, y_i)|i = 1, 2, \cdots, n\}$. Let U be an input space and V be an output spaces. Let $U_j, j = 1, 2, \cdots, t$, and $V_k, k = 1, 2, \cdots, l$ be crisp intervals in U and V, respectively. If sample X gives information gain at E_{jk}, by formulas (2.8) and (2.9), to Cartesian space point (U_j, V_k), then matrix

$$E = \begin{array}{c} \\ U_1 \\ U_2 \\ \cdots \\ U_t \end{array} \begin{array}{c} V_1 \quad V_2 \quad \cdots \quad V_l \\ \left(\begin{array}{cccc} E_{11} & E_{12} & \cdots & E_{1l} \\ E_{21} & E_{22} & \cdots & E_{2l} \\ \cdots & \cdots & \cdots & \cdots \\ E_{t1} & E_{t2} & \cdots & E_{tl} \end{array} \right) \end{array}, \tag{2.10}$$

is called the *crisp-interval information matrix* of X on $U \times V$.

For example, a frequency histogram (Subsection 3.4.3 gives the definition) is a crisp-interval information matrix with one column.

Obviously, E is always complete because $\forall (x_i, y_i) \in X, \exists (U_j, V_k) \in \mathbf{U} \times \mathbf{V}$ such that $\chi_{jk}(x_i, y_i) = 1$.

Example 2.2 We have a batch of medium and strong earthquake data (see Appendix 2.A) including 134 seismic records observed in China from 1900 to 1975 with magnitudes in the range 4.25-8.5 and epicentral intensity in the range VI-XII degrees.

For earthquake magnitude, we take intervals

$$
\begin{aligned}
\mathbf{U} &= \{U_1, U_2, \cdots, U_{11}\} \\
&= \{[4.2, 4.6[, [4.6, 5.0[, [5.0, 5.4[, [5.4, 5.8[, [5.8, 6.2[, [6.2, 6.6[, [6.6, 7.0[, \\
&\quad [7.0, 7.4[, [7.4, 7.8[, [7.8, 8.2[, [8.2, 8.6[\}
\end{aligned}
$$

For epicentral intensity, intervals are

$$
\begin{aligned}
\mathbf{V} &= \{V_1, V_2, \cdots, V_7\} \\
&= \{\{VI^-, VI, VI^+\}, \{VII^-, VII, VII^+\}, \{VIII^-, VIII, VIII^+\}, \\
&\quad \{IX^-, IX, IX^+\}, \{X^-, X, X^+\}, \{XI^-, XI, XI^+\}, \\
&\quad \{XII^-, XII, XII^+\}\}
\end{aligned}
$$

In Appendix 2.A, there are two observations that are special in epicentral intensity VI—VII: No.62 and No.82. Their epicentral intensity belong to both of interval V_1 and V_2. For convenience, we make VI—VII belong to V_2.

For the first record of Appendix 2.A, considering magnitude as input and epicentral intensity as output, we obtain an observation $(x_1, y_1) = (5.75, VII)$. Because $(5.75, VII)$ is a point of 2-cube $U_4 \times V_2$, where $U_4 = [5.4, 5.8[, V_2 = \{VII^-, VII, VII^+\}$, we can obtain fallen information $\chi_{42}(x_1, y_1) = 1$. For other j, k, $\chi_{jk}(x_1, y_1) = 0$. With the method, we can calculate all of fallen information $\chi_{jk}(x_i, y_i), i = 1, 2, \cdots, 134; j = 1, 2, \cdots, 11; k = 1, 2, \cdots, 7$.

Then, employing the bumping formula in (2.9), we can obtain an information matrix

$$
E_{[\cdot, \cdot[} =
\begin{array}{cc}
\begin{array}{l}
U_1 \; ([4.2, 4.6[) \\
U_2 \; ([4.6, 5.0[) \\
U_3 \; ([5.0, 5.4[) \\
U_4 \; ([5.4, 5.8[) \\
U_5 \; ([5.8, 6.2[) \\
U_6 \; ([6.2, 6.6[) \\
U_7 \; ([6.6, 7.0[) \\
U_8 \; ([7.0, 7.4[) \\
U_9 \; ([7.4, 7.8[) \\
U_{10} \; ([7.8, 8.2[) \\
U_{11} \; ([8.2, 8.6[)
\end{array}
&
\begin{array}{ccccccc}
V_1 & V_2 & V_3 & V_4 & V_5 & V_6 & V_7 \\
1 & 0 & 0 & 0 & 0 & 0 & 0 \\
5 & 2 & 0 & 0 & 0 & 0 & 0 \\
16 & 7 & 0 & 0 & 0 & 0 & 0 \\
6 & 25 & 2 & 0 & 0 & 0 & 0 \\
0 & 11 & 10 & 0 & 0 & 0 & 0 \\
0 & 5 & 10 & 6 & 0 & 0 & 0 \\
0 & 0 & 2 & 8 & 0 & 0 & 0 \\
0 & 0 & 2 & 5 & 3 & 0 & 0 \\
0 & 0 & 0 & 2 & 1 & 0 & 0 \\
0 & 0 & 0 & 0 & 2 & 2 & 0 \\
0 & 0 & 0 & 0 & 0 & 0 & 1
\end{array}
\end{array},
$$

where $[\cdot, \cdot[$ indicates that we chosen left-closed and right-open intervals, in earthquake magnitude domain of the mapping, to construct the information matrix. Obviously, there are distinct flaws in E to show the relationship between magnitude, M, and epicentral intensity, I_0. It cannot show all gradients for V_k along $U_1, U_2, \cdots, U_{11}$, or for U_j along $V_1, V_2, \cdots, V_7$. Some neighboring elements in the matrix are same. For example, element E_{53} and E_{63} are same, both are 10. The worst of it is that, if we choose the bounds of the intervals to be left open and right closed, the information matrix will be changed significantly. The new information matrix is

$$
E_{],]} =
\begin{array}{c}
\\
U_1' \ (]4.2, 4.6]) \\
U_2' \ (]4.6, 5.0]) \\
U_3' \ (]5.0, 5.4]) \\
U_4' \ (]5.4, 5.8]) \\
U_5' \ (]5.8, 6.2]) \\
U_6' \ (]6.2, 6.6]) \\
U_7' \ (]6.6, 7.0]) \\
U_8' \ (]7.0, 7.4]) \\
U_9' \ (]7.4, 7.8]) \\
U_{10}' \ (]7.8, 8.2]) \\
U_{11}' \ (]8.2, 8.6])
\end{array}
\begin{array}{ccccccc}
V_1 & V_2 & V_3 & V_4 & V_5 & V_6 & V_7 \\
\left(\begin{array}{ccccccc}
2 & 0 & 0 & 0 & 0 & 0 & 0 \\
15 & 4 & 0 & 0 & 0 & 0 & 0 \\
5 & 5 & 0 & 0 & 0 & 0 & 0 \\
6 & 25 & 2 & 0 & 0 & 0 & 0 \\
0 & 15 & 11 & 0 & 0 & 0 & 0 \\
0 & 1 & 10 & 6 & 0 & 0 & 0 \\
0 & 0 & 3 & 10 & 0 & 0 & 0 \\
0 & 0 & 0 & 3 & 3 & 0 & 0 \\
0 & 0 & 0 & 2 & 1 & 0 & 0 \\
0 & 0 & 0 & 0 & 2 & 2 & 0 \\
0 & 0 & 0 & 0 & 0 & 0 & 1
\end{array}\right)
\end{array}.
$$

Where, maximum element of the first column is in 2nd row instead of 3th row.

The reason why a little change in the bounds makes a strong change in the information matrix is that, in the above model, all seismic records falling into a same interval are considered to play a same role. It neglects the difference between them. In fact, the records may occupy different positions in the interval. For example, in Appendix 2.A, 8th and 14th records are

| 8 | Yunnan | Jan. 12,1934 | 23.7° | 102.7° | 6.0 | VIII |
| 14 | Yunnan | Sept. 13,1950 | 23.5° | 103.1° | 5.8 | VIII |

We obtain two observations, $(x_8, y_8) = (6.0, VIII)$ and $(x_{14}, y_{14}) = (5.8, VIII)$. For epicentral intensity, they are same. For magnitude, although both of the records belong to interval $U_5 = [5.8, 6.2[$, their positions in U_5 are different. The former is located at the center of the interval, and the latter is located at the boundary. Neglecting the position difference implies that we throw away some information. It is easy to pick up the information about the differences, if we employ fuzzy bounds instead of crisp bounds of the intervals in the matrix.

2.4 Information Matrix on Fuzzy Intervals

We start from a very general concept, namely the so-called fuzzy quantity[9]. Imposing supplementary conditions leads successively to the notion of a fuzzy interval, a fuzzy closed interval and finally a fuzzy number.

(1) A fuzzy quantity is a fuzzy set on the real line $\mathbb{R}$. It's well-known that a fuzzy quantity Z may be interpreted as a possibility distribution; $Z(x)$ expresses the possibility that the value of the fuzzily known quantity Z is equal to x. For every $\alpha \in]0,1]$ we define the α-cut of Z as $Z_\alpha = \{x \mid x \in \mathbb{R} \text{ and } Z(x) \geq \alpha\}$.
The 1-cut of Z is called the kernel of Z:

$$\ker(Z) = \{x \mid x \in \mathbb{R} \text{ and } Z(x) = 1\}.$$

For every $\alpha \in [0,1[$ we may define the strong α-cut (or strong α-cut) of Z as $Z_{\bar{\alpha}} = \{x \mid x \in \mathbb{R} \text{ and } Z(x) > \alpha\}$. The strong 0-cut of Z is called the support of Z:

$$\operatorname{supp} Z = \{x \mid x \in \mathbb{R} \text{ and } Z(x) > 0\}.$$

(2) A fuzzy quantity Z satisfying ker $(Z) \neq \emptyset$ is called a normalized fuzzy quantity. Every element of ker(Z) is called a modal value of Z.

(3) A convex fuzzy quantity Z is called a fuzzy interval; it satisfies the convexity condition

$$(\forall (x_1, x_2) \in R^2)(\forall \alpha \in [0,1])(Z(\alpha x_1 + (1-\alpha)x_2) \geq \min(Z(x_1), Z(x_2))).$$

(4) An upper-semicontinuous fuzzy interval is called a fuzzy closed interval. Let (X, τ) be a topological space and $\mathbb{R}$ being equipped with the euclidean topology $\tau_{|\cdot|}$ derived from the absolute value mapping $|\cdot|$. It's wellknown that a subbase for $\tau_{|\cdot|}$ is given by $\{]\alpha, +\infty[\mid \alpha \in \mathbb{R}\} \cup \{]-\infty, \alpha[\mid \alpha \in \mathbb{R}\}$. The general form for continuity of an $X - \mathbb{R}$ mapping f is given as :

$$f \text{ is continuous} \iff (\forall O_2 \in \tau_{|\cdot|})(f^{-1}(O_2) \in \tau)$$
$$\iff (\forall F_2 \in \tau'_{|\cdot|})(f^{-1}(F_2) \in \tau')$$

where $\tau', \tau'_{|\cdot|}$ denote the class of closed sets corresponding with $\tau, \tau_{|\cdot|}$. Using the subbase one easily obtains the following equivalent forms :

$$
\begin{aligned}
f \text{ is continuous} \quad &\iff \quad (\forall \alpha \in \mathbb{R})(f^{-1}(]\alpha, +\infty[) \in \tau) \\
&\text{and} \quad (\forall \alpha \in \mathbb{R})(f^{-1}(]-\infty, \alpha[) \in \tau) \\
&\iff \quad (\forall \alpha \in \mathbb{R})(f^{-1}([\alpha, +\infty[) \in \tau') \\
&\text{and} \quad (\forall \alpha \in \mathbb{R})(f^{-1}(]-\infty, \alpha]) \in \tau').
\end{aligned}
$$

Splitting these conjunctions leads to the notions of upper and lower-semicontinuity:

$$\begin{aligned}
f \text{ is upper} - \text{semicontinuous} \quad &\Longleftrightarrow \quad (\forall \alpha \in \mathbb{R})(f^{-1}(] - \infty, \alpha[) \in \tau) \\
&\Longleftrightarrow \quad (\forall \alpha \in \mathbb{R})(f^{-1}([\alpha, +\infty[) \in \tau') \\
f \text{ is lower} - \text{semicontinuous} \quad &\Longleftrightarrow \quad (\forall \alpha \in \mathbb{R})(f^{-}(]\alpha, +\infty[) \in \tau) \\
&\Longleftrightarrow \quad (\forall \alpha \in \mathbb{R})(f^{-1}(] - \infty, \alpha]) \in \tau').
\end{aligned}$$

In particular, if $X = \mathbb{R}$ and range $(f) \subseteq [0, 1]$, i.e. if we restrict ourselves to fuzzy quantities Z, we obtain:

$$\begin{aligned}
Z \text{ is upper} - \text{semicontinuous} \quad &\Longleftrightarrow \quad (\forall \alpha \in]0, 1])(Z_\alpha \text{is closed}) \\
Z \text{ is lower} - \text{semicontinuous} \quad &\Longleftrightarrow \quad (\forall \alpha \in [0, 1[)(Z_{\overline{\alpha}} \text{ is open}).
\end{aligned}$$

(5) A fuzzy real number or shortly a fuzzy number is a fuzzy closed interval with a unique modal value and bounded support. ($\forall x \in \mathbb{R}$ is called a modal value of fuzzy set A if and only if $\mu_A(x) = 1$).

Triangular numbers, trapezoidal numbers and L-R numbers of Dubois and Prade are special cases of fuzzy intervals.

A triangular fuzzy number N is defined[4] by three numbers $a_1 < a_2 < a_3$ where the graph of the membership function $\mu_N(x)$ is a triangle with base on the interval $[a_1, a_3]$ and vertex at $x = a_2$. We specify N as $(a_1/a_2/a_3)$. A triangular shaped fuzzy number N is partially defined by three number $a_1 < a_2 < a_3$ where: (1) the graph of $\mu_M(x)$ is continuous and monotonically increasing from zero to one on $[a_1, a_2]$; (2) $\mu_M(a_2) = 1$; and (3) the graph of $\mu_M(x)$ is continuous and monotonically decreasing from one to zero on $[a_2, a_3]$. An example is shown in Fig. 2.1.

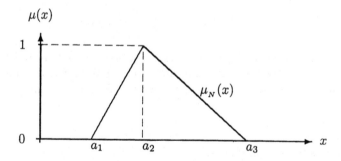

Fig. 2.1 Membership function of a triangular shaped fuzzy number N

A triangular shaped fuzzy set is built for each interval extreme, which is the fuzzy set modal value. Each fuzzy set covers two adjacent intervals, with the exception of the first and last fuzzy sets which cover one interval.

The membership function of a triangular shaped fuzzy set N is

$$\mu_N(x) = \begin{cases} 0, & \text{if } x < a_1; \\ \frac{x-a_1}{a_2-a_1}, & \text{if } a_1 \le x \le a_2; \\ \frac{x-a_3}{a_2-a_3}, & \text{if } a_2 \le x \le a_3; \\ 0, & \text{if } a_3 < x. \end{cases}$$

A trapezoidal fuzzy interval A, as shown in Fig. 2.2, is a fuzzy set that is defined[1] by a pair of linear functions L (left) and R (right):

$$\mu_A(x) = \begin{cases} 0, & \text{if } x < L(0); \\ \frac{x-L(0)}{L(1)-L(0)}, & \text{if} L(0) \le x < L(1); \\ 1, & \text{if } L(1) \le x \le R(1); \\ \frac{x-R(0)}{R(1)-R(0)}, & \text{if } R(1) \le x \le R(0); \\ 0, & \text{if } R(0) < x. \end{cases}$$

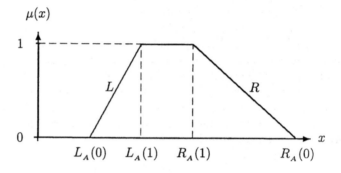

Fig. 2.2 A trapezoidal fuzzy interval A

The concept of the L-R numbers was suggested by Dubois and Prade in 1978 (see [3]). Firstly they gave the definition of the reference function: a function, usually denoted L or R, is a reference function of numbers if and only if
(1) $L(x) = L(-x)$;
(2) $L(0) = 1$;
(3) L is nonincreasing on $[0, +\infty]$.
For instance, $L(x) = 1$ for $x \in [-1, +1]$ and 0 outside; $L(x) = max(0, 1-|x|^p)$, $p \ge 0$; $L(x) = e^{-|x|^p}$, $p \ge 0$; $L(x) = 1/(1+|x|^p)$, $p \ge 0$.
A fuzzy number M is said to be an L-R type fuzzy number if and only if

$$\mu_M(x) = \begin{cases} L((m-x)/\alpha), & \text{for } x \le m, \ \alpha > 0; \\ R((x-m)/\beta), & \text{for } x \ge m, \ \beta > 0. \end{cases}$$

L is for left and R for right reference. m is the *mean value* of M. α and β are call *left* and *right spreads*, respectively. When the spreads are zero, M is a nonfuzzy number by convention. As the spread increase, M becomes fuzzier and fuzzier. Symbolically, they write

$$M = (m, \alpha, \beta)_{L-R}.$$

The simplest fuzzy interval is symmetrical triangle fuzzy number.

Definition 2.4 Let u be an arbitrary real number. Then the membership function of symmetrical triangle fuzzy number $I_{(x_0, \Delta)}$ is

$$\mu_I(u) = \begin{cases} 1 - |x_0 - u|/\Delta, & \text{if } |x_0 - u| \le \Delta, \\ 0, & \text{if } |x_0 - u| > \Delta. \end{cases} \qquad (2.11)$$

where x_0 and Δ are the central number of I and the fuzzy degree of I respectively.

A symmetrical triangle fuzzy number $I_{(x_0, \Delta)}$ can be shown in Fig. 2.3.

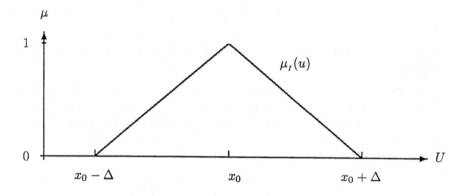

Fig. 2.3 Membership function of a symmetrical triangle fuzzy number $I_{(x_0, \Delta)}$

A symmetrical triangle fuzzy number is approximate to fuzzy concept "around x_0". Its boundary is fuzzy. Employing it to be fuzzy intervals, we can construct a good quality information matrix.

Let u_j be the center point of interval U_j in (2.7). Taking u_j be the central number of a fuzzy interval and h_x, width of U_j, to be the fuzzy degree respectively, we can obtain a fuzzy interval, in the domain of the mapping, as

$$\mu_{A_j}(u) = \begin{cases} 1 - |u_j - u|/h_x, & \text{if } |u_j - u| \le h_x, \\ 0, & \text{if } |u_j - u| > h_x. \end{cases} \qquad (2.12)$$

For V_k, we also can obtain a fuzzy interval, in the range of the mapping, as

$$\mu_{B_k}(v) = \begin{cases} 1 - |v_j - v|/h_y, & \text{if } |v_k - v| \le h_y, \\ 0, & \text{if } |v_k - v| > h_y. \end{cases} \qquad (2.13)$$

The symbol $A(u_j, h_x)$ and $B(v_k, h_y)$ are used for indicating to denote fuzzy interval A_j and B_k, with membership function $\mu_{A_j}(u)$ and $\mu_{B_k}(v)$, respectively. Let $\mathcal{U} = \{A_j | j = 1, 2, \cdots, t\}$ and $\mathcal{V} = \{B_k | k = 1, 2, \cdots, l\}$. Then, we define so-called soft Cartesian space as the following

$$\mathcal{U} \times \mathcal{V} = \{(A_j, B_k)|j = 1, 2, \cdots, t; k = 1, 2, \cdots, l\}. \tag{2.14}$$

Corresponding to the falling formula in (2.3) and the characteristic falling formula in (2.8), we construct so-called distributing formula.

Given an observation $(x_i, y_i) \in X$ and Cartesian space point $(A_j, B_k) \in \mathcal{U} \times \mathcal{V}$, we can calculate distributed information of (x_i, y_i) on (A_j, B_k) by

$$\mu_{A_j B_k}(x_i, y_i)$$
$$= \begin{cases} (1 - \frac{|u_j - x_i|}{h_x})(1 - \frac{|v_k - y_i|}{h_y}), & \text{if } |u_j - x_i| \le h_x \text{ and } |v_k - y_i| \le h_y, \\ 0, & \text{otherwise.} \end{cases}$$
$$\tag{2.15}$$

$\mu_{jk}(x_i, y_i)$ is short for $\mu_{A_j B_k}(x_i, y_i)$. We call (2.15) *distributing formula*. Let

$$\mathcal{Q}_{jk} = \sum_{i=1}^{n} \mu_{jk}(x_i, y_i), \tag{2.16}$$

we say that X distributes its information, in $\mathcal{Q}_{jk}$, to Cartesian space point (A_j, B_k).

Definition 2.5 Given a sample $X = \{(x_i, y_i)|i = 1, 2, \cdots, n\}$. Let U be input space and V be output spaces. Let $A_j, j = 1, 2, \cdots, t$, and $B_k, k = 1, 2, \cdots, l$ be fuzzy intervals in U and V, respectively. If sample X gives information gain at $\mathcal{Q}_{jk}$, by formulas (2.15) and (2.16), to Cartesian space point (A_j, B_k), then matrix

$$\mathcal{Q} = \begin{array}{c} \\ A_1 \\ A_2 \\ \cdots \\ A_t \end{array} \begin{pmatrix} B_1 & B_2 & \cdots & B_l \\ \mathcal{Q}_{11} & \mathcal{Q}_{12} & \cdots & \mathcal{Q}_{1l} \\ \mathcal{Q}_{21} & \mathcal{Q}_{22} & \cdots & \mathcal{Q}_{2l} \\ \cdots & \cdots & \cdots & \cdots \\ \mathcal{Q}_{t1} & \mathcal{Q}_{t2} & \cdots & \mathcal{Q}_{tl} \end{pmatrix}, \tag{2.17}$$

is called the *fuzzy-interval information matrix* of X on $U \times V$.

Obviously, $\mathcal{Q}$ also always is complete because $\forall (x_i, y_i) \in X, \exists (A_j, B_k) \in \mathcal{U} \times \mathcal{V}$ such that $|u_j - u| \le h_x$ and $|v_k - y_i| \le h_y$.

Example 2.3 We consider the data in Appendix 2.A again. For earthquake magnitude, we take fuzzy intervals

$$\mathcal{U} = \{A_1, A_2, \cdots, A_{11}\}$$

where $A_j, j = 1, 2, \cdots, 11$, is defined by expression (2.12), and whose central numbers consist a discrete universe of discourse as

$$U = \{u_1, u_2, \cdots, u_{11}\}$$
$$= \{4.4, 4.8, 5.2, 5.6, 6.0, 6.4, 6.8, 7.2, 7.6, 8.0, 8.4\} \tag{2.18}$$

The fuzzy degree of these fuzzy intervals is $h_x = 0.4$. That is, $A_j = (u_j, 0.4), j = 1, 2, \cdots, 11$.

For epicentral intensity, firstly, according to engineer experience, we define a mathematical mapping f from earthquake intensity into the real line $\mathbb{R}$:

$$
f: \begin{array}{rcl}
I_0 & \to & \mathbb{R} \\
VI & \mapsto & 6 \\
VII & \mapsto & 7 \\
VIII & \mapsto & 8 \\
IX & \mapsto & 9 \\
X & \mapsto & 10 \\
XI & \mapsto & 11 \\
XII & \mapsto & 12
\end{array}
\qquad (2.19)
$$

Then, we extend the domain of the mapping f to include VI^-, VI^+, $\cdots$, XII^-, XII^+ and I_0'—I_0'' with a definition as

$$
f(I_0^-) = f(I_0) - 0.2, \quad f(I_0^+) = f(I_0) + 0.2, \quad f(I_0'\!-\!I_0'') = \frac{f(I_0') + f(I_0'')}{2}
$$
$$(2.20)$$

We take $6, 7, \cdots, 12$ as central number and $h_y = 1$ as fuzzy degree. Then, $B_k = B(k+5, 1), k = 1, 2, \cdots, 7$.

For example, from 21th record of Appendix 2.A, considering the mapping, we obtain an observation $(x_{21}, y_{21}) = (6.0, f(VII^-)) = (6.0, 6.8)$. In nonzero membership, it belongs to 2-dimension fuzzy intervals $A_5 \times B_1$ and $A_5 \times B_2$, where $A_5 = A(6.0, 0.4)$, $B_1 = B(6, 1)$ and $B_2 = B(7, 1)$. Distributing (x_{21}, y_{21}) to these 2-dimension fuzzy intervals, we have

$$
\begin{aligned}
\mu_{51}(x_{21}, y_{21}) &= (1 - \tfrac{|u_5 - x_{21}|}{h_x})(1 - \tfrac{|v_1 - y_{21}|}{h_y}) \\
&= (1 - \tfrac{|6.0 - 6.0|}{0.4})(1 - \tfrac{|6 - 6.8|}{1}) \\
&= 0.2
\end{aligned}
$$

$$
\begin{aligned}
\mu_{52}(x_{21}, y_{21}) &= (1 - \tfrac{|u_5 - x_{21}|}{h_x})(1 - \tfrac{|v_2 - y_{21}|}{h_y}) \\
&= (1 - \tfrac{|6.0 - 6.0|}{0.4})(1 - \tfrac{|7 - 6.8|}{1}) \\
&= 0.8
\end{aligned}
$$

For other j, k, $\mu_{jk}(x_{21}, y_{21}) = 0$.

Distributing all observations of X to all 2-dimension fuzzy intervals and summing them for every 2-dimension fuzzy intervals, we can obtain an information matrix shown in Eq. (2.21).

Now we have shown three kinds of information matrixes, respectively, on discrete universes of discourse, crisp intervals and fuzzy intervals. The first kind may be incomplete, i.e., there are some observations which cannot be illustrated on matrix; the second one cannot fully show all gradients along domain of the mapping or along the range of the mapping even if the given sample has provided enough information. Comparing Q in Example 2.3 and E in Example 2.2, we know that the third kind has overcome the problems we have met. That is, (1) All observations can be illustrated on Q; (2) The

$$
Q = \begin{array}{c} \\ A_1 \\ A_2 \\ A_3 \\ A_4 \\ A_5 \\ A_6 \\ A_7 \\ A_8 \\ A_9 \\ A_{10} \\ A_{11} \end{array}
\begin{array}{ccccccc}
B_1 & B_2 & B_3 & B_4 & B_5 & B_6 & B_7 \\
1.22 & 0.12 & 0 & 0 & 0 & 0 & 0 \\
9.40 & 3.28 & 0.15 & 0 & 0 & 0 & 0 \\
11.28 & 10.72 & 0.75 & 0 & 0 & 0 & 0 \\
5.13 & 19.88 & 3.12 & 0 & 0 & 0 & 0 \\
0.40 & 11.25 & 10.85 & 0.50 & 0 & 0 & 0 \\
0 & 2.95 & 8.58 & 4.72 & 0 & 0 & 0 \\
0 & 0 & 3.45 & 9.47 & 0.20 & 0 & 0 \\
0 & 0 & 0.90 & 4.10 & 2.87 & 0 & 0 \\
0 & 0 & 0 & 2 & 1.88 & 0 & 0 \\
0 & 0 & 0 & 0 & 1.25 & 2 & 0 \\
0 & 0 & 0 & 0 & 0 & 0 & 0.75
\end{array}
\qquad (2.21)
$$

position difference between the observations, in a same interval with crisp bounds, can be shown. The difference is regarded as an extra information.

2.5 Mechanism of Information Matrix

When we use a matrix framework to illustrate a given sample for its information structure, the framework plays a role to collect the observation's information by some means. An observation can be regarded as a small "ball". The information of the ball is a striking of the ball on the sensors located on the framework when the ball falls.

The model on discrete universes of discourse is as a special framework consisting of a number of nodes (u, v) with sensors, where $(u, v) \in U \times V$ (U and V is shown in (2.5)). Because the balls (observations) are small, some of them may escape from the framework when they are falling on it. Therefore, the model, in many cases, cannot collect all information carried by a given sample.

A crisp-interval model can be regarded as a group of boxes, with length h_x and width h_y described in section 2.3. The cover of a box is formed by four nodes (u, v) of $U \times V$. Suppose that a box has an inverted cone bottom, and their covers have been taken off. We place these boxes in order, one by one without any empty to fill in the square area encircled by the framework. Then, for each box, we set a sensor at the center of the bottom. Certainly, the box-matrix can catch all balls (observations) and every ball can touch one of the sensors located in the bottoms. However, the sensors cannot identify the first position where a ball falls. Hence, the second model cannot totally show all gradients. In other words, some information cannot be grabbed by the model.

The last model, so-called fuzzy-interval matrix, is a more intelligent architecture. Firstly, to each node $(u, v) \in U \times V$, we set a sensor at it. Then, for a group of four sensors, we image a thin wood board, with length h_x and width h_y, covering on them. We can put $(t-1) \times (l-1)$ boards on all groups. These boards are independent.

Now, when a ball (observation) falls, it must strike one of the boards. If its first position is not just in one of the supporting nodes, all four sensors under the board can detect it. However, the messages received by these sensors are not as strong as that when a ball (observation) falls straightly on one of sensors. In other words, the striking force caused by a ball (observation) is shared by the four supporting nodes, and, at the same time, the information carried by a ball (observation) is distributed by the four sensors. Particularly, when a ball (observation) falls straightly on one of sensors, other three sensors receive zero message. By this way, not only no ball (observation) can escape, but also we can identify the first positions by four sensors.

Obviously, the value, Q_{jk}, in an information matrix indicates the cause-effect connection strength between input A_j and output B_k. For example, from Q in (2.21) we know that if an earthquake in magnitude around $M = 5.2$, in A_3, occurs, then the possibility of that the epicentral intensity produced is about $I_0 = VII$, in B_2, is more larger than about $I_0 = VIII$, in B_3, because $Q_{32} = 10.72 > Q_{33} = 0.75$.

The trace consists of the points with the strongest connection, and may embody the relationship between input and output in a system.

In fact, at least, there exist four approaches to describe or produce relationships in systems, which are equations of mathematical physics, regression, neural networks, and fuzzy graph [27,30]. Let us review these approaches for identifying the special aptitude of the information matrix.

2.6 Some Approaches Describing or Producing Relationships

The reason why we are interested in illustrating a given small-sample by an information matrix is that we want to, with the given sample, know how the input variable x effects the output variable y, i.e., we want to know the relationship between x and y. The relationships in systems are so important that there are many approaches developed to describe or produce them. Equations of mathematical physics would be the most ideal, regression models in terms of mathematical statistics would be the most common, artificial neural networks could be considered as the latest fashion, and the fuzzy graphs might be the most visual. Relatively speaking, the approach based on the information matrix is more visual and precise than the fuzzy graphs.

2.6.1 Equations of mathematical physics

In the 18th century, researchers found that numerous phenomena of physics and mechanics can be described by boundary value problems for differential equations. Then, equations of mathematical physics were developed into a popular approach to describe the relationships in physical systems.

For example, the equation of oscillations

$$\rho \frac{\partial^2 u}{\partial t^2} = \text{div}(p \text{ grad } u) - qu + f(x, t)$$

describes the small vibrations of strings, membranes, and acoustic and electromagnetic oscillations. Where, the space variables $x = (x_1, x_2, \cdots, x_n)$ vary in a region $G \in R^n$, $n = 1, 2, 3$, in which the physical process under consideration evolves. For $\rho = 1$, $p = a^2$ =const. and $q = 0$, it becomes the wave equation

$$\rho \frac{\partial^2 u}{\partial t^2} = a^2 \triangle u + f(x, t),$$

where $\triangle$ is the Laplace operator:

$$\triangle = \frac{\partial^2}{\partial x_1^2} + \frac{\partial^2}{\partial x_2^2} + \cdots + \frac{\partial^2}{\partial x_n^2}.$$

If in the wave equation the external perturbation f is periodic with frequency ω:

$$f(x, t) = a^2 f(x) e^{i\omega t},$$

then the amplitude $u(x)$ of a periodic solution with the same frequency ω,

$$u(x, t) = u(x) e^{i\omega t},$$

satisfies the Helmholtz equation[6]

$$\triangle u + k^2 u = -f(x), \quad k^2 = \frac{\omega^2}{a^2}.$$

For a complete description of the oscillatory process it is necessary to give the initial perturbation and initial velocity:

$$u|_{t=0} = u_0(x), \quad \frac{\partial u}{\partial t}|_{t=0} = u_1(x).$$

The formulation of the boundary value problems discussed above assumes that the solutions are sufficiently regular in the interior of the region as well as up to the boundary. Such formulations of boundary value problems are termed classical. However, in many problems of physical interest one must relinquish such regularity requirements. Inside the region the solution may be a generalized function and satisfy the equation in the sense of generalized functions, while the boundary value conditions may be fulfilled in some generalized sense. Such formulations are called generalized, and the corresponding solutions are called generalized solutions.

Laying aside non-regularity, we must meet the following natural requirements:

(1) a solution must be exist;
(2) the solution must be unique;
(3) the solution must depend continuously on the data of the problem.

It is clear that equations of mathematical physics can describe physical phenomena with functions (i.e., relationships) in terms of partial differential equations, presuming that the basic laws of physics are known. Even then, the requirement (3) is imposed in connection with the fact that, as a rule, the data of physical problems are determined experimentally only approximately, and hence it is necessary to be sure that the solution of the problem does not depend essentially on the measurement errors of these data.

2.6.2 Regression

Although the term "regression" denotes a broad collection of statistical methods, nowadays the term much more generally means the description of the nature of the relationship between two or more variables.

First, a basic question: Given two presumably dependent random variables, how should we go about studying their joint behavior? One approach would be to consider the conditional probability density function of Y given x as a function of x — that is, make a graph of $f_{Y|x}(y)$ versus x. While such graphs would show the relationship between X and Y, their use and interpretation are difficult. A better solution is to replace $f_{Y|x}$ with one of its numerical descriptors — say $E(Y|x)$ — and graph $E(Y|x)$ versus x.

For example, let X and Y have the following density:

$$f_{X,Y}(x,y) = y^2 e^{-y(x+1)}, \quad x \geq 0, y \geq 0.$$

Hence, the marginal probability density function for X is

$$f_X(x) = \int_{-\infty}^{+\infty} f_{X,Y}(x,y)dy = \frac{2}{(x+1)^3}.$$

Therefore, the conditional probability density function of Y given x is

$$f_{Y|x}(y) = \int_{-\infty}^{y} \frac{f_{X,Y}(x,u)}{f_X(x)}du = \frac{y^2 e^{-y(x+1)}}{\frac{2}{(x+1)^3}} = \frac{1}{2}(x+1)^3 y^2 e^{-y(x+1)}.$$

Then

$$E(Y|x) = \int_0^{+\infty} y \cdot \frac{1}{2}(x+1)^3 y^2 e^{-y(x+1)}dy = \frac{3}{x+1}.$$

In more general terminology, the search for a relationship between $E(Y|x)$ and x is known as the *regression problem* (and the function $h(x) = E(Y|x)$ is referred to as the regression curve of Y on x). Experimentally, problems of this short arise when a researcher records a set of (x_i, y_i) pair, pots the points on a scatter diagram, and then asks for an question that adequately describes the functional relationship between the two variables. What the equation is seeking to approximate, of course, is $E(Y|x)$ versus x.

It may be argued that, regression with respect to a given sample is to estimate a conditional expectation $E(Y|x)$ versus x with the given sample $\{(x_i, y_i)|i = 1, 2 \cdots, n\}$. Particularly, when the population from which observations are taken is normal, i.e., the joint density function is

$$f(x, y) = \frac{1}{2\pi \sigma_X \sigma_Y \sqrt{1 - \rho^2}} e^{Q(x,y)},$$

where

$$Q(x, y) = -\frac{1}{2(1 - \rho^2)} \left[\left(\frac{x - \mu_X}{\sigma_X} \right)^2 - 2\rho \left(\frac{x - \mu_X}{\sigma_X} \right) \left(\frac{y - \mu_Y}{\sigma_Y} \right) + \left(\frac{y - \mu_Y}{\sigma_Y} \right)^2 \right],$$

then, the regression curve of Y on x is

$$y = \bar{y} + \hat{\rho} \frac{\hat{\sigma}_Y}{\hat{\sigma}_X} (x - \bar{x}), \tag{2.22}$$

where

$$\bar{x} = \frac{1}{n} \sum_{i=1}^{n} x_i, \quad \bar{y} = \frac{1}{n} \sum_{i=1}^{n} y_i,$$

$$\hat{\sigma}_X^2 = \frac{1}{n} \sum_{i=1}^{n} (x_i - \bar{x})^2, \quad \hat{\sigma}_Y^2 = \frac{1}{n} \sum_{i=1}^{n} (y_i - \bar{y})^2,$$

$$\hat{\rho} = \frac{1}{n \hat{\sigma}_X \hat{\sigma}_Y} \sum_{i=1}^{n} [(x_i - \bar{x})(y_i - \bar{y})].$$

In other words, given a sample $\{(x_i, y_i)|i = 1, 2 \cdots, n\}$, on the condition of normal assumption, we can, with the regression method, obtain an estimated relationship between input x and output y, which is a linear function.

In principle, for any population, we can calculate the conditional probability density function showing the relationship between two or more variables. Let x and y be input variable (also called independent variable) and output variable (also called dependent variable) respectively. Where, x may be a vector. We can employ the method of least squares to estimate the relationship with a given sample.

Suppose that the desired relationship can be written as a function $h(x)$. The function we will term "best" is the one whose coefficients minimize the square-sum function L, where

$$L = \sum_{i=1}^{n} [y_i - h(x_i)]^2.$$

For example, let $y = h(x) = a + bx$ then

$$L = \sum_{i=1}^{n} [y_i - (a + bx_i)]^2.$$

By the first step, we get

$$\frac{\partial L}{\partial a} = \sum_{i=1}^{n} (-2)[y_i - (a + bx_i)],$$

and

$$\frac{\partial L}{\partial b} = \sum_{i=1}^{n} (-2)x_i[y_i - (a + bx_i)].$$

Now, set the right-hand sides of $\partial L/\partial a$ and $\partial L/\partial b$ equal to 0 and simplify. This gives

$$na + \left(\sum_{i=1}^{n} x_i\right) b = \sum_{i=1}^{n} y_i,$$

and

$$\left(\sum_{i=1}^{n} x_i\right) a + \left(\sum_{i=1}^{n} x_i^2\right) b = \sum_{i=1}^{n} x_i y_i,$$

Then, we obtain

$$b = \left[n \sum_{i=1}^{n} x_i y_i - \left(\sum_{i=1}^{n} x_i\right)\left(\sum_{i=1}^{n} y_i\right)\right] / \left[n\left(\sum_{i=1}^{n} x_i^2\right) - \left(\sum_{i=1}^{n} x_i\right)^2\right],$$

and

$$a = \frac{1}{n}\left(\sum_{i=1}^{n} y_i - b \sum_{i=1}^{n} x_i\right).$$

It can be verify that the result is just as same as the formula (2.22).

It is clear that regression results based on a given sample can describe the relationship between input and output, presuming that the type of the population from which observations are taken is known and the size of the given sample is sufficiently large.

For a small sample without any information about the population type, it is very difficult to obtain a reasonable regression result.

2.6.3 Neural networks

Artificial neural networks (ANN) have received extensive attention during the last two decades. ANN is well known as a tool to solve many practical problems as pattern recognition, function approximation, system identification, time series forecasting, etc. [22,18,17]. A neural network can be understood[11] as a mapping $f : R^p \rightarrow R^q$, defined by

$$y = f(x) = \varphi(Wx)$$

where $x \in R^p$ is the input vector, $y \in R^q$ is the output vector. The weight matrix W is a $p \times q$ matrix and φ is a nonlinear function that is often referred to as the activation function. The typical activation function is the Sigmoid function

$$\varphi(x) = \frac{1}{1 + e^{-\alpha x}}, \quad \alpha > 0$$

The mapping f can be decomposed into a composition of mappings; the result is a multi-layer network,

$$R^p \to R^m \to \cdots \to R^n \to R^q$$

The algorithm for computing W is often called the training algorithm. The most popular neural networks are the multi-layer back-propagation networks whose training algorithm is the well-known gradient descent method. Relationships between variables are most often recognized by learning neural networks with data or patterns collected. The approach is also called adaptive pattern recognition[16]. For the majority of cases, the applied neural networks, from a statistical point of view, solve conditional estimation problems[10]. The celebrated back propagation error algorithm used for training feed forward neural networks is shown to be a special case of gradient optimization in the sense of mean squared error[21]. Feed forward neural networks are analyzed in paper [25] for consistent estimation of conditional expectation functions, which optimize expected squared error.

In the learning phase of training such a network, we present the pattern $\underline{x}_p = \{i_{pi}\}$ as input and ask that the network adjust the set of weights in all the connecting links and also all the thresholds in the nodes such that the desired outputs t_{pk} are obtained at the output notes. Once this adjustment is made, we present another pair of $\underline{x}_p = \{i_{pi}\}$ and t_{pk}, and ask that the network learn that association also. In fact, we ask that the network find a single set of weights and biases that will satisfy all the (input, output) pairs presented to do it. This process can pose a very strenuous learning task and is not always readily accomplished.

In general, the outputs $\{o_{pk}\}$ of the network will not be the same as the target or desired values $\{t_{pk}\}$. For each pattern, the square of the error is

$$E_p = \frac{1}{2} \sum_k (t_{pk} - o_{pk})^2,$$

and the average system error is

$$E = \frac{1}{2P} \sum_p \sum_k (t_{pk} - o_{pk})^2, \tag{2.23}$$

where P is the sample size and the one-half scaling factor is inserted for mathematical convenience. A true gradient search for minimum system error should be based on the minimization of expression (2.23).

A number of authors have discussed the property of universal approximation with respect to neural networks. For example, Cybenko (1989)(see [2]) and Funahashi (1989) (see [5]) show that any continuous function can be approximated by a neural network with one internal hidden layer using sigmoidal nonlinearities. Hornik, Stinchcombe, and White (1989) (see [7]) prove that multilayer networks using arbitrary squashing functions can approximate any continuous function to any degree of accuracy, provided enough hidden units are available. Wray and Green (1995) (see [26]) prove that, due to the fact that networks are implemented on computers, the property of universal approximation (to any degree of accuracy) does not hold in practice.

It is clear that the approximate function described by a trained neural network can be regarded as the estimation for the relationship we want to know. However, when a trained neural network is performing as a mapping from input space to output space, it is a black box. This means it is not possible to understand how a neural system works, and it is very hard to incorporate human a priori knowledge into a neural network.

Furthermore, the well-known back-propagation algorithm [20] has the problem of becoming trapped in local minima, which may lead to failure in finding a global optimal solution [14]. Besides, the convergence rate of back-propagation is still too slow even if learning can be achieved.

2.6.4 Fuzzy graphs

The fuzzy graphs we will discuss differ from extended ones based on traditional graph theory[23].

A traditional graph is a pair $G = (V, E)$ where V is a finite set of vertices and E a nonfuzzy relation on $V \times V$, i.e., a set of ordered pairs of vertices; these pairs are the edges of G. It plays an important role in the modeling of structures, especially in operations research. Extending the traditional concept, Rosenfeld (1975), see [19], given so-called fuzzy graph $\widetilde{G}$ which is a pair $(\widetilde{V}, \widetilde{E})$ where $\widetilde{V}$ is a fuzzy set on V and $\widetilde{E}$ is a fuzzy relation on $V \times V$ such that

$$\mu_{\widetilde{E}}(v, v') \leq \min(\mu_{\widetilde{V}}(v), \mu_{\widetilde{V}}(v')).$$

It may be helpful for representing soft or ill-defined structures, for instance, in humanistic system.

However, in Zadeh's definition, the central idea of a fuzzy graph is to represent a function with means of fuzzy logic instead of mathematical equations, see [29]. The domain of the output parameter is described through individual membership functions. The mapping from input to output is defined by a set of fuzzy points with individual membership functions[8]. Well-known techniques like the center-of-gravity calculation allow us to approximate real-valued functions based on this fuzzy graph.

Let the membership function of A be $\mu_A(x)$. The center of gravity of A can be calculated by

$$CG(A) = \frac{\int_{a_1}^{a_2} x\mu_A(x)dx}{\int_{a_1}^{a_2} \mu_A(x)dx}.$$

The main advantage of the fuzzy graph concept is the very compact and easy to understand representation of a function with if-then rules.

From the point of view of fuzzy if-then rules, a fuzzy graph can be considered as the integration of some fuzzy patches. A fuzzy patch is visually defined by a fuzzy subset of the product space $X \times Y$. The rules define fuzzy patches in the input-output space $X \times Y$. The fuzzy system $F: X \to Y$ approximates a function $f: R^p \to R^q$. All centroidal fuzzy systems F compute a conditional mean: $f(x) = E[y|X = x]$. An additive fuzzy system[12] splits this global conditional mean into a convex sum of local conditional means.

Zadeh [28] first observed that planar rule patches can cover both functions and relations. He did not observe that resulting exponential rule explosion or the optimality of rule patches that cover extrema or show how to covert a set of abstract rules or rule patches into a well-defined function $f: R^p \to R^q$ with output $f(x)$ from each input x.

The patch structure of a rule suggests that a neural or statistical clustering algorithm can learn rules from data. Data clusters can define or center the first set of rules. The simplest scheme centers a rule patch at a sample vector value and in practice centers fuzzy sets at the vector components. Then new data can tune the rules and thus move them in the product space or change their shape. Sparse or noisy data tend to give large patches and thus less certain rules. Dense or filtered data tend to give small patches and thus more certain rules.

An expert might give the rules approximate the relationship we want to know. This gives the first-cut fuzzy approximation to the relationship with a fuzzy graph. Much of fuzzy engineering deals with tuning this rules and perhaps adding new rules and deleting old ones.

It is clear that a fuzzy graph can be regarded as the estimation for the relationship we want to know. And, some fuzzy rule generators from training data by neural networks is more automatic. However, if the first-cut fuzzy approximation is far away the real relationship, the training data can do nothing. The method does not ensure that we can always find the rules with a given pool of experts or with a fixed set of data.

Another problem is that fuzzy graphs suffer from the curse of dimensionality: rule explosion. They need too many rules to approximate most function. The number of rules grows exponentially with the number of input and output variables.

2.7 Conclusion and Discussion

In this chapter, we give three kinds of information matrixes of a sample $X = \{(x_1, y_1), (x_2, y_2), \cdots, (x_n, y_n)\}$ on a Cartesian space $U \times V$, where $U = \{u_1, u_2, \cdots, u_t\}$, $V = \{v_1, v_2, \cdots, v_l\}$.

1. Simple information matrix $Q = \{Q_{jk}\}_{t \times l}$, calculated by

$$Q_{jk} = \sum_{i=1}^{n} q_{u_j v_k}(x_i, y_i),$$

and

$$q_{u_j v_k}(x_i, y_i) = \begin{cases} 1, & \text{if } x_i = u_j \text{ and } y_i = v_k, \\ 0, & \text{otherwise.} \end{cases}$$

2. Crisp-interval information matrix $E = \{E_{jk}\}_{(t-1) \times (l-1)}$, calculated by

$$E_{jk} = \sum_{i=1}^{n} \chi_{U_j V_k}(x_i, y_i),$$

and

$$\chi_{U_j V_k}(x_i, y_i) = \begin{cases} 1, & \text{if } x_i \in U_j \text{ and } y_i \in V_k, \\ 0, & \text{otherwise.} \end{cases}$$

where $U_j = u_{j+1} - u_j, j = 1, 2, \cdots, t-1$; $V_k = v_{k+1} - v_k, k = 1, 2, \cdots, l-1$.

3. Fuzzy-interval information matrix $\mathcal{Q} = \{\mathcal{Q}_{jk}\}_{t \times l}$, calculated by

$$\mathcal{Q}_{jk} = \sum_{i=1}^{n} \mu_{jk}(x_i, y_i),$$

and

$$\mu_{jk}(x_i, y_i)$$
$$= \begin{cases} (1 - \frac{|u_j - x_i|}{h_x})(1 - \frac{|v_k - y_i|}{h_y}), & \text{if } |u_j - x_i| \leq h_x \text{ and } |v_k - y_i| \leq h_y, \\ 0, & \text{otherwise.} \end{cases}$$

where $h_x \equiv u_{j+1} - u_j, j = 1, 2, \cdots, t-1$; $h_y \equiv v_{k+1} - u_k, k = 1, 2, \cdots, l-1$.

We use the matrixes to illustrate the information structure of a sample and serve as the discovery of relations between objects from observing, experiments and data. A fuzzy-interval information matrix is better to illustrate the information structure than a simple information matrix and a crisp-interval information matrix.

In all common existing approaches on relationship identification, we cannot obtain a fitness function for the relationship in the population with respect to the given sample. Equations of mathematical physics mainly suffer from presuming that the basic laws of physics are known. Regression bases on that the type of the population from which observations are taken has been known and the size of the given sample is sufficient large. A neural network is a black box and we might meet the problem of getting trapped in local minima. Fuzzy graphs may lead to a wrong result if the expert fails to confirm the first-cut fuzzy approximation under the acceptable form.

In an information matrix, the trace consists of the points with the strongest connection may embody the relationship between input and output in a system. To produce the matrix, we do not use any assumption in the basic laws of physics, the type of the population. And, it is unnecessary to derive any first-cut approximation from experts. Certainly, the trace of the strongest points in an information matrix is visual and we cannot meet any convergence problem as in neural networks.

In general, any arrangement of numbers, figures, or signs in a square made up of rows can be called a matrix. If it shows some information, it is an information matrix. For example, see [24],the Fisher' s information matrix for the network is an inner product of the form:

$$F = X_{aug}^T W X_{aug},$$

where X_{aug} is referred to as the augmented design matrix and is an augmentation of the design matrix with additional columns that arise due to the hidden units. Another example, see [15], the plain α-information matrix is provided in order to explain deriving α-weighted EM (α-weighted Expectation and Maximization) algorithms for neural networks of module mixtures.

In the book, the concept of information matrix is supposed to more directly illustrate a given small-sample for showing its information structure. Employing the information matrixes, we might find a new approach which is much more effective in relationship recognition from a given small-sample.

A primary information matrix cannot be directly used as a relationship for logical inference or prediction. We can employ more general fuzzy sets instead of symmetrical triangle fuzzy numbers as the fuzzy intervals to construct an information matrix. And, an information matrix is allowed to be 1-dimension or $n - dimension$. We will introduce the concept of information distribution and corresponding methods to discuss them.

References

1. Cross, V., and Setnes, M. (1998), A Generalized Model for Ranking Fuzzy Sets. Proceedings of FUZZ-IEEE'98, Anchorage, USA, pp.773-778
2. Cybenko, G. (1989), Approximation by superpositions of a sigmoidal function. Mathematics of Control, Signals, and Systems, $2(4)$, pp.303-314
3. Dobois, D., and Prade, H. (1978), Operations of fuzzy numbers. International Journal of Systems Science, $9(6)$, pp.613-626
4. Feuring, T., James, J., and Hayashi, Y.(1998), A Gradient Descent Learning Algorithm for Fuzzy Neural Networks. Proceedings of FUZZ-IEEE'98, Anchorage, USA, pp.1136-1141
5. Funahashi, K. (1989), On the approximate realization of continuous mappings by neural networks. Neural Networks, $2(3)$, pp.183-192
6. Hazewinkel, M. (Ed.) (1995), Encyclopaedia of mathematics. Kluwer Academic Publishers, Singapore
7. Hornik, K., Stinchcombe, M., and White, H. (1989), Multilayer feedforward networks are universal approximators. Neural Networks, $2(5)$, pp.359-366
8. Huber, K.-P. and Berthold, M.(1998), application of fuzzy graphs for meta-modeling. Proceedings of FUZZ-IEEE'98, Anchorage, USA, pp.640-644
9. Kerre, E. (1999), Fuzzy Sets and Approximate Reasoning. Xian Jiaotong University Press, Xian, China
10. Kulczycki, P.(1998), Estimating conditional distributions by neural networks. Proceedings of IJCNN'98, Anchorage, USA, pp.1344-1349
11. Ky Van Ha (1998), Hierarchical radial basis function networks. Proceedings of IJCNN'98, Anchorage, USA, pp.1893-1898
12. Kosko, B.(1987), Foundations of Fuzzy Estimation Theory. Ph.D. dissertation, Department of Electrical Engineering, University of California at Irvine, June 1987; Order Number 8801936, University Microfilms International, 300 N. Zeeb Road, Ann Arbor, MI 48106
13. Liu, Z.R., Huang, C.F., Kong, Q.Z., and Yin, X.F. (1987), A fuzzy quantitative study on the effect of active fault distribution on isoseismal area in Yunnan. Journal of seismology, 1, pp.9-16
14. Marco Gori and Alberto Tesi (1992), On the problem of local minima in back-propagation. IEEE Transactions on Pattern Analysis and Machine Intelligence, $14(1)$, pp.76-86
15. Matsuyama, Y., Furukawa, S., Takeda, N., and Ikeda, T. (1998), Fast α-weighted EM learning for neural networks of module mixtures. Proceedings of IJCNN'98, Anchorage, USA, pp.2306-2311
16. Pao, Y.H.(1989), Adaptive Pattern Recognition and Neural Networks. Addison-Wesley, Reading, Massachusetts
17. Patrick, K. Simpson (1990), Artificial Neural Systems Foundations, Paradigms, Applications, and Implementations. McGraw-Hill, New York
18. Ripley, B. D. (1996) , Pattern Recognition and Neural Networks. Cambridge University Press, Cambridge
19. Rosenfeld, A. (1975), Fuzzy graphs. Zadeh, L.A., Fu, K.S. Tanaka, K., and Shimura, M. (eds): Fuzzy Sets and Applications to Cognitive and Decision Processes. Academic Press, New York, pp.77-95
20. Rumelhart, D. E., Hinton, G. E. and Williams, R. J.(1986), Learning internal representations by error propagation. Rumelhart, D.E. and McClelland, J. L. (eds): Parallel Distributed Processing: Explorations in the Microstructure of Cognitions; Vol. 1: Foundations. MIT Press, Cambridge, MA, pp.318-362
21. Rumelhart, D.E. and McClelland J. L. (1986). Parallel Distributed Processing. MIT Press, Cambridge, MA

22. Simon Haykin (1994), Neural Networks: A Comprehensive Foundation. Prentice-Hall, Inc., Englewood Cliffs, New Jersey
23. Skiena, S. (1990), Implementing Discrete Mathematics Combinatorics and Graph Theory with Mathematics. Addison-Wesley, Reading, MA
24. Warner, B. A. and Misra, M. (1998), Iteratively reweighted least squares based learning. Proceedings of IJCNN'98, Anchorage, USA, pp.1327–1331
25. White, H. (1990), Connectionist nonparametric regression: multilayer feedforward network can learn arbitrary mappings. Neural Networks, 3(5), pp.535-549
26. Wray, J. and Green, G.G.R.(1995), Neural networks, approximation theory, and finite precision computation. Neural Networks, 8(1), pp.3l-37
27. Zadeh, L.A.(1974), On the analysis of large scale system. in: Gottinger H. Ed., Systems Approaches and Environment Problems, Vandenhoeck and Ruprecht, Gottingen, pp. 23-37
28. Zadeh, L.A.(1987), Fuzzy Sets and Applications: Selected Papers by Zadeh, L.A., Yager, R. R., Ovchinnikov, S., Tong, R.M., and Nguyen, H.T. (eds), John Wiley and Sons, New York
29. Zadeh, L.A. (1994), Soft computing and fuzzy logic. IEEE Software, 11(6), pp.48-56
30. Zadeh, L.A. (1995), Fuzzy control, fuzzy graphs, and fuzzy inference. in: Yam, Y. and Leung, K.S. Eds., Future Directions of Fuzzy Theory and Systems (World Scientific, Singapore), pp.1-9

Appendix 2.A: Some Earthquake Data

134 data of middle and strong earthquakes observed in China from 1900 to 1975

No.	Region	Date	North Latitude	East Longitude	Richter Magnitude	Epicentral Intensity
1	Jianxi	Sept. 21,1941	(25.1°	115.8°)	$5\frac{3}{4}$	VII
2	Guangxi	Sept. 25,1958	22.6°	109.5°	$5\frac{3}{4}$	VII
3	Guangdong	Mar. 19,1962	23.7°	114.7°	6.4	VIII
4	Guangdong	July 26,1969	21.8°	111.8°	6.4	VIII
5	Yunnan	Dec. 31,1913	24.2°	102.5°	6.5	IX
6	Yunnan	July 31, 1917	(28.0°	104.0°)	6.5	IX
7	Yunnan	Mar. 15,1927	(25.4°	103.1°)	6.0	VIII
8	Yunnan	Jan. 12,1934	23.7°	102.7°	6.0	VIII
9	Yunnan	May 14, 1938	22.5°	100.0°	6.0	VII
10	Yunnan	Dec. 26, 1941	22.7°	99.9°	7.0	VIII
11	Yunnan	May 16, 1941	23.6°	99.8°	7.0	IX
12	Yunnan	June 27,1948	(26.6°	99.6°)	$6\frac{1}{4}$	VIII
13	Yunnan	Oct. 9,1948	(27.2°	104.0°)	$5\frac{3}{4}$	VIII
14	Yunnan	Sept. 13,1950	23.5°	103.1°	5.8	VIII
15	Yunnan	Dec. 21,1951	26.7°	100.0°	6.3	IX
16	Yunnan	June 19, 1952	22.7°	99.8°	6.5	VIII
17	Yunnan	Dec. 8,1952	22.9°	99.7°	5.8	VII$^+$
18	Yunnan	May 4,1953	24.2°	103.2°	5.0	VII$^-$
19	Yunnan	May 21,1953	(26.5°	99.9°)	5.0	VI
20	Yunnan	May 27,1955	25.5°	105.0°	5.0	VI$^+$
21	Yunnan	June 7,1955	26.5°	101.1°	6.0	VII$^-$
22	Yunnan	Aug. 24,1956	27.0°	101.5°	4.8	VI
23	Yunnan	Sept. 2,1960	28.9°	98.5°	5.5	VI
24	Yunnan	June 12,1961	24.9°	98.7°	5.8	VIII
25	Yunnan	June 27,1961	27.9°	99.7°	6.0	VIII
26	Yunnan	Apr. 25,1962	23.6°	106.1°	5.5	VII
27	Yunnan	June 24,1962	25.2°	101.2°	6.2	VIII$^-$
28	Yunnan	Apr. 23,1963	25.8°	99.5°	6.0	VII$^-$
29	Yunnan	Feb. 13,1964	25.6°	100.6°	5.4	VII
30	Yunnan	May 24,1965	24.1°	102.6°	5.2	VI$^+$

31	Yunnan	July 3,1965	22.4°	101.6°	6.1	VII+
32	Yunnan	Jan. 31,1966	27.8°	99.7°	5.1	VII+
33	Yunnan	Feb. 5,1966	26.2°	103.2°	6.5	IX−
34	Yunnan	Feb. 13,1966	26.1°	103.1°	6.2	VII
35	Yunnan	Feb. 18,1966	26.0°	103.2°	5.2	VI+
36	Yunnan	Sept. 19,1966	23.8°	97.9°	5.4	VII+
37	Yunnan	Sept. 21,1966	23.8°	97.9°	5.2	VII
38	Yunnan	Sept. 23,1966	26.1°	104.5°	5.0	VI+
39	Yunnan	Sept. 28,1966	27.5°	100.0°	6.4	IX
40	Yunnan	Oct. 11,1966	28.2°	103.7°	5.2	VI
41	Yunnan	Jan. 5,1970	24.1°	102.6°	7.8	X
42	Yunnan	Jan. 5,1970	23.9°	103.6°	5.3	VI
43	Yunnan	Feb. 5,1970	24.2°	102.2°	5.5	VI+
44	Yunnan	Feb. 7,1970	22.9°	100.8°	5.5	VII+
45	Yunnan	Feb. 5,1971	24.9°	99.2°	5.5	VII+
46	Yunnan	Apr. 28,1971	22.5°	101.2°	6.5	VIII
47	Yunnan	Jan. 23,1972	23.5°	102.5°	5.5	VII
48	Yunnan	Mar. 22,1973	22.1°	100.9°	5.5	VII
49	Yunnan	Apr. 22,1973	22.7°	104.0°	5.0	VI
50	Yunnan	June 1,1973	25.0°	98.7°	5.0	VI
51	Yunnan	June 2,1973	25.0°	98.7°	5.0	VI
52	Yunnan	June 2,1973	25.0°	98.7°	$4\frac{3}{4}$	VI−
53	Yunnan	Aug. 2,1973	27.9°	104.6°	5.4	VI
54	Yunnan	Aug. 16,1973	22.9°	101.1°	6.3	VIII
55	Sichuan	Aug. 25,1933	32.0°	103.7°	7.5	X
56	Sichuan	Dec. 18,1935	28.6°	103.7°	6.0	VIII
57	Sichuan	Apr. 27,1936	(28.7°	103.7°)	6.8	IX
58	Sichuan	Oct. 8,1941	(32.1°	103.3°)	6.0	VIII
59	Sichuan	June 7,1947	(26.7°	102.9°)	5.5	VII
60	Sichuan	May 25,1948	(29.7°	100.3°)	$7\frac{1}{4}$	X
61	Sichuan	June 18,1948	28.7°	101.4°	$5\frac{3}{4}$	VII
62	Sichuan	Nov. 13,1949	(29.5°	102.0°)	$5\frac{1}{2}$	VI—VII
63	Sichuan	Sept. 30,1952	(28.4°	102.2°)	6.8	IX
64	Sichuan	July 21,1954	(27.7°	101.1°)	5.3	VII
65	Sichuan	Apr. 14,1955	(30.0°	101.8°)	7.5	IX
66	Sichuan	Sept. 23,1955	(26.4°	101.9°)	6.8	IX
67	Sichuan	Sept. 28,1955	26.6°	101.3°	5.5	VII
68	Sichuan	Oct. 1,1955	29.9°	101.4°	5.8	VII
69	Sichuan	Feb. 8,1958	(31.8°	104.4°)	6.2	VII
70	Sichuan	Nov. 9,1960	32.8°	103.7°	6.8	IX

71	Sichuan	Feb. 27,1962	27.1°	101.8°	5.5	VII
72	Sichuan	Jan. 24,1967	(30.3°	104.1°)	5.5	VII
73	Sichuan	Aug. 30,1967	(31.6°	100.3°)	6.8	IX
74	Sichuan	Feb. 24,1970	(30.6°	103.2°)	6.3	IX
75	Sichuan	Aug. 16,1971	28.9°	103.6°	5.8	VII$^+$
76	Sichuan	Sept. 27,1972	30.1°	101.6°	5.8	VII
77	Sichuan	Feb. 6,1973	31.1°	100.1°	7.9	X
78	Ningxia	Dec. 16,1920	(36.5°	105.7°)	8.5	XII
79	Gansu	May 23,1927	(33.6°	102.6°)	8.0	XI
80	Gansu	Aug. 1,1936	34.2°	105.7°	6.0	VIII
81	Gansu	Feb. 11,1954	39.0°	101.3°	7.3	X
82	Shaanxi	Aug. 11,1959	35.5°	110.6°	5.4	VI—VII
83	Gansu	Feb. 3,1960	33.8°	104.5°	5.3	VI
84	Gansu	Oct. 1,1961	34.3°	104.8°	5.7	VII
85	Ningxia	Dec. 7,1962	(38.1°	106.3°)	5.4	VII
86	Gansu	Dec. 11,1962	34.8°	105.1°	5.0	VI$^+$
87	Qinghai	Jan. 11,1963	37.5°	101.6°	4.8	VII
88	Qinghai	Apr. 19,1963	35.7°	97.0°	7.0	VIII$^+$
89	Gansu	Aug. 20,1967	32.7°	106.8°	5.0	VI
90	Gansu	Oct. 16,1967	(30.8°	105.1°)	4.8	VI
91	Ningxia	Dec. 3,1970	35.9°	105.6°	5.5	VII
92	Ningxia	Mar. 24,1971	35.5°	98.0°	6.8	VIII
93	Xinjian	Aug. 5,1914	43.5°	91.5°	7.5	IX
94	Xinjian	Aug. 11,1931	47.1°	89.8°	8.0	XI
95	Xinjian	Mar. 10,1944	42.5°	82.5°	$7\frac{1}{4}$	IX
96	Xinjian	Feb. 24,1949	42.0°	84.0°	$7\frac{1}{4}$	IX
97	Xinjian	Apr. 15,1955	39.9°	74.6°	7.0	IX
98	Xinjian	Apr. 14,1961	(39.9°	77.8°)	6.8	IX
99	Xinjian	Aug. 20,1962	44.7°	81.6°	6.4	VIII
100	Xinjian	Aug. 29,1963	39.8°	74.3°	6.5	VIII
101	Xinjian	Nov. 13,1965	(43.6°	88.3°)	6.6	VIII
102	Xinjian	Feb. 12,1969	(41.5°	79.3°)	6.5	VIII
103	Xinjian	Sept. 14,1969	39.7°	74.8°	5.5	VII
104	Xinjian	July 29,1970	(39.9°	77.7°)	5.8	VII
105	Xinjian	Mar. 23,1971	(41.5°	79.3°)	6.0	VIII
106	Xinjian	June 16,1971	41.4°	79.3°	5.8	VII
107	Xinjian	July 26,1971	39.9°	77.3°	5.6	VII
108	Xinjian	Aug. 1,1971	43.9°	82.4°	4.8	VI
109	Xinjian	Jan. 16,1972	40.2°	78.9°	6.2	VII
110	Xinjian	Apr. 9,1972	42.2°	84.6°	5.6	VII

111	Xinjian	June 3,1973	44.4°	83.5°	6.0	VII
112	Shandong	Aug. 1,1937	(35.2°	115.4°)	$6\frac{3}{4}$	IX
113	Shanxi	Oct. 8,1952	(38.9°	112.8°)	5.5	VIII
114	Hebei	Feb. 16,1954	(37.6°	115.7°)	$4\frac{3}{4}$	VII
115	Shanxi	Aug. 19,1956	(37.9°	113.9°)	5.0	VII
116	Hebei	Jan. 1,1957	(40.4°	115.3°)	5.0	VI
117	Shanxi	June 6,1957	(37.6°	112.5°)	5.0	VI
118	Shanxi	June 11,1957	(37.8°	112.5°)	5.0	VI
119	Jilin	Apr. 13,1960	(44.7°	121.0°)	$5\frac{3}{4}$	VII
120	Shanxi	Jan. 13,1965	(35.1°	111.6°)	5.5	VII$^+$
121	Hebei	May 7,1965	/	/	$4\frac{1}{4}$	VI
122	Hebei	Mar. 6,1966	37.5°	115.0°	5.2	VII
123	Hebei	Mar. 8,1966	37.4°	114.9°	6.8	IX$^+$
124	Hebei	Mar. 20,1966	37.3°	115.0°	5.6	VI
125	Hebei	Mar. 22,1966	37.5°	115.1°	7.2	X
126	Hebei	Mar. 26,1966	37.6°	115.3°	6.2	VII$^+$
127	Hebei	Mar. 29,1966	37.5°	114.9°	6.0	VIII
128	Jilin	Oct. 2,1966	(43.8°	125.0°)	5.2	VII
129	Hebei	Mar. 27,1967	38.5°	116.5°	6.3	VII
130	Hebei	July 28,1967	(40.7°	115.8°)	5.5	VI
131	Hebei	Dec. 3,1967	37.6°	115.2°	5.7	VII
132	Shanxi	Dec. 18,1967	36.5°	110.8°	5.4	VI$^+$
133	Shanxi	Apr. 24,1969	39.3°	113.3°	4.6	VI
134	Liaoning	Feb. 4,1975	40.7°	122.8°	7.3	IX

3. Some Concepts From Probability and Statistics

This chapter reviews some preliminary concepts from probability and statistics. They are necessary to make preparations for the following chapters. In section 3.1, the information matrix is introduced into probability estimation field. Section 3.2 reviews some preliminary concepts from probability. Section 3.3 describes some continuous distributions we will use in this book. Section 3.4 reviews some necessary concepts from statistics, and describes some traditional estimation methods we likely to compare them with the new methods. Section 3.5 introduces Monte Carlo methods and gives some Fortran programs to generate samples for computer simulation experiments.

3.1 Introduction

As we know regression results based on a sample can describe the relationship between input X and output Y. From the point of view of the statistics, the regression curve is the conditional probability density function of Y given x. In the final analysis, the accuracy of the function approximation in terms of the regression must be determined by the corresponding probability distribution estimation.

It is easy to develop an information matrix on crisp intervals into an estimator about probability distribution.

Let $E = \{E_{jk}|j = 1, 2, \cdots, t; k = 1, 2, \cdots, l\}$ be an information matrix of the given sample $X = \{x_1, x_2, \cdots, x_n\}$ on the Cartesian space $\mathbf{U} \times \mathbf{V} = \{(U_j, V_k)|j = 1, 2, \cdots, t; k = 1, 2, \cdots, l\}$, where x_i is 2-dimension, and U_j and V_k are crisp intervals in input space X and output space Y, respectively. According to the definition of the information matrix on crisp intervals, we know that the number of x_i in the small box $U_j \times V_k$ is E_{jk}. It implies that the probability that x_i falls into the box is E_{jk}/n, because the information matrix must be complete. Therefore, $\widehat{P} = \{E_{jk}/n|j = 1, 2, \cdots, t; k = 1, 2, \cdots, l\}$ is an estimator about probability distribution on $X \times Y$. Calculating its conditional probability distribution, we can obtain a rough relationship between input and output.

Particularly, if x_i is 1-dimension, the information matrix reduces to the histogram in terms of number of x_i in the same interval.

In the probability estimation field, the estimation methods in terms of continuous distributions are categorized into two groups: the parametric estimation and the nonparametric estimation. The parametric estimation makes specific distributional assumptions to be identified, but there are no guarantee that the assumptions are always satisfied. The nonparametric estimation, in contrast, makes no assumptions other than imposes a certain degree of smoothness. Despite that the nonparametric density estimation is computationally intensive, many research problems regarding this approach have been extensively studied because of the availability of the fast computating facilities[4]. The kernel density estimators[8],[12] are a possible class of generative models which have particular theoretical advantages.

In the kernel method the performance of the estimated density function depends on the choice of kernel function and the bandwidth. Smyth (1994, see [10]) applied four different kernel functions (Gaussian kernel, Cauchy kernel, Triangular kernel and Epanechnikov Kernel) to a fault diagnosis problem and the vowel data set described in Ng and Lippmann [7] and compared the performances of kernels. He reported that the Epanechnikov kernel was the optimal kernel in terms of minimizing the mean integrated square error between the estimator and the true density, but the Gaussian kernel has been found to give the most consistent results since the tail behavior is consistent without oversmoothing the main part of the density function. When the bandwidth b is too small, the estimated probability density function becomes "spiky" and when the bandwidth is too large the density function smooths out the detail. The worst case occurs when a given sample is small the kernel method can do nothing.

It is well known that the so-called fuzzy-interval matrix has an intelligent architecture. Can we use the approach to improve probability distribution estimation? Or, can we get a more accurate relationship from a fuzzy-interval matrix? The answer is positive. The method of information distribution is derived from it. The benefit of the information distribution can be shown in results from computer simulation experiment based on Monte Carlo methods.

Before presenting the method, we first review some necessary concepts.

3.2 Probability

3.2.1 Sample spaces, outcomes, and events

Let us now be more precise about our terminology and define some necessary terms that lead to the concept of a probability distribution function. Consistent with standard textbooks, we will refer to the physical or mathematical process as an *experiment*, and this experiment has a (possibly infinite) number of *outcomes*, to which we will assign probabilities. The *sample space* of the experiment is the collection of all possible outcomes. Thus if the experiment is carried out, its outcome is assured to be in the sample space. We will also describe one realization of the experiment as a *trial,* and by definition a trial will have an outcome in the sample space. The experiment may result in the occurrence of a specific *event.* An event may be viewed as a consequence of the outcome (or outcomes) of the experiment. Let us now illustrate these concepts with a simple example.

Example 3.1 An illustration of an experiment and the terminology used in the experiment.

The experiment consists of one roll of a normal die (with faces labeled 1, 2, 3, 4, 5, and 6) and observing the top face of the die. The outcomes are the six faces, and the sample space consists of these six outcomes, since every realization of the experiment (i.e., each trial) results in one of these faces being the top face. (We will assume that the die will not balance on an edge or corner.) Events can then be defined in terms of the possible outcomes. Possible events that may be defined in terms of the six unique outcomes are:

A_1: Result is an even number,

A_2: Result is larger than 4,

A_3: Result is equal to 2 (hence the event is one of the outcomes).

3.2.2 Probability

Since this book is not intended to be a complete and rigorous treatment of probability, we will avoid the formal theory of probability and instead present a functional description. This should be sufficient preparation to understand the concept of a probability density function, which is the goal of this section.

To an event A_i we will assign a probability p_i, which is also denoted $P(A_i)$, or "probability of event." The quantity p_i must satisfy the properties given in Table 3-1 to be a legitimate probability.

Table 3-1. Properties of a valid probability p_i

(1) $0 \leq p_i \leq 1$
(2) If A_i is to occur, $p_i=1$
If A_i does not occur, $p_i=0$
(3) If events A_i and A_j are mutually exclusive, then
$P(A_i \text{ and } A_j) = 0$
$P(A_i \text{ or } A_j) = p_i + p_j$
(4) If events A_i, $i = 1, 2, \cdots, n$, are mutually exclusive and exhaustive (one of the n events A_i is assured to occur), then
$\sum_{i=1}^{n} p_i = 1$

3.2.3 Joint, marginal, and conditional probabilities

We now consider an experiment that consists of two parts, and each part leads to the occurrence of specified events. Let us define events arising from the first part of the experiment by A_i with probability f_i and events from the second part by B_j with probability g_j. The combination of events A_i and B_j may be called a composite event, denoted by the ordered pair $C_{ij} = (A_i, B_j)$. We wish to generalize the definition of probability to apply to the composite event C_{ij}. The joint probability p_{ij} is defined to be the probability that the first part of the experiment led to event A_i and the second part of the experiment led to event B_j. Thus, the joint probability p_{ij} is the probability that the composite event C_{ij} occurred (i.e., the probability that both events A_i and B_j occur).

Any joint probability can be factored into the product of a marginal probability and a conditional probability:

$$p_{ij} = p(i)p(j|i), \tag{3.1}$$

where p_{ij} is the joint probability, $p(i)$ is the marginal probability (the probability that event A_i occurs regardless of event B_j), and $p(j|i)$ is the conditional probability (the probability that event B_j occurs given that event A_i occurs). Note that the marginal probability for event A_i to occur is simply the probability that the event A_i occurs, or $p(i) = f_i$. Let us now assume that there are J mutually-exclusive events B_j, $j = 1, 2, \cdots, J$ and the following identity is evident:

$$p(i) = \sum_{k=1}^{J} p_{ik}. \tag{3.2}$$

Using Eq. (3.2), we easily manipulate Eq. (3.1) to obtain the following expression for the joint probability

$$p_{ij} = p_{ij}\left(\frac{\sum_{k=1}^{J} p_{ik}}{\sum_{k=1}^{J} p_{ik}}\right) = p(i)\left(\frac{p_{ij}}{\sum_{k=1}^{J} p_{ik}}\right). \tag{3.3}$$

Using Eq. (3.1), Eq. (3.3) leads to the following expression for the conditional probability:

$$p(j|i) = \frac{p_{ij}}{\sum_{k=1}^{J} p_{ik}}. \tag{3.4}$$

3.2.4 Random variables

We now define the concept of a random variable, a key definition in probability and statistics. We define a random variable as a real number x that is assigned to an event A. It is random because the event A is random, and it is variable because the assignment of the value may vary over the real axis. We will use "r.v." as an abbreviation for "random variable."

Example 3.2 Associating a random variable with a roll of a die.
Assign the number $10n$ to each face n of a die. When face n appears, the r.v. is $10n$.

3.2.5 Expectation value, variance, functions of random variables

Now that we have assigned a number to the outcome of an event, we can define an "average" value for the r.v. over the possible events. This average value is called the *expected value* for the random variable x, and has the following definition:

$$\text{expected value (or mean)} \equiv \text{E}(x) = \bar{x} = \sum_i p_i x_i. \tag{3.5}$$

One can define a unique, real-valued function of a r.v., which will also be a r.v. That is, given a r.v. x, then the real-valued function $g(x)$ is also a r.v. and we can define the expected value of $g(x)$:

$$\text{E}[g(x)] = \bar{g} = \sum_i p_i g(x_i). \tag{3.6}$$

The expected value of a linear combination of r.v.'s is simply the linear combination of their respective expected values;

$$\text{E}[ag(x) + bh(x)] = a\text{E}[g(x)] + b\text{E}[h(x)]. \tag{3.7}$$

The expected value is simply the "first moment" of the r.v., meaning that one is finding the average of the r.v. itself, rather than its square or cube or square root. Thus the mean is the average value of the first moment of the r.v., and one might ask whether or not averages of the higher moments have any significance. In fact, the average of the square of the r.v. does lead to an important quantity, the variance, and we will now define the higher moments of a r.v. x as follows:

$$\text{E}(x^n) = \overline{x^n}. \tag{3.8}$$

We also define "central" moments that express the variation of a r.v. about its mean, hence "corrected for the mean":

$$n^{th}\text{central moment} = \overline{(x - \bar{x})^n}. \tag{3.9}$$

The first central moment is zero. The second central moment is the *variance*:

$$\text{variance} \equiv \text{Var}(x) \equiv \sigma^2(x) = \overline{(x - \bar{x})^2} = \sum_i p_i(x_i - \bar{x})^2. \tag{3.10}$$

It is straightforward to show the following important identity:

$$\sigma^2 = \overline{x^2} - \bar{x}^2. \tag{3.11}$$

We will also find useful the square root of the variance, which is the standard deviation,

$$\text{standard deviation} = \sigma(x) = [\text{Var}(x)]^{1/2}. \tag{3.12}$$

3.2.6 Continuous random variables

So far we have considered only discrete r.v.'s, that is, a specific number x_i is assigned to the event A_i, but what if the events cannot be enumerated by integers, such as the angle of scattering for an electron scattering off a gold nucleus or the time to failure for a computer chip? The above definitions for discrete r.v.'s can be easily generalized to the continuous case.

First of all, if there is a continuous range of values, such as an angle between 0 and 2π, then the probability of getting exactly a specific angle is zero, because there are an infinite number of angles to choose from, and it would be impossible to choose exactly the correct angle. For example, the probability of choosing the angle $\theta = 1.34$ radians must be zero, since there are an infinite number of alternative angles. In fact, there are an infinite number of angles between 1.33 and 1.35 radians or between 1.335 and 1.345 radians, hence the probability of a given angle must be zero. However, we can talk about the probability of a r.v. taking on a value *within* a given interval, e.g., an angle θ between 1.33 and 1.35 radians. To do this, we define a probability density function.

3.2.7 Probability density function

The significance of the probability density function $p(x)$ is that $p(x)dx$ is the probability that the r.v. is in the interval $(x, x + dx)$, written as:

$$Prob(x \leq x' \leq x + dx) \equiv P(x \leq x' \leq x + dx) = p(x)dx. \qquad (3.13)$$

This is an operational definition of $p(x)$. Since $p(x)dx$ is unitless (i.e., it is a probability), then $p(x)$ has units of inverse r.v. units, e.g., 1/cm or 1/s, depending on the units of x. We will use "PDF" as an abbreviation for "probability density function."

We can also determine the probability of finding the r.v. somewhere in the finite interval $[a, b]$:

$$Prob(a \leq x \leq b) \equiv P(a \leq x \leq b) = \int_a^b p(x)dx, \qquad (3.14)$$

which, of course, is the area under the curve $p(x)$ from $x = a$ to $x = b$.

As with the definition of discrete probability distributions, there are some restrictions on the PDF. Since $p(x)$ is a probability density, it must be positive for all values of the r.v. x. Furthermore, the probability of finding the r.v. somewhere on the real axis must be unity. As it turns out, these two conditions are the only necessary conditions for $p(x)$ to be a legitimate PDF, and are summarized below.

$$p(x) \geq 0, \quad -\infty < x < \infty, \qquad (3.15)$$

$$\int_{-\infty}^{\infty} p(x)dx = 1. \qquad (3.16)$$

3.2.8 Cumulative distribution function

The cumulative distribution function (cdf) $F(x)$ gives the probability that the r.v. x' is less than or equal to x:

$$F(x) \equiv Prob(x' \leq x) = \int_{-\infty}^{x} p(x')dx'. \tag{3.17}$$

Note that since $p(x) \geq 0$, and the integral of $p(x)$ is normalized to unity, $F(x)$ obeys the following conditions:
  (1) $F(x)$ is monotone increasing,
  (2) $F(-\infty) = 0$,
  (3) $F(+\infty) = 1$.
Since $F(x)$ is the indefinite integral of $p(x)$, $p(x) = F'(x)$.

Theorem 3.1 Let X have cumulative distribution function F. Suppose that F is continuous and strictly increasing. If U is uniform on (0,1), then $F^{-1}(U)$ has distribution F.

Proof Because U is uniform on (0,1), $P[U \leq x] = x$, $0 \leq x \leq 1$. Hence $P[F^{-1}(U) \leq t] = P[U \leq F(t)] = F(t)$.

$\square$

3.3 Some Probability Density Functions

3.3.1 Uniform distribution

$$p(x) = \frac{1}{b-a}, \quad a < b, \quad a \leq x \leq b. \tag{3.18}$$

This distribution is based on a constant PDF. From this standpoint it is therefore the simplest of all statistical distributions. This distribution is also called the *rectangular distribution*. We also use symbol $U(a, b)$ to denote the distribution. The mean and variance for the uniform distribution are given by:

$$\mu = \frac{a+b}{2}, \quad \sigma^2 = (b-a)^2/12. \tag{3.19}$$

Fig. 3.1 illustrates the uniform distribution.

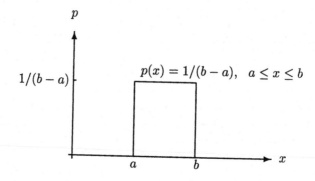

Fig. 3.1 Uniform distribution $U(a, b)$

3.3.2 Normal distribution

The most important PDF in probability and statistics is perhaps the normal distribution:

$$p(x) = \frac{1}{\sigma\sqrt{2\pi}} \exp[-\frac{(x - \mu)^2}{2\sigma^2}], \quad -\infty < x < \infty. \tag{3.20}$$

This distribution is also called the Gaussian distribution. This is a two-parameter (σ and μ) distribution, and it can be shown that μ is the mean of the distribution and σ^2 is the variance. Fig. 3.2 illustrates the normal distribution PDF. We also use symbol $N(\mu, \sigma^2)$ to denote the normal distribution. The standard normal distribution is the special case $\mu = 0$, $\sigma^2 = 1$. In the case where x is standard normal, the random variable $z = \sigma x + \mu$ is normal with mean μ and variance σ^2. Thus, for purposes of random number generation there is no loss in generality when one specializes to the standard normal.

Let us calculate the probability that a sample from the normal distribution will fall within a single standard deviation σ of the mean μ:

$$P(\mu - \sigma \leq x \leq \mu + \sigma) = 0.6826.$$

Similarly, the probability that the sample is within two standard deviations (within "2σ") of the mean is

$$P(\mu - 2\sigma \leq x \leq \mu + 2\sigma) = 0.9544.$$

Hence 68% of the samples will, on average, fall within one σ, and over 95% of the samples will fall within two σ of the mean μ.

The normal distribution will be encountered frequently in this book, not only because it is a fundamental PDF for many physical and mathematical applications, but also because it plays a central role in the estimation of errors with Monte Carlo simulation.

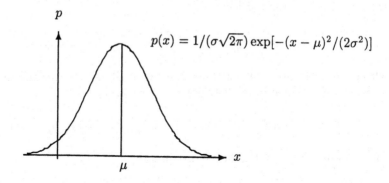

$$p(x) = 1/(\sigma\sqrt{2\pi})\exp[-(x-\mu)^2/(2\sigma^2)]$$

Fig. 3.2 Normal distribution

3.3.3 Exponential distribution

$$p(x) = \lambda e^{-\lambda x}, \quad x \geq 0, \quad \lambda > 0. \tag{3.21}$$

This distribution can be described as a number of physical phenomena, such as the time t for a radioactive nucleus to decay, or the time x for a component to fail, or the distance z a photon travels in the atmosphere before colliding with a water molecule. The exponential distribution is characterized by the single parameter λ, and one can easily show that the mean and variance for the exponential distribution are given by:

$$\mu = \frac{1}{\lambda}, \quad \sigma^2 = (\frac{1}{\lambda})^2. \tag{3.22}$$

Fig. 3.3 graphically illustrates the exponential distribution.

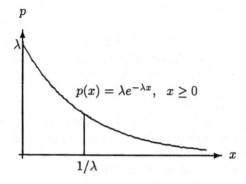

$$p(x) = \lambda e^{-\lambda x}, \quad x \geq 0$$

Fig. 3.3 Exponential distribution

3.3.4 Lognormal distribution

A r.v. x is lognormal if $\log x$ is normal. The PDF for the lognormal distribution with parameters μ and σ is

$$p(x) = \frac{1}{x\sigma\sqrt{2\pi}} \exp[-\frac{(\log x - \mu)^2}{2\sigma^2}], \quad 0 < x < \infty, \tag{3.23}$$

where "log" is natural logarithm. One can easily show that the mean and variance for the lognormal distribution are given by:

$$E(x) = \exp(\mu + \frac{\sigma^2}{2}), \quad Var(x) = [\exp(\sigma^2) - 1]\exp(2\mu + \sigma^2).$$

Fig. 3.4 illustrates the lognormal distribution.

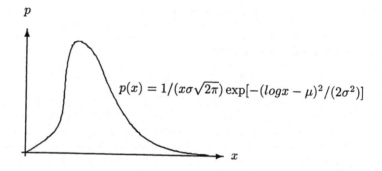

$$p(x) = 1/(x\sigma\sqrt{2\pi})\exp[-(\log x - \mu)^2/(2\sigma^2)]$$

Fig. 3.4 Lognormal distribution

3.4 Statistics and Some Traditional Estimation Methods

3.4.1 Statistics

The word *statistics* has two meanings[6]. In the more common usage, statistics refers to numerical facts. The second meaning refers to the field or discipline of study. In this sense of the word as in this book, statistics is defined as a group of methods that are used to collect, analyze, present, and interpret data and to make decisions.

In statistics, the collection of all elements of interest is called a *population*. The selection of a few elements from this population is called a *sample*. The elements in a sample is also called *objects* or *sample points*. The number of elements in a sample is called the *sample size*.

A sample drawn in such a way that each element of the population has some chance of being selected is called a *random sample*. If the chance of being selected is the same for each element of the population, it is called a *simple random sample*.

A *variable* is a characteristic under study that assumes different values for different elements. The value of a variable for an element is called an *observation*. A *data set* is a collection of observations on one or more variables.

We let $\mathbb{R}$ stand for the real line, and R^r for r-dimension real space.

Definition 3.1 If the observations of a random sample are in R^r, $r > 1$, the random sample is called a *r-dimension random sample*.

For a r-dimension random sample $X = \{x_1, x_2, \cdots, x_n\}$, we say that $\forall x \in X$ we have $x \in R^r$. It implies that an element and its observation are regarded identically.

A major portion of statistics deals with estimating probability distributions of the populations based on some given samples. As stated in section 3.1, the estimation methods in terms of continuous distributions are categorized into two groups: the parametric estimation and the nonparametric estimation. The former includes the point estimation and the interval estimation. The histogram and the kernel method belong to the nonparametric estimation. The interval estimation was discussed in Chapter 2. In the point estimation, the maximum likelihood principle is very often used for selecting an estimator. Now, we briefly review the maximum likelihood estimate, the histogram and the kernel method which will be frequently used in the following chapters.

3.4.2 Maximum likelihood estimate

Let X be an r.v. with PDF $f(\cdot; \theta)$ of known functional form but depending upon an unknown r-dimension constant vector $\theta = (\theta_1, \theta_2, \cdots, \theta_r)'$ which is called a *parameter*. We let Θ stand for the set of all possible values of θ and called it the *parameter space*. So $\theta \in \Theta \subseteq R^r$, $r \geq 1$.

If θ is known, we can calculate, in principle, all probabilities in which we might be interested. In practice, however, θ is generally unknown. Therefore the problem of estimating θ arise; or more generally, we might be interested in estimating some function of θ, $g(\theta)$, say, where g is (measurable and) usually a real-valued function. We now proceed to define what we mean by an estimator and an estimate of $g(\theta)$. Let $X_1, X_2, \cdots, X_n$ be i.i.d. (independent and identical distribution) r.v.'s with PDF $f(\cdot; \theta)$.

Definition 3.2 Any function computed from the values of a random sample, such as the sample mean, is called a *statistic*.

Definition 3.3 Any statistic $T = T(X_1, X_2, \cdots, X_n)$ which used for estimating the unknown quantity $g(\theta)$ is called an *estimator* of $g(\theta)$. The value $T = T(x_1, x_2, \cdots, x_n)$ of T for the observed values of X's is called an *estimate* of $g(\theta)$.

For simplicity, the terms estimator and estimate are often used interchangeably.

Let us consider the joint PDF of the X's $f(x_1; \theta) \cdots f(x_1; \theta)$. Treating the $x_1, x_2, \cdots, x_n$ as if they were constants and looking at this joint PDF as a function of θ, we denote it by $L(\theta | x_1, x_2, \cdots, x_n)$ and call it the *likelihood function*.

Definition 3.4 The estimate $\hat{\theta} = \hat{\theta}(x_1, x_2, \cdots, x_n)$ is called a *maximum likelihood estimate* (MLE) of θ if

$$L(\hat{\theta} | x_1, x_2, \cdots, x_n) = \max[L(\theta | x_1, x_2, \cdots, x_n); \theta \in \Theta]. \qquad (3.24)$$

$\hat{\theta}(X_1, X_2, \cdots, X_n)$ is called an MLE of θ.

Since the function $y = \log x$, $x > 0$ is strictly increasing, in order to maximize (with respect to θ) $L(\theta | x_1, x_2, \cdots, x_n)$ in the case that $\Theta \subseteq \mathbb{R}$, it suffices to maximize $\log L(\theta | x_1, x_2, \cdots, x_n)$. This is much more convenient to work with, as will become apparent from examples to be discussed below.

Example 3.3 Let $X_1, X_2, \cdots, X_n$ be i.i.d. r.v.'s from an exponential distribution with the unknown parameter λ:

$$f(\lambda; x) = \lambda e^{-\lambda x}.$$

Then

$$L(\lambda | x_1, x_2, \cdots, x_n) = \lambda e^{-\lambda x_1} \cdot \lambda e^{-\lambda x_2} \cdots \lambda e^{-\lambda x_n} = \lambda^n e^{-\lambda(x_1 + x_2 + \cdots + x_n)},$$

and hence

$$\log L(\lambda | x_1, x_2, \cdots, x_n) = n \log \lambda - \lambda(x_1 + x_2 + \cdots + x_n).$$

Therefore the likelihood equation

$$\frac{\partial}{\partial \lambda} \log L(\lambda | x_1, x_2, \cdots, x_n) = 0 \quad \text{become} \quad \frac{n}{\lambda} - (x_1 + x_2 + \cdots + x_n) = 0,$$

which gives

$$\lambda = \frac{n}{x_1 + x_2 + \cdots + x_n} = \frac{1}{\bar{x}}.$$

Example 3.4 Let $X_1, X_2, \cdots, X_n$ be i.i.d. r.v.'s from $N(\mu, \sigma^2)$ with parameter $\theta = (\mu, \sigma^2)'$. Then

$$L(\theta | x_1, x_2, \cdots, x_n) = \left(\frac{1}{\sqrt{2\pi\sigma^2}} \right)^n \exp\left[-\frac{1}{2\sigma^2} \sum_{i=1}^{n} (x_i - \mu)^2 \right],$$

so that

$$\log L(\theta | x_1, x_2, \cdots, x_n) = -n \log \sqrt{2\pi} - n \log \sqrt{\sigma^2} - \frac{1}{2\sigma^2} \sum_{i=1}^{n} (x_i - \mu)^2.$$

Differentiating with respect to μ and σ^2 and equating the resulting expressions to zero, we obtain

$$\frac{\partial}{\partial \mu} \log L(\theta|x_1, x_2, \cdots, x_n) = \frac{n}{\sigma^2}(\bar{x} - \mu) = 0,$$

$$\frac{\partial}{\partial \sigma^2} L(\theta|x_1, x_2, \cdots, x_n) = -\frac{n}{2\sigma^2} + \frac{1}{2\sigma^4}\sum_{i=1}^{n}(x_i - \mu)^2 = 0.$$

Then

$$\tilde{\mu} = \bar{x} \quad \text{and} \quad \tilde{\sigma}^2 = \frac{1}{n}\sum_{i=1}^{n}(x_i - \bar{x})^2$$

are the roots of these equations. It can be proven that $\tilde{\mu}$ and $\tilde{\sigma}^2$ actually maximize the likelihood function and therefore

$$\hat{\mu} = \bar{x} \quad \text{and} \quad \hat{\sigma}^2 = \frac{1}{n}\sum_{i=1}^{n}(x_i - \bar{x})^2$$

are the MLE's of μ and σ^2, respectively.

3.4.3 Histogram

A *histogram* is a graph of grouped (binned) data in which the number of values in each bin is represented by the area of a rectangular box.

A *frequency histogram* is a bar graph constructed in such a way that the area of each bar is proportional to the number of observations in the category that it represents. We read the actual number of observations per class from a frequency histogram.

A *relative frequency histogram* is a bar graph constructed in such a way that the area of each bar is proportional to the fraction of observations in the category that it represents.

According to the definition of the PDF, we can develop a relative frequency histogram by dividing the width of the bin to be an estimator about probability distribution.

Example 3.5 Appendix 2.A includes 24 seismic records with epicentral intensity $I_0 = VIII$. Table 3-2 shows their Richter magnitudes, where we renumber them according to their order in Appendix 2.A. For example, x_3 is 7th record in the appendix but it is the third one with epicentral intensity $I_0 = VIII$, and its Richter magnitude is 6.0.

Table 3-2. Magnitude records with epicentral intensity $I_0 = VIII$

i	1	2	3	4	5	6	7	8	9	10	11	12
x_i	6.4	6.4	6.0	6.0	7.0	6.25	5.75	5.8	6.5	5.8	6.0	6.5
i	13	14	15	16	17	18	19	20	21	22	23	24
x_i	6.3	6.0	6.0	6.0	6.8	6.4	6.5	6.6	6.5	6.0	5.5	6.0

Given an origin $x_0 = 5.4$, a bin width $h = 0.4$, for positive integers $m = 5$, we can construct the histogram on the intervals:

$$I_1 = [5.4, 5.8[, \; I_2 = [5.8, 6.2[, \; I_3 = [6.2, 6.6[, \; I_4 = [6.6, 7.0[, \; I_5 = [7.0, 7.4[$$

We now count the observations falling into each interval. The observations that fall into interval $I_1 = [5.4, 5.8[$ are $x_{23} = 5.5$ and $x_7 = 5.75$. There are two observations falling this interval. We do the same for each interval. The result of these counts are recorded in Table 3-3 in the column labeled frequency.

Table 3-3. Frequency distribution of the magnitude records

Interval	Midpoint	Frequency
[5.4, 5.8[	5.6	2
[5.8, 6.2[	6.0	10
[6.2, 6.6[	6.4	9
[6.6, 7.0[	6.8	2
[7.0, 7.4[	7.2	1

From this frequency table, we can construct the frequency histogram. The histogram, shown in Fig. 3.5, consists of five vertical bars each of width 0.4, each centered at its respective interval mark. The height of each bar is equal to the interval frequency, the number of observations per interval.

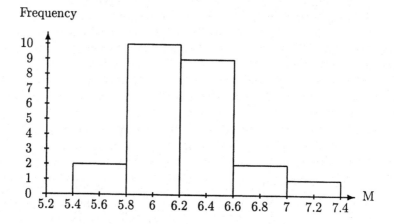

Fig. 3.5 Frequency histogram for a sample of observations on magnitude.

Dividing the number of observations per interval by the total number 24 of observations, we obtain a relative frequency histogram shown in Fig. 3.6. Note that the vertical axes is labeled 0 to 1.0.

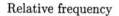

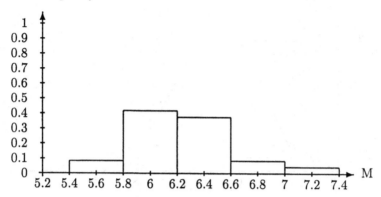

Fig. 3.6 Relative frequency histogram for a sample of observations on magnitude.

We use $u_1, u_2, \cdots, u_m$ to denote the midpoints of the intervals in the histogram. They are the standard points of the intervals. Dividing the relative frequency per interval by the bin width, we written the result by $\hat{p}(u_i)$, $i = 1, 2, \cdots, m$. Plotting $(u_i, \hat{p}(u_i))$, $i = 1, 2, \cdots, m$ and linking them, we obtain an estimate of probability distribution $p(x)$ of the population. For simplicity, $(u_i, \hat{p}(u_i))$, $i = 1, 2, \cdots, m$ and the curve produced from them are often considered to be same.

Definition 3.5 Let $X = \{x_1, x_2, \cdots, x_n\}$ be a given sample drawn from a population with PDF $p(x)$. Given an *origin* x_0 and a *bin width* h, we define the *bins* of the histogram to be the intervals $[x_0 + mh, x_0 + (m + 1)h[$ for positive and negative integers m. The intervals are chosen to be closed on the left and open on the right.

$$\hat{p}(x) = \frac{1}{nh}(\text{number of } x_i \text{ in the same bin as } x). \qquad (3.25)$$

is called a *histogram estimate* (HE) of $p(x)$.

From the data in Table 3-2, we can obtain a histogram estimate of $p(x)$ that an earthquake in magnitude x causes the damage measured by epicentral intensity $I_0 = VIII$. The estimate is shown in Fig. 3.7.

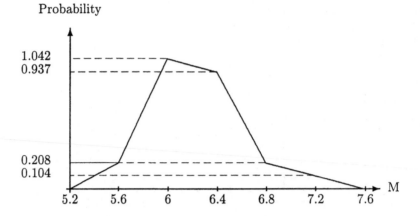

Fig. 3.7 Histogram estimate of the probability distribution to $I_0 = VIII$.

3.4.4 Kernel method

From the definition of a probability density, if the random variable X has density p, then

$$p(x) = \lim_{h \to 0} \frac{1}{2h} P(x - h < X < x + h).$$

For any given h, we can of course estimate $P(x - h < X < x + h)$ by the proportion of the sample falling in the interval $(x - h, x + h)$. Thus, for a chosen small number h, a natural estimator $\hat{p}$ of the density is given by

$$\hat{p}(x) = \frac{1}{2hn}[\text{number of } x_1, \cdots, x_n \text{ falling in } (x - h, x + h)], \qquad (3.26)$$

and is called the naive estimator[9] of p.

To express the estimator more clearly, define the weight function w by

$$w(x) = \begin{cases} \frac{1}{2}, & \text{if } |x| < 1, \\ 0, & \text{otherwise.} \end{cases} \qquad (3.27)$$

Then it is easy to see that the naive estimator can be written as

$$\hat{p}(x) = \frac{1}{nh} \sum_{i=1}^{n} w\left(\frac{x - x_i}{h}\right). \qquad (3.28)$$

It follows from (3.27) that the estimate is constructed by placing a 'box' of width $2h$ and height $(2nh)^{-1}$ on each observation and sum up the corresponding terms.

The naive estimator can be seen as an attempt to construct a histogram where every point is the centre of a sampling interval, thus freeing the histogram from a particular choice of bin positions. The choice of bin width is still necessary and is governed by the parameter h, which controls the amount by which the data are smoothened to produce the estimate.

Analogous to the definition of the naive estimator, the *kernel estimator*[8] with kernel K is defined by

$$\hat{p}(x) = \frac{1}{nh} \sum_{i=1}^{n} K\left(\frac{x - x_i}{h}\right), \qquad (3.29)$$

where h is the *window width* , also called the *smoothing parameter* or *bandwidth,* and the weight function w of the naive estimator is replaced by a *kernel function K* which satisfies the condition

$$\int_{-\infty}^{\infty} K(x)dx = 1.$$

Usually, K is a symmetric PDF, or the weight function w used in the definition of the naive estimator.

Just like the naive estimator which can be considered as a sum of 'boxes centred on the observations, the kernel estimate is a sum of 'bumps' placed at the observations. The kernel function K determines the shape of the bumps while the window width h determines their widths.

There is a theorem[8],[9],[11] which states that the optimal choice of the kernel K with $\int_{-\infty}^{\infty} x^2 K(x)dx = 1$ is

$$K_{opt}(x) = \begin{cases} \frac{3}{4\sqrt{5}}(1 - \frac{x^2}{5}), & \text{for } |x| \leq \sqrt{5}, \\ 0, & \text{otherwise;} \end{cases} \qquad (3.30)$$

(called *the Epanechnikov kernel*) and the ideal window width is

$$h_{opt} = 0.7687 \left\{ \int_{-\infty}^{\infty} p''(x)dx \right\}^{-1/5} n^{-1/5}. \qquad (3.31)$$

However, the problem of choosing h is still of crucial importance in density estimation. In fact, h_{opt} is invalid when the population density $p(x)$ is unknown. A very easy and natural approach is to use a standard family of distributions to assign a value to the term $\int_{-\infty}^{\infty} p''(x)dx$ in the expression (3.31) for the ideal window width. If the Gaussian kernel,

$$K_{Gauss}(x) = \frac{1}{\sqrt{2\pi}} \exp[-\frac{1}{2}x^2], \qquad (3.32)$$

is being used, then the window width obtained would be

$$h_{opt} = 1.06\sigma n^{-1/5}. \qquad (3.33)$$

There are some arguments in favour of using kernels which take negative as well as positive values although there is a small improvement in the order of magnitude of the best achievable mean integrated square error, from $n^{-4/5}$ to $n^{-8/9}$. The kernel estimate does suffer from a drawback when applied to small samples because there is insufficient information for choosing a better kernel function and the window width. Although the nearest neighbour method[9] and the adaptive kernel method have been developed to improve the kernel estimate, using the variable window width instead of h destroys the density property. In fact, we have $\int_{-\infty}^{\infty} \hat{p}(x)dx = \infty$ for all n.

Example 3.6 Using the Gaussian kernel to the data in Table 3-2, we obtain

$$
\begin{aligned}
h_{opt} &= 1.06\sigma n^{-1/5} \\
&= 1.06\sqrt{\tfrac{1}{n}\sum_{i=1}^{n}(x_i - \bar{x})^2}\, n^{-1/5} \\
&= 1.06\sqrt{\tfrac{1}{n}\sum_{i=1}^{24}(x_i - 6.208)^2}\, 24^{-1/5} \\
&= 1.06 \times 0.353 \times 0.530 \\
&= 0.198
\end{aligned}
$$

Then, the kernel estimate is

$$
\begin{aligned}
\hat{p}(x) &= \tfrac{1}{24\times 0.198}\sum_{i=1}^{24}\tfrac{1}{\sqrt{2\pi}}\exp[-\tfrac{1}{2}(\tfrac{x-x_i}{0.198})^2] \\
&= 0.084\sum_{i=1}^{24}\exp[-12.75(x - x_i)^2].
\end{aligned}
$$

which is shown in Fig. 3.8.

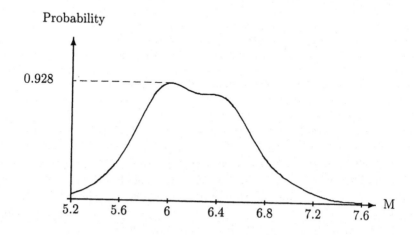

Fig. 3.8 Kernel estimate of the probability distribution to $I_0 = VIII$.

3.5 Monte Carlo Methods

Monte Carlo (MC for short) methods are statistical techniques for numerically approximating the solution of a mathematical problem by studying the distribution of some random variable, often generated by a computer. They are also called statistical simulation methods, where statistical simulation is defined in quite general terms to be any method that utilizes sequences of random numbers to perform the simulation. Monte Carlo methods have been used for centuries, but only in the past several decades has the technique gained the status of a full-fledged numerical method capable of addressing the most complex applications. (The name "Monte Carlo" was coined during the Manhattan Project of World War II, because of the similarity of statistical simulation to games of chance, and because the capital of Monaco was a center for gambling and similar pursuits.) MC is now used routinely in many diverse fields, from the simulation of complex physical phenomena to the mundane.

MC calculation of π is the simplest type of MC method to understand. We first draw a unit circle circumscribed by a square. The circle is shaded. Then, we could examine this problem in terms of the full circle and square, but it's easier to examine just one quadrant of the circle. If you are a very poor dart player, it is easy to imagine throwing darts randomly at the quadrant of the circle, and it should be apparent that of the total number of darts that hit within the square, the number of darts that hit the shaded part (circle quadrant) is proportional to the area of that part. In other words,

$$\frac{\text{darts hitting shaded area}}{\text{darts hitting inside square}} = \frac{\text{area of shaded area}}{\text{area of square}}$$

If you remember your geometry, it's easy to show that

$$\frac{\text{darts hitting shaded area}}{\text{darts hitting inside square}} = \frac{\frac{1}{4}\pi r^2}{r^2} = \frac{\pi}{4}$$

or

$$\pi = 4\frac{\text{darts hitting shaded area}}{\text{darts hitting inside square}}$$

If each dart thrown lands somewhere inside the square, the ratio of "hits" (in the shaded area) to "throws" will be one-fourth the value of π. If you actually do this experiment, you will soon realize that it takes a very large number of throws to get a decent value of π...well over 1,000. To make things easy on ourselves, we can have computers to generate random numbers.

Today, the expression "Monte Carlo method" is more general. Any method based on the use of random numbers and probability statistics to investigate problems can be called a MC method. In this book, we will use a MC method to judge which probability estimate is better. In contrast, a conventional numerical solution approach would start with some strict mathematical assumptions, discussing every situations and then solving a set of algebraic equations for the limit state. Relatively speaking, the simulation results from a MC method are rough but the conclusion based on these results usually is correct. Meanwhile, a theory result depend on the strict assumptions may be accurate but the corresponding conclusion to a real system might be wrong.

For now, we will assume that the behavior of a system can be described by PDF's. Once the PDF's are known, the MC simulation can proceed by random sampling from the PDF's.

3.5.1 Pseudo-random numbers

Real random numbers are non-deterministic. A generator needs to collect data from natural stochastic processes. A common generator is an electronic geiger counter which generates a pulse every time it detects a radioactive decay. The time between decays has a pure random component. To produce a usable pure random sequence, latent patterns must be removed. The known exponential distribution is easily removed mathematically; however, the mean time between decays in background radiation rises in the daytime and falls at night. There may also be other non-random patterns lying hidden in the data which could potentially be exploited by an analyst attempting to reduce his uncertainty for the next number in the sequence.

A pseudo-random number generator is a deterministic algorithm for producing a series of numbers that have many of the same properties as a series of purely random numbers. The sequences can mirror true random numbers in frequency distribution, clustering of runs, and the absence of readily discernible patterns. However, unlike a pure sequence, a pseudo-random sequence is 100% predictable when its algorithm and seed are known.

Computer-generated "random" numbers are more properly referred to as pseudo-random numbers, and pseudo-random sequences of such numbers. A variety of clever algorithms have been developed that generate sequences of numbers that pass every statistical test used to distinguish random sequences from those containing some pattern or internal order.

A deterministic computer program cannot generate true random numbers. For most purposes, pseudo-random numbers suffice. A good pseudo-random sequence often works fine for applications such as statistical sampling or simulation models. In these applications, the generator need be of only moderate sophistication so that its inherent patterns do not resonate with those occurring in nature.

In this book, we use pseudo-random numbers for our simulation models to show the benefit of the principle of information distribution. Where, pseudo-random numbers are called random numbers for short.

3.5.2 Uniform random numbers

In simulation field[2], the most commonly used generators are linear congruent generators proposed by Lehmer in 1948 (see [5]). They Compute the ith inter X_i in the random sequence from X_{I-1} by the recursion

$$X_i = (aX_{i-1} + c) \bmod m. \tag{3.34}$$

That parameters a, c, and m determine the statistical quality of the generator.

Symbol "mod" is one important arithmetical operator called the modulus or remainder operator. The mod operator returns the remainder of two numbers. For instance 10 mod 3 is 1 because 10 divided by 3 leaves a remainder of 1. Here are more examples:

$$1 \bmod 3 = 1$$

$$2 \bmod 3 = 2$$

$$3 \bmod 3 = 0$$

$$4 \bmod 3 = 1$$

$$5 \bmod 3 = 2$$

$$6 \bmod 3 = 0$$

This seemingly simple operator is actually a very powerful tool for mathematicians, especially those who specialize in Number Theory. Mod can be used just as you might use any other more common operator like + or -.

If c in (3.34) is zero, then the resulting generator is called a "multiplicative congruent" generator. The multiplicative generators with $m = 2^{31} - 1 = 2147483647$ are widely used. Looking at related measures of goodness such as one-step serial correlation and cycle length, we suggest Program 3-1 to generate uniform random numbers X(I), I=1,2,$\cdots$, N, based on seed number SEED (i.e., we let X_0 =SEED)

Program 3-1   Generator of Random Numbers Obeying
Uniform Distribution $U(0, 1)$

```
PROGRAM MAIN
INTEGER N,SEED
REAL X(10)
READ(*,*)N,SEED
IX=SEED
DO 10 I=1,N
K1=IX/60466
IX=35515*(IX-K1*60466)-K1*33657
IF(IX.LT.0)IX=IX+2147483647
K1=IX/102657
```

```
      IX=20919*(IX-K1*102657)-K1*1864
      IF(IX.LT.0)IX=IX+2147483647
      RANUN=FLOAT(IX)/2.147483647e9
      X(I)=RANUN
10    CONTINUE
      WRITE(*,20)(X(I),I=1,N)
20    FORMAT(1X,F12.4)
      STOP
      END
```

In which we employ

$$X_{i+1} = 742938285 * X_i \text{ MOD } 2147483647,$$

and

$$U_{i+1} = X_{i+1}/M, \quad M = 2147483647.$$

They are used in GPSS/H simulation language[3],[13].

In order to avoid overflow when we do integer calculation. Based on Number Theory we can get an equivalent algorithm.

$$742938285 = 35515 * 20919,$$

$$2147483647 = 35515 * 60466 + 33657,$$

$$2147483647 = 20919 * 102657 + 1864,$$

and

$$a * b * X \text{ MOD } c = a * (b * X \text{ MOD } c) \text{ MOD } c,$$

$$(X + Y) \text{ MOD } c = ((X \text{ MOD } c) + (Y \text{ MOD } c)) \text{ MOD } c.$$

Then we will get the algorithm which is given in the Fortran program above.

The uniform will be frequently used when we generate some other distributions. For that, we rewrite the program to be a function Program 3-2.

Program 3-2   Uniform Function Generating Random Numbers
              Obeying Uniform Distribution $U(0,1)$

```
      FUNCTION RANUN(IX)
      K1=IX/60466
      IX=35515*(IX-K1*60466)-K1*33657
      IF(IX.LT.0)IX=IX+2147483647
      K1=IX/102657
      IX=20919*(IX-K1*102657)-K1*1864
      IF(IX.LT.0)IX=IX+2147483647
      RANUN=FLOAT(IX)/2.147483647e9
      RETURN
      END
```

3.5.3 Normal random numbers

Box and Muller (1958, see[1]) show that one can generate two independent standard normal variables X and Y by setting

$$X = \cos(2\pi U_{i+1})\sqrt{-2\log(U_i)},$$
$$Y = \sin(2\pi U_{i+1})\sqrt{-2\log(U_i)},$$

where U_{i+1} and U_i are supposed uniform on $(0,1)$. This method is superficially akin to inversion. We know that there is no exact method using a single uniform random variate and inversion to generate a single normal variate. Being suitably bold we decide to try to generate two simultaneously. This, surprisingly, is possible. The method is not one-to-one, but rather two-to-two. We suggest Program 3-3 to generate normal random numbers $X(I)$, $I=1,2,\cdots$, N, based on seed number SEED and uniform function RANUN(SEED).

Program 3-3   Generator of Random Numbers Obeying
Normal Distribution $N(\text{MU},\text{SIGMA}^2)$

```
      PROGRAM MAIN
      INTEGER N,SEED
      REAL MU,SIGMA,X(1000),U1,U2,Y
      READ(*,*)MU,SIGMA,N,SEED
      DO 10 I=1,N
      U1=RANUN(SEED)
      U2=RANUN(SEED)
      Y=SQRT(-2.*ALOG(U2))*COS(6.283*U1)
      X(I)=MU+SIGMA*Y
10    CONTINUE
      WRITE(*,20)(X(I),I=1,N)
20    FORMAT(1X,F12.4)
      STOP
      END
```

The normal will be used when we generate lognormal distribution. For that, we rewrite the program to be a subprogram Program 3-4.

Program 3-4   Normal Subprogram Generating Random Numbers
Obeying Normal Distribution $N(\text{MU},\text{SIGMA}^2)$

```
      SUBROUTINE NORMAL(MU,SIGMA,SEED,Y)
      INTEGER SEED
      REAL MU,SIGMA,U1,U2,X,Y
      U1=RANUN(SEED)
      U2=RANUN(SEED)
      X=SQRT(-2.*ALOG(U2))*COS(6.283*U1)
      Y=MU+SIGMA*X
      RETURN
      END
```

3.5.4 Exponential random numbers

The exponential distribution with parameter λ has cdf

$$y = F(x) = 1 - e^{-\lambda x}, \quad x \geq 0. \tag{3.35}$$

Its inverse function is

$$x = F^{-1}(y) = -\frac{1}{\lambda} \log(1 - y).$$

Let U be uniform on (0,1) and set

$$X = -\frac{1}{\lambda} \log(1 - U).$$

According to Theorem 3.1, we know that X is exponential with parameter λ. Because $1 - U$ has the same distribution, we recommend using the simpler transformation

$$X = -\frac{1}{\lambda} \log(U). \tag{3.36}$$

We suggest Program 3-5 to generate exponential random numbers X(I), I=1,2,$\cdots$, N, with parameter LAMBDA, based on seed number SEED and uniform function RANUN($\cdot$).

Program 3-5   Generator of Random Numbers Obeying
               Exponential Distribution with parameter λ

```
      PROGRAM MAIN
      INTEGER N,SEED
      REAL LAMBDA,X(1000),U
      READ(*,*)LAMBDA,N,SEED
      DO 10 I=1,N
      U=RANUN(SEED)
      X(I)=-1.0/LAMBDA*ALOG(U)
10    CONTINUE
      WRITE(*,20)(X(I),I=1,N)
20    FORMAT(1X,F12.4)
      STOP
      END
```

3.5.5 Lognormal random numbers

If log$x$ is normal with mean μ and variance σ^2 then x is lognormal with mean $\exp[\mu + \sigma^2/2]$ and variance $[\exp(\sigma^2) - 1]\exp[2\mu + \sigma^2]$ (see subsection 3.3.4), we know that if Y is normal with mean μ and variance σ^2 then $X = \exp Y$ is lognormal with mean $\exp[\mu + \sigma^2/2]$ and variance $[\exp(\sigma^2) - 1]\exp[2\mu + \sigma^2]$. Therefore Program 3-6 can be employed to generate lognormal random numbers X(I), I=1,2,$\cdots$, N, with parameters MU and SIGMA, based on seed number SEED and normal subprogram NORMAL(MU,SIGMA,SEED,Y).

Program 3-6   Generator of Random Numbers Obeying
                Lognormal Distribution with parameters μ and σ

```
        PROGRAM MAIN
        INTEGER N,SEED
        REAL MU,SIGMA,X(1000),Y
        READ(*,*)MU,SIGMA,N,SEED
        DO 10 I=1,N
        CALL NORMAL(MU,SIGMA,SEED,Y)
        X(I)=EXP(Y)
10      CONTINUE
        WRITE(*,20)(X(I),I=1,N)
20      FORMAT(1X,F12.4)
        STOP
        END
```

References

1. Box, G. E. P. and Muller, M. E. (1985), A note on the generation of random normal deviates. Annals of Mathematical Statistics,**29**, pp.610-611
2. Bratley, P., Fox, B. L., and Schrage, L. E. (1987), A Guide to Simulation. Springer-Verlag, New York
3. Fishman, G. S., Moore, L. R. (1986), An exhaustive ananlysis of multiplicative congruential random number generators with modulus $2^{31} - 1$. SIAM Journ. Sci. Stat. Comp. Vol.7, pp. 24-45
4. Jizenman, A. (1991), Recent developments in nonparametric density estimation. J. of American Statistical Association, Vol. 86, No. 413, pp. 205-224
5. Lehmer, D. H. (1949), Mathematical methods in large-scale computing units. Proceedings of 2nd Sympos. on Large-Scale Digital Calculating Machinery, Cambridge, MA, pp.141-146
6. Mann, P. S. (1998), Introductory Statistics. John Wiley & Sons, New York
7. Ng, K. and Lippmann, R. (1991), A Comparative study of the practical characteristics of neural networks and conventional pattern classifiers. Advances in Neural Information Processing Systems, Vol. 3, pp. 970-976
8. Parzen, E. (1962), On estimation of a probability density function and mode. Annals of Mathematical Statistics,**33**, pp.1065-1076
9. Silverman, B. W. (1986), Density Estimation for Statistics and Data Analysis. Chapman and Hall, London
10. Smyth, P. (1994), Probability density estimation and local basis function networks. in: Hanson, S. J., Petsche, T., Kearns, M., and Rivest, R. (eds), Computational Learning Theory and Natural Learning Systems 2, MIT Press, Cambridge, MA, pp. 233-248
11. Wertz, W. (1978), Statistical Density Estimation: a Survey. Vandenhoeck & Ruprecht, Göttingen
12. White, H. (1990), Connectionist nonparametric regression: multi layer feedforward network can learn arbitrary mappings. Neural Networks, **3**, pp. 535-549
13. Zhang, Z., Zou, Z., and Feng, Y. (1995), Selection of random number generators for simulation. Journal of Xidian University, Vol.22, Sup. pp26-33

4. Information Distribution

This chapter presents the concept of information distribution. Through information distribution, we can change the crisp observations of a given sample into fuzzy sets. Hence, fuzzy sets are employed to describe the fuzzy transition information in a small sample. It is useful to improve the estimation of the probability distribution. Based on this estimation, we can construct fuzzy relationships, directly, without any assumptions. In detail, we discuss the method of 1-dimension linear-information-distribution. Computer simulation shows the work efficiency of the new method is about 23% higher than the histogram method for the estimation of a probability distribution. The chapter is organized as follows: in section 4.1, we introduce the concept of information distribution. In section 4.2, we give the mathematical definition of information distribution. Section 4.3 gives the method of 1-dimension linear-information-distribution. Section 4.4 demonstrates the benefit of information distribution for probability distribution estimation. In section 4.5, we construct a fuzzy relation matrixes with the method of information distribution. In section 4.6, we discuss approximate inference based on information distribution.

4.1 Introduction

Let us recall the main difference between the crisp-interval and fuzzy-interval information matrixes.

From Chapter 2 we know that, the information matrixes are constructed by using a sample $X = \{(x_1, y_1), (x_2, y_2), \cdots, (x_n, y_n)\}$ on a Cartesian space $U \times V$, where $U = \{u_1, u_2, \cdots, u_t\}$, $V = \{v_1, v_2, \cdots, v_l\}$. In general, we let $h_x \equiv u_{j+1} - u_j$, $j = 1, 2, \cdots, t - 1$ and $h_y \equiv v_{k+1} - u_k$, $k = 1, 2, \cdots, l - 1$.

Let $U_j = [u_j, u_{j+1}[, j = 1, 2, \cdots, t - 1$ and $V_k = [v_k, v_{k+1}[, k = 1, 2, \cdots, l - 1$. A small Cartesian space $U_j \times V_k$ is called a *box* in the Cartesian space $U \times V$. The length of a box is h_x and width h_y.

In the crisp model, the element E_{jk} of a crisp-interval information matrix $E = \{E_{jk}\}_{(t-1) \times (l-1)}$ is calculated by counting the observations falling into the box $U_j \times V_k$. Here, an observation fall to only one box. The model neglects the difference between the observations in same box, who may occupy different positions in the box. It leads to that a crisp-interval information matrix cannot illustrate all information.

It is interesting to note that, when $X \in \mathbb{R}$ and $U = V$, a box reduces to an interval and a crisp-interval information matrix reduces to a frequency histogram. In this case, formula (2.8) will reduce to

$$\chi_{U_j}(x_i) = \begin{cases} 1, & \text{if } x_i \in U_j, \\ 0, & \text{otherwise.} \end{cases} \tag{4.1}$$

and formula (2.9) becomes

$$E_j = \sum_{i=1}^{n} \chi_{U_j}(x_i). \tag{4.2}$$

In the fuzzy model, the element Q_{jk} of a fuzzy-interval information matrix $Q = \{Q_{jk}\}_{t \times l}$ is calculated by weighted counting the observations being near the node (u_j, v_k). The nearer the node, the larger the weight. The weight of observation (x_i, y_i) with respect to node (u_j, v_k) is $\mu_{jk}(x_i, y_i)$, calculated by

$$\mu_{jk}(x_i, y_i) = \begin{cases} (1 - \frac{|u_j - x_i|}{h_x})(1 - \frac{|v_k - y_i|}{h_y}), & \text{if } |u_j - x_i| \le h_x \text{ and } |v_k - y_i| \le h_y, \\ 0, & \text{otherwise.} \end{cases}$$

The weighted counting is executed by

$$Q_{jk} = \sum_{i=1}^{n} \mu_{jk}(x_i, y_i).$$

The main advantage of the fuzzy model lies in that it can distinguish the difference between the observations in same box. Therefore, a fuzzy-interval information matrix can illustrate more information.

In the fuzzy model, for 1-dimension X, formula (2.15) reduces to

$$\mu_{A_j}(x_i) = \begin{cases} 1 - \frac{|u_j - x_i|}{h_x}, & \text{if } |u_j - x_i| \le h_x, \\ 0, & \text{otherwise.} \end{cases} \tag{4.3}$$

and formula (2.16) becomes

$$Q_j = \sum_{i=1}^{n} \mu_{A_j}(x_i), \tag{4.4}$$

One easily finds that the distribution formula (4.3) divides x_i into two parts if x_i is not equal any u_j. If $u_j \le x_i \le u_{j+1}$, we say that x_i is shared by fuzzy intervals A_j and A_{j+1}. Because X must include some knowledge in which we are interested, we say that any observation x_i carries some information about the knowledge. Therefore, the information carried by x_i is shared by the two fuzzy intervals. So, (4.3) is also called the *information distribution formula*.

Using this formula, we can map any pair (x_i, u_j) to a value in $[0,1]$. If $x_i = u_j$, the mapping value $=1$. If $|x_i - u_j| > h_x$, the value $=0$. The nearer x_i is to u_j, the larger the value is. Total of the values in terms of x_i is 1.

Generalizing the model, we obtain the definition of information distribution.

4.2 Definition of Information Distribution

Let $X = \{x_1, x_2, \cdots, x_n\}$ be a sample observed from an experiment, and $U = \{u_1, u_2, \cdots, u_m\}$ be a discrete universe of X for monitoring. This experiment has a number (possibly infinite) *outcomes*. The *population* of the experiment is the collection of all possible outcomes. We assume that X is a *random sample* to guarantee that every element of the population has the same chance of being selected. We also assume that $x_i (i = 1, 2, \cdots, n)$ are i.i.d. r.v.'s.

For example, the 2000 Census Bureau Employees collected all ages of the citizens in a town. The age data forms a sample. We can choose one of $U = \{5, 10, \cdots, 120\}$ and $\mathbf{U} = \{[0, 5[, [5, 10[, \cdots, [115, 120]\}$ to monitor age. U is a discrete universe of age, but $\mathbf{U}$ is not. In this case, the population is the anticipated 275 million people across the United States.

Definition 4.1 A mapping from $X \times U$ to $[0,1]$

$$\begin{aligned} \mu: \quad & X \times U \quad \to [0,1] \\ & (x, u) \quad \mapsto \mu(x, u), \forall (x, u) \in X \times U \end{aligned}$$

is called an *information distribution* of X on U, if $\mu(x, u)$ has the following properties :

(1) $\forall x \in X$, if $\exists u \in U$, such that $x = u$, then $\mu(x, u) = 1$, i.e., μ is *reflexive*;
(2) For $x \in X$, $\forall u', u'' \in U$, if $||u' - x|| \leq ||u'' - x||$ then $\mu(x, u') \geq \mu(x, u'')$, i.e., μ is *decreasing* when $|| x - u ||$ is increasing;
(3) $\sum_{j=1}^{m} \mu(x_i, u_j) = 1$, $i = 1, 2, \cdots, n$, i.e., information *conserved*.

For example, if John is 23 years old, he must appear to be between 20 years old and 25 years old. A mapping to assign the possibilities of being 20 and 25 is an information distribution.

Example 4.1 A surveyor wants to know an outline of the age for a special class. However, when he arrives at the campus he only meets three students, John, Mary and Smith who are 16, 20 and 23 years old, respectively. Then, the given sample is $X = \{16, 20, 23\}$. The outline he needs is a frequency distribution on the three fuzzy concepts defined by fuzzy numbers,

$$A_1 \equiv \text{ about 15 years old, } \mu_{A_1} = I_{(15,5)},$$

$$A_2 \equiv \text{ about 20 years old, } \mu_{A_2} = I_{(20,5)},$$

$$A_3 \equiv \text{ about 25 years old, } \mu_{A_3} = I_{(25,5)}.$$

Let the center points of the three fuzzy sets be the monitor points In which he is interested. Then, he obtain a discrete universe of age $U = \{15, 20, 25\}$. By formula (4.3) with $h_x = 5$, this maps $X \times U$ into [0,1] as shown in Table 4-1.

Table 4-1   An information distribution of an age sample on the center points of fuzzy sets in terms of age

$\mu(x_i, u_j)$	$u_1 = 15$	$u_2 = 20$	$u_3 = 25$
$x_1 = 16$	0.8	0.2	0
$x_2 = 20$	0	1	0
$x_3 = 23$	0	0.4	0.6
Σ	0.8	1.6	0.6
%	27	53	20

Table 4-1 shows the age-information structure of X on U. Naturally, his reference is that, in the class, 27% students are about 15 years old, 53% are about 20, and 20% are about 25.

In Definition 4.1, $u_j, j = 1, 2, \cdots, m$ are called the *controlling points*. μ is called a *distribution function* of X on U. We say that observation x_i gives information, gain at $q_{ij} = \mu(x_i, u_j)$, to controlling point u_j. q_{ij} is called *distributed information* on u_j from x_i. U is also called the *framework space*. An information distribution illustrates a sample on a chosen framework space.

Let $Q_j = \sum_{i=1}^{n} q_{ij}, j = 1, 2, \cdots, m$. We say that sample X provides information in the total gain Q_j to controlling point u_j. Q_j is also called the *total distributed information* to controlling point u_j. $Q = (Q_1, Q_2, \cdots, Q_m)$ is called the *primary information distribution* of X on U.

The simplest distribution function is the 1-dimension linear-information-distribution.

4.3 1-Dimension Linear-Information-Distribution

Definition 4.2 Let $X = \{x_1, x_2, \cdots, x_n\}$ be a 1-dimension random sample, and $U = \{u_1, u_2, \cdots, u_m\}$ be the chosen framework space with $u_j - u_{j-1} \equiv \Delta$, $j = 2, 3, \cdots, m$. For any $x \in X$, and $u \in U$, the following formula is called *1-dimension linear-information-distribution*:

$$\mu(x,u) = \begin{cases} 1 - \frac{|x-u|}{\Delta}, & \text{for } |x-u| \le \Delta, \\ 0, & \text{otherwise,} \end{cases} \qquad (4.5)$$

where Δ is called *step length*. In short, this μ is called *linear distribution*.

Theorem 4.1 The linear distribution μ is reflexive, decreasing and conserved.
Proof:
(1) $\forall x \in X$, if $\exists u \in U$, such that $x = u$, then $|x - u| = 0 < \Delta$, therefore $\mu(x,u) = 1 - |x - u|/\Delta = 1 - 0 = 1$, i.e., μ is reflexive;
(2) For $x \in X$, $\forall u', u'' \in U$
    (i) If $\Delta \le |u' - x| \le |u'' - x|$, then $\mu(x,u') = \mu(x,u'') = 0$;
    (ii) If $|u' - x| \le \Delta \le |u'' - x|$, then $\mu(x,u') \ge 0, \mu(x,u'') = 0$, i.e., $\mu(x,u') \ge \mu(x,u'')$
    (iii) If $|u' - x| \le |u'' - x| \le \Delta$, then $\mu(x,u') = 1 - |x - u'|/\Delta \ge 1 - |x - u''|/\Delta = \mu(x,u'')$, hence $\mu(x,u') \ge \mu(x,u'')$.
    Therefor, μ is decreasing when $\| x - u \|$ is increasing;
(3) For an arbitrary element x_i of X, there is no loss in generality when we suppose $u_j \le x_i \le u_{j+1}$. In the case, we know that $\forall u_k \in U$, if $k \ne j, j+1$, then $\mu(x_i, u_k) = 0$. Therefore

$$\sum_{t=1}^{m} \mu(x_i, u_t)$$

$$= \sum_{k \ne j, j+1} \mu(x_i, u_k) + \mu(x_i, u_j) + \mu(x_i, u_{j+1})$$

$$= 0 + 1 - |x_i - u_j|/\Delta + 1 - |x_i - u_{j+1}|/\Delta$$

$$= 1 - (x_i - u_j)/\Delta + 1 - (u_{j+1} - x_i)/\Delta$$

$$= 2 - \frac{x_i - u_j + u_{j+1} - x_i}{\Delta}$$

$$= 2 - \frac{u_{j+1} - u_j}{\Delta}$$

$$= 2 - 1 \quad (\text{because } u_{j+1} - u_j = \Delta)$$

$$= 1,$$

i.e., the information is conserved.

$\qquad\qquad\qquad\qquad\qquad\qquad\qquad\qquad\qquad\qquad\qquad\qquad\qquad\quad \square$

In fact, with μ, we have changed an observation x_i into a fuzzy set on U. For x_i, the fuzzy set can be written as μ_{x_i}. If $u_j \le x_i \le u_{j+1}$, The fuzzy set is the following:

$$\underset{\sim}{x_i} = \frac{0}{u_1} + \frac{0}{u_2} + \cdots + \frac{q_{ij}}{u_j} + \frac{q_{ij+1}}{u_{j+1}} + \cdots + \frac{0}{u_m}, \qquad (4.6)$$

where $q_{ij} = \mu(x_i, u_j)$ and $q_{ij+1} = \mu(x_i, u_{j+1})$.

An information distribution acts as a singleton fuzzifier[9] mapping the crisp measurement x_i into a fuzzy set $\underset{\sim}{x_i}$ whose membership function μ_{x_i} is described by a distribution function.

A fuzzy set on $\mathbb{R}$ is called a fuzzy singleton if it takes the value 0 for all points x in $\mathbb{R}$ except one. The point in which a fuzzy singleton takes the nonzero value is called the support of the fuzzy singleton.

Using the concept of a fuzzy singleton, we know that x_i is a special fuzzy set on $\mathbb{R}$. If $\exists\, u \in U$, such that $x_i = u$, we can change x_i into a general fuzzy set on U by the method of information distribution.

Using the linear distribution, we can obtain an estimator of the probability about the population with respect to X, which is a new histogram.

For doing that, firstly, let us return to analyze the classical frequency histogram presented in subsection 3.4.3.

We know that, if the intervals in Fig. 3.5 are combined into a large one $[5.4, 7.4[$, the resulting histogram becomes one bar, which cannot display any gradient. In other words, if the intervals in a histogram are too large, we can learn nothing from the given sample. In fact, the larger the intervals are, the rougher the estimation of probability distribution is. Oppositely, if we take smaller intervals, what will happen?

In Fig. 3.5, the domain is divided into five intervals. Now, we divide it into six intervals

$$I_1 = [5.40, 5.73[, \quad I_2 = [5.73, 6.07[, \quad I_3 = [6.07, 6.40[,$$

$$I_4 = [6.40, 6.73[, \quad I_5 = [6.73, 7.07[, \quad I_6 = [7.07, 7.40[,$$

and obtain a corresponding histogram shown in Fig. 4.1.

It is waving with two crests (at I_2, I_4) and one trough (at I_3), which also cannot show any statistical law.

This follows from the fact that, in the model of classical histogram, all observations falling into a same interval are considered to play a same role. It neglects the difference between them. In fact, the records may occupy different positions in the interval. For example, both $x_1 = 6.4$ and $x_9 = 6.5$ belong to interval $I_4 = [6.40, 6.73[$, but their positions in I_4 are different. The former is located at the boundary, and the latter is near the center of the interval. Neglecting the position difference implies that we throw away some information.

If the sample size is large, by the central limit theorem, the neglected information is insignificant. However, for a small sample, we have to pay attention to each observation. Any observation may be near the center or boundary of an interval. A little random disturbance in the experiment may make the observations near the boundaries move other intervals. The information showing position is called the *transition information*. Due to fuzziness of a small sample, we also called it the *fuzzy transition information*.

Frequency

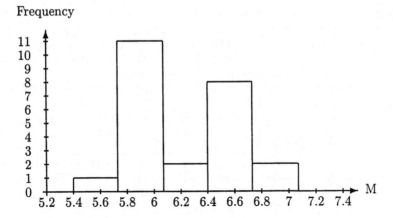

Fig. 4.1 A classical frequency histogram with smaller intervals.

It is easy to pick up the information, if we change observations as single-tons into general fuzzy sets. The mechanism of the performance is to use fuzzy bounds instead of crisp bounds of the intervals in the classical histogram. The new histogram is called the soft frequency histogram.

Definition 4.3 Suppose a classical frequency histogram be constructed on m intervals $I_1, I_2, \cdots, I_m$ with width h. Let u_j be the center point of interval I_j. We use $U = \{u_1, u_2, \cdots, u_m\}$ to be the framework space with step length $\Delta = h$. Let

$\tilde{H}(x)$ = total provided information to controlling point in same interval as x.
$$(4.7)$$

which is called a *soft frequency histogram* of X on $\mathbb{R}$.

When we have got Q of X on U, the soft histogram, in fact, is

$$\forall x \in I_j, \quad \tilde{H}(x) = Q_j, \tag{4.8}$$

Example 4.2 Let us reconstruct the classical histogram in Fig. 4.1 into a soft histogram. X is given in Table 3-2, $n = 24$, $m = 6$, $\Delta = h = (7.4 - 5.4)/6 = 0.3\dot{3}$, and the corresponding set of controlling points is

$$\begin{aligned} U &= \{u_1, u_2, u_3, u_4, u_5, u_6\} \\ &= \{5.57, 5.90, 6.24, 6.57, 6.90, 7.24\}. \end{aligned}$$

Using the linear distribution formula in (4.5), we obtain all distributed information q_{ij} shown in Table 4-2. For instance, for $x_3 = 6$, because

$$|x_3 - u_2| = |6 - 5.9| = 0.1 < \Delta, \text{ and } |x_3 - u_3| = |6 - 6.24| = 0.24 < \Delta,$$

and

$$|x_3 - u_1|, |x_3 - u_4|, |x_3 - u_5|, |x_3 - u_6| > \Delta,$$

we have

$$q_{32} = \mu(x_3, u_2) = 1 - \frac{|x_3 - u_2|}{\Delta} = 1 - \frac{|6 - 5.9|}{0.33} = 1 - \frac{0.1}{0.33} = 0.697,$$

$$q_{33} = \mu(x_3, u_3) = 1 - \frac{|x_3 - u_3|}{\Delta} = 1 - \frac{|6 - 6.24|}{0.33} = 1 - \frac{0.24}{0.33} = 0.273,$$

$$q_{3j} = \mu(x_3, u_j) = 0, j = 1, 4, 5, 6.$$

Table 4-2 Distributed information on u_j from x_i

q_{ij}	u_1 5.57	u_2 5.90	u_3 6.24	u_4 6.57	u_5 6.90	u_6 7.24
$x_1 = 6.4$	0	0	0.515	0.485	0	0
$x_2 = 6.4$	0	0	0.515	0.485	0	0
$x_3 = 6.0$	0	0.697	0.273	0	0	0
$x_4 = 6.0$	0	0.697	0.273	0	0	0
$x_5 = 7.0$	0	0	0	0	0.697	0.273
$x_6 = 6.25$	0	0	0.970	0.030	0	0
$x_7 = 5.75$	0.455	0.545	0	0	0	0
$x_8 = 5.8$	0.303	0.697	0	0	0	0
$x_9 = 6.5$	0	0	0.212	0.788	0	0
$x_{10} = 5.8$	0.303	0.697	0	0	0	0
$x_{11} = 6.0$	0	0.697	0.273	0	0	0
$x_{12} = 6.5$	0	0	0.212	0.788	0	0
$x_{13} = 6.3$	0	0	0.818	0.182	0	0
$x_{14} = 6.0$	0	0.697	0.273	0	0	0
$x_{15} = 6.0$	0	0.697	0.273	0	0	0
$x_{16} = 6.0$	0	0.697	0.273	0	0	0
$x_{17} = 6.8$	0	0	0	0.303	0.697	0
$x_{18} = 6.4$	0	0	0.515	0.485	0	0
$x_{19} = 6.5$	0	0	0.212	0.788	0	0
$x_{20} = 6.6$	0	0	0	0.909	0.091	0
$x_{21} = 6.5$	0	0	0.212	0.788	0	0
$x_{22} = 6.0$	0	0.697	0.273	0	0	0
$x_{23} = 5.5$	0.788	0	0	0	0	0
$x_{24} = 6.0$	0	0.697	0.273	0	0	0
$\sum_{i=1}^{24} q_{ij}$	1.848	7.515	6.364	6.030	1.485	0.273

Then, the primary information distribution of X on U is

$$\begin{aligned} Q &= (Q_1, Q_2, Q_3, Q_4, Q_5, Q_6) \\ &= (1.848, 7.515, 6.364, 6.030, 1.485, 0.273). \end{aligned}$$

By formula (4.8), we obtain a soft frequency histogram shown in Fig. 4.2.

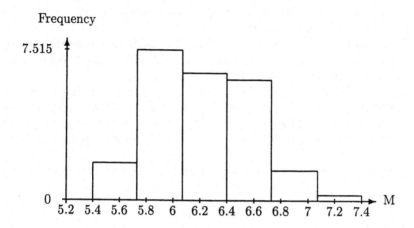

Fig. 4.2 A soft frequency histogram for a sample of observations on magnitude.

Obviously, the new histogram is more regular than the classical one because the bars show some statistical law. It will be useful to estimate a probability distribution with the following expression.

Definition 4.4 Let $X = \{x_1, x_2, \cdots, x_n\}$ be a given sample drawn from a population with PDF $p(x)$. Given m intervals $I_1, I_2, \cdots, I_m$ with width Δ. Let u_j of the center point of I_j be a controlling point. If the distributed information is calculated by linear distribution,

$$\tilde{p}(x) = \frac{1}{n\Delta}(\text{total information to controlling point in same interval as } x).$$
(4.9)

is called a *soft histogram estimate* (SHE) of $p(x)$.

With computer simulation experiments, we can prove that SHE $\tilde{p}(x)$ is better than HE $\hat{p}(x)$ (see Definition 3.5 in subsection 3.4.3).

4.4 Demonstration of Benefit for Probability Estimation

In this section, we use computer simulation to prove that SHE is better than HE when a given sample X is small.

4.4.1 Model description

Consider the following case: In an experiment group there are three researchers:

(1) A computer scientist who draws a sample X consisting of n random numbers $x_i, i = 1, 2, \cdots, n$. He knows that the sample is drawn from a population with density $p(x), x \in \mathbb{R}$;

(2) A statistician who is good at extracting statistical laws through the histogram method. He does not know where the given sample comes from. Studying the sample, he obtain $\hat{p}(x)$ to estimate $p(x)$;

(3) A fuzzy engineer who is interested in mining fuzzy information carried by a small sample and good at constructing a soft histogram by the method of information distribution. He also does not know where the given sample comes from. Analyzing the same sample, the engineer gives $\tilde{p}(x)$ to estimate $p(x)$.

The computer scientist might use the Kullback-Leibler divergence[4] $\hat{\varepsilon}$ and $\tilde{\varepsilon}$:

$$\hat{\varepsilon} = \int p(x) \log \frac{\hat{p}(x)}{p(x)} dx,$$

$$\tilde{\varepsilon} = \int p(x) \log \frac{\tilde{p}(x)}{p(x)} dx,$$

as errors, to judge which method is better. The smaller the error is, the better the method is.

In the case that the given sample is small, sometimes, there exists x_t in the domain of $p(x)$ such that $\hat{p}(x_t) = 0$ or $\tilde{p}(x_t) = 0$. Therefore, Kullback-Leibler divergence cannot be directly employed to judge the methods. In this case, we suggest the *non-log-divergence* defined by (4.10), as errors, to make the comparison.

$$\hat{\rho} = \int p(x)|\hat{p}(x) - p(x)|dx, \tag{4.10a}$$

$$\tilde{\rho} = \int p(x)|\tilde{p}(x) - p(x)|dx. \tag{4.10b}$$

For $x \in I_j$, we know that $\hat{p}(x)$ and $\tilde{p}(x)$ are constant. Hence, the non-log-divergence can be simplified to

$$\hat{\rho} = \frac{1}{m} \sum_{j=1}^{m} p(u_j)|\hat{p}(u_j) - p(u_j)|, \tag{4.11a}$$

$$\tilde{\rho} = \frac{1}{m} \sum_{j=1}^{m} p(u_j)|\tilde{p}(u_j) - p(u_j)|. \tag{4.11b}$$

We use $\hat{\rho}$ and $\tilde{\rho}$ to denote the errors with respect to HE and SHE, respectively. If $\tilde{\rho} \leq \hat{\rho}$, the computer scientist concludes that $\tilde{p}(x)$ is better than $\hat{p}(x)$ to approximate true distribution $p(x)$, and vice versa. In this section, we will calculate the errors and obtain the answer.

Random numbers $x_i, i = 1, 2, \cdots, n$, are produced from a computer program with a seed number selected randomly. For that, firstly, based on the Monte Carlo method, the computer scientist produces n random numbers

z_i, $i = 1, 2, \cdots, n$, belonging to uniform distribution $U(0,1)$. Program 3-1 in subsection 3.5.2 can be used for doing that.

Then, employing some formula in mathematical statistics, the computer scientist can obtain random numbers x_i, $i = 1, 2, \cdots, n$.

A simulation experiment includes producing $X = \{x_1, x_2, \cdots, x_n\}$, calculating $\hat{p}(x)$, $\tilde{p}(x)$, and comparing $\tilde{\rho}$ and $\hat{\rho}$. Obviously, there is randomness in the result from a simulation experiment, because the random numbers in an experiment are pseudo-random sequence which is 100% predictable when its algorithm and seed are known.

To reduce the randomness of the simulation experiment, we must do many simulations with different seed numbers given randomly. The average divergence of estimators is a better index to show the quality of a method. Suppose we have done N simulations, and $\rho(i)$ is the result from the i-th simulation, the average divergence is then defined by

$$\rho = \frac{1}{N} \sum_{i=1}^{N} \rho(i). \tag{4.12}$$

There are many kinds of function curves in terms of the probability distributions. However they have performance between two shape functions: the normal distribution and exponential distribution. The former is a symmetry curve; the latter is a monotone decreasing curve. In other words, if $\tilde{\rho} \leq \hat{\rho}$ holds for both of the normal distribution and exponential distribution, then the computer scientist will say that the method of information distribution is better than the histogram method. Therefore, it is enough to simulate the two distributions. As a transition curve, in this section, the lognormal distribution is also selected to be studied in our simulation experiment.

4.4.2 Normal experiment

Let

$$p_1(x) = \frac{1}{0.372\sqrt{2\pi}} \exp[-\frac{(x - 6.86)^2}{2 \times 0.372^2}], -\infty < x < \infty, \tag{4.13}$$

x is a random variable obeying normal distribution $N(6.86, 0.372^2)$.

Running Program 3-3 in subsection 3.5.3 with MU=6.86, SIGMA=0.327, N=19, SEED=6459, we obtain 19 random number:

$$\begin{aligned} X = \ & \{x_1, x_2, \cdots, x_{19}\} \\ = \ & \{6.74, 6.52, 7.04, 6.55, 7.57, 7.11, 7.29, 6.73, 6.89, 6.67, \qquad (4.14) \\ & 6.49, 7.27, 6.78, 6.15, 6.80, 6.94, 6.55, 6.98, 7.20\}. \end{aligned}$$

Given an origin $x_0 = 5.74$, a bin width $h = 0.2$, for positive integer $m = 11$, we obtain 11 intervals for making a classical frequency histogram

$[5.74, 5.94[, [5.94, 6.14[, [6.14, 6.34[, [6.34, 6.54[, [6.54, 6.74[, [6.74, 6.94[,$
$[6.94, 7.14[, [7.14, 7.34[, [7.34, 7.54[, [7.54, 7.74[, [7.74, 7.94[.$

We use their center points to construct a framework space U consists of 11 controlling points,

$$
\begin{aligned}
U &= \{u_1, u_2, \cdots, u_{11}\} \\
&= \{5.84, 6.04, 6.24, 6.44, 6.64, 6.84, 7.04, 7.24, 7.44, 7.64, 7.84\}.
\end{aligned}
$$

Its step length is $\Delta = h = 0.2$.

Employing formula (3.25) with $h = 0.2$ and X in (4.14) to estimate density $p(x)$, we obtain

$$
\begin{aligned}
\widehat{P}_1 &= \{\hat{p}(u_1), \hat{p}(u_2), \cdots, \hat{p}(u_{11})\} \\
&= \{0, 0, 0.26, 0.53, 1.05, 1.32, 0.79, 0.79, 0, 0.26, 0\}.
\end{aligned}
$$

Because the true probability distribution calculated by formula (4.13), on discrete universe of discourse U, is

$$
\begin{aligned}
P_1 &= \{p(u_1), p(u_2), \cdots, p(u_{11})\} \\
&= \{0.02, 0.09, 0.27, 0.57, 0.90, 1.07, 0.95, 0.64, 0.32, 0.12, 0.03\}
\end{aligned}
$$

Therefore, we obtained the non-log-divergence $\hat{\rho} = 0.073$.

Now, employing formula (4.9) with $\Delta = 0.2$ and the same X, we obtain

$$
\begin{aligned}
\widetilde{P}_1 &= \{\tilde{p}(u_1), \tilde{p}(u_2), \cdots, \tilde{p}(u_{11})\} \\
&= \{0, 0.12, 0.14, 0.59, 1.09, 1.10, 0.86, 0.73, 0.20, 0.17, 0\},
\end{aligned}
$$

and its non-log-divergence is $\tilde{\rho} = 0.041$.

That is:

Divergence	Classical histogram	Soft histogram
ρ	0.073	0.041

Fig. 4.3 shows probability density estimators by the classical histogram

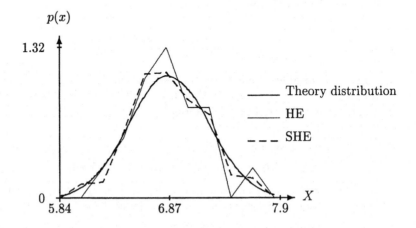

Fig. 4.3   Probability density estimators by HE and SHE
with respect to a normal distribution.

and the soft histogram with respect to a normal distribution. Table 4-3 shows the result of the simulation experiment with $N = 90$ and sample size $n = 10, 12, \cdots, 30$. We use random seed number for different experiment. The table also shows the relative error between HE and SHE defined as:

$$e = \frac{\hat{\rho} - \tilde{\rho}}{\hat{\rho}} \tag{4.15}$$

Table 4-3 Average non-log-divergence ρ and the relative error e for $p_1(x)$

n	10	12	14	16	18	20	22	24	26	28	30
$\hat{\rho}$	0.195	0.189	0.172	0.154	0.150	0.144	0.135	0.126	0.119	0.113	0.112
$\tilde{\rho}$	0.157	0.148	0.140	0.128	0.122	0.114	0.109	0.103	0.097	0.091	0.087
e	0.195	0.217	0.186	0.169	0.187	0.208	0.193	0.183	0.185	0.195	0.223

From Table 4-3 we know that the:

(1) method of information distribution is better than histogram method to estimate a normal distribution, because, for any n, $\tilde{\rho} < \hat{\rho}$.

(2) method of information distribution can improve a histogram estimator to reduce the error about 19.5% because the average relative error in the experiment is 0.195.

(3) method of information distribution can raise the work efficiency, i.e., a small sample acts as a larger one for estimating.

We use ρ_n to denote the average non-log-divergence with n observations. If $\hat{\rho}_m \approx \tilde{\rho}_n$ and $t = \frac{m-n}{m}$, we say that the *work efficiency* of the method for $\tilde{\rho}$ is t higher than one for $\hat{\rho}$ is.

From Table 4-3 we obtain

$$\hat{\rho}_{18} = 0.150 \approx 0.148 = \tilde{\rho}_{12}, \quad t_1 = \frac{18 - 12}{18} = 0.33,$$

$$\hat{\rho}_{24} = 0.126 \approx 0.128 = \tilde{\rho}_{16}, \quad t_2 = \frac{24 - 16}{24} = 0.33,$$

$$\hat{\rho}_{28} = 0.113 \approx 0.114 = \tilde{\rho}_{20}, \quad t_3 = \frac{28 - 20}{28} = 0.29,$$

$$T_1 = \frac{1}{3} \sum_{i=1}^{3} t_i = \frac{0.33 + 0.33 + 0.29}{3} = 0.32.$$

Therefore, the work efficiency of the method of information distribution is 32% higher than histogram method. In other words, using a small sample, less 32% size than one used in the histogram method, we can obtain a same accurate result by the method of information distribution. For example, if a statistician needs 30 observations to estimate a normal distribution by HE, then a fuzzy engineer only needs 20 observations (because 30-30×32% ≈ 20) to do it by SHE.

For other parameters in a normal distribution, the results are similar. To verify it, let us simulate the standard normal distribution $N(0, 1)$ and one with standard deviation $\sigma = 0.2$. The simulation results are shown in Table 4-4 and Table 4-5, respectively.

Table 4-4 Average non-log-divergence ρ and the relative error e for $N(0,1)$

n	10	12	14	16	18	20	22	24	26	28	30
$\hat{\rho}$	0.028	0.026	0.025	0.022	0.021	0.020	0.019	0.018	0.017	0.016	0.015
$\tilde{\rho}$	0.021	0.020	0.019	0.017	0.016	0.015	0.015	0.014	0.013	0.012	0.012
e	0.239	0.237	0.238	0.207	0.216	0.226	0.226	0.222	0.223	0.226	0.226

Table 4-5 Average non-log-divergence ρ and the relative error e for $N(0, 0.2^2)$

n	10	12	14	16	18	20	22	24	26	28	30
$\hat{\rho}$	0.688	0.671	0.630	0.561	0.537	0.515	0.480	0.458	0.430	0.407	0.386
$\tilde{\rho}$	0.525	0.503	0.471	0.433	0.411	0.389	0.365	0.347	0.326	0.306	0.290
e	0.237	0.250	0.253	0.229	0.236	0.246	0.240	0.241	0.241	0.250	0.248

4.4.3 Exponential experiment

Let

$$p_2(x) = 1.5e^{-1.5x}, \qquad x \geq 0. \tag{4.16}$$

Running Program 3-5 in subsection 3.5.4 with LAMBDA=1.5, N=19, SEED= 876905, we obtain 19 random number:

$$
\begin{aligned}
X &= \{x_1, x_2, \cdots, x_{19}\} \\
&= \{0.03, 0.21, 0.07, 0.05, 0.19, 0.78, 1.27, 0.12, 0.1, 1.38, \\
&\qquad 0.32, 0.67, 1.68, 1.94, 1.05, 0.52, 0.28, 0.16, 0.08\}.
\end{aligned} \tag{4.17}
$$

Given an origin $x_0 = 0$, a bin width $h = 0.18$, for positive integer $m = 11$, we obtain 11 intervals for making a classical frequency histogram

$$[0.00, 0.18[, [0.18, 0.36[, [0.36, 0.54[, [0.54, 0.72[, [0.72, 0.90[, [0.90, 1.08[,$$
$$[1.08, 1.26[, [1.26, 1.44[, [1.44, 1.62[, [1.62, 1.80[, [1.80, 1.98[.$$

We use their center points to construct a framework space U consists of 11 controlling points,

$$
\begin{aligned}
U &= \{u_1, u_2, \cdots, u_{11}\} \\
&= \{0.09, 0.27, 0.45, 0.62, 0.81, 0.99, 1.17, 1.35, 1.53, 1.71, 1.89\}.
\end{aligned}
$$

Its step length is $\Delta = h = 0.18$.

Employing formula (3.25) with $h = 0.18$ and X in (4.17) to estimate density $p(x)$, we obtain

$$
\begin{aligned}
\widehat{P}_2 &= \{\hat{p}(u_1), \hat{p}(u_2), \cdots, \hat{p}(u_{11})\} \\
&= \{2.047, 1.170, 0.292, 0.292, 0.292, 0.292, 0.000, 0.585, 0.000, 0.292, 0.292\}.
\end{aligned}
$$

Because the true probability distribution calculated by formula (4.16), on discrete universe of discourse U, is

$$
\begin{aligned}
P_2 &= \{p(u_1), p(u_2), \cdots, p(u_{11})\} \\
&= \{1.311, 1.000, 0.764, 0.592, 0.445, 0.340, 0.259, 0.198, 0.151, 0.115, 0.088\}
\end{aligned}
$$

Therefore, we obtained the non-log-divergence $\hat{\rho} = 0.178$.

Now, employing formula (4.9) with $\Delta = 0.18$ and the same X, we obtain

$$\tilde{P}_2 = \{\tilde{p}(u_1), \tilde{p}(u_2), \cdots, \tilde{p}(u_{11})\}$$
$$= \{1.995, 1.084, 0.292, 0.396, 0.327, 0.206, 0.241, 0.430, 0.103, 0.258, 0.224\},$$

and its non-log-divergence is $\tilde{\rho} = 0.149$. Fig. 4.4 shows the estimators.

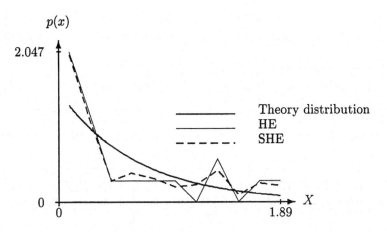

Fig. 4.4   Probability density estimators by HE and SHE
with respect to an exponential distribution.

Running 90 simulation experiments for $n = 10, 12, \cdots, 30$, we obtain Table 4-6 which shows average non-log-divergence ρ and the relative error e related to exponential distribution $p_2(x)$.

Table 4-6 Average non-log-divergence ρ and the relative error e for $p_2(x)$

n	10	12	14	16	18	20	22	24	26	28	30
$\hat{\rho}$	0.334	0.257	0.211	0.161	0.151	0.141	0.129	0.119	0.100	0.094	0.083
$\tilde{\rho}$	0.273	0.213	0.174	0.138	0.127	0.117	0.106	0.098	0.084	0.079	0.072
e	0.184	0.171	0.176	0.140	0.158	0.169	0.173	0.178	0.162	0.155	0.128

It proves that SHE is also better than HE with respect to an exponential distribution, although the reduction of the error is only 16.3%.

From Table 4-6 we obtain

$$\hat{\rho}_{20} = 0.141 \approx 0.138 = \tilde{\rho}_{16}, \quad t_1 = \frac{20 - 16}{20} = 0.2,$$

$$\hat{\rho}_{22} = 0.129 \approx 0.127 = \tilde{\rho}_{18}, \quad t_2 = \frac{22 - 18}{22} = 0.18,$$

$$\hat{\rho}_{30} = 0.083 \approx 0.084 = \tilde{\rho}_{26}, \quad t_3 = \frac{30 - 26}{30} = 0.13,$$

$$T_2 = \frac{1}{3}\sum_{i=1}^{3} t_i = \frac{0.2 + 0.18 + 0.13}{3} = 0.17.$$

Therefore, the work efficiency of method of information distribution is 17% higher than histogram method is.

For other parameters in an exponential distribution, the results are similar. To verify it, let a bigger $\lambda = 3$ and a smaller $\lambda = 0.9$, we obtain Table 4-7 and Table 4-8, respectively.

Table 4-7 Average non-log-divergence and relative error to estimate $p(x) = 3e^{-3x}$

n	10	12	14	16	18	20	22	24	26	28	30
$\hat{\rho}$	1.419	1.047	0.857	0.653	0.600	0.559	0.511	0.476	0.405	0.375	0.340
$\tilde{\rho}$	1.146	0.884	0.713	0.565	0.511	0.468	0.425	0.398	0.341	0.319	0.297
e	0.192	0.155	0.168	0.135	0.149	0.162	0.168	0.163	0.158	0.148	0.128

Table 4-8 Average non-log-divergence and relative error to estimate $p(x) = 0.9e^{-0.9x}$.

n	10	12	14	16	18	20	22	24	26	28	30
$\hat{\rho}$	0.120	0.090	0.076	0.059	0.054	0.050	0.046	0.043	0.036	0.033	0.030
$\tilde{\rho}$	0.098	0.076	0.062	0.049	0.045	0.042	0.038	0.035	0.030	0.028	0.026
e	0.182	0.163	0.183	0.165	0.161	0.168	0.173	0.176	0.158	0.150	0.127

4.4.4 Lognormal experiment

Let

$$p_3(x) = \frac{1}{0.4x\sqrt{2\pi}} \exp[-\frac{(\log x - 3)^2}{2 \times 0.4^2}], \quad 0 < x < \infty, \qquad (4.18)$$

Running Program 3-6 in subsection 3.5.5 with MU=3, SIGMA=0.4, N=19, SEED=6459, we obtain 19 random number:

$$
\begin{aligned}
X &= \{x_1, x_2, \cdots, x_{19}\} \\
&= \{17.70, 13.93, 24.27, 14.46, 43.11, 26.35, 31.76, 17.48, 20.79, 16.38, \\
&\quad 13.51, 31.32, 18.48, 9.32, 18.81, 21.85, 14.39, 22.84, 28.89\}.
\end{aligned}
$$

$$(4.19)$$

Given an origin $x_0 = 0$, a bin width $h = 4.1$, for positive integer $m = 11$, we obtain 11 intervals for making a classical frequency histogram

$[0, 4.1[, [4.1, 8.2[, [8.2, 12.3[, [12.3, 16.4[, [16.4, 20.5[, [20.5, 24.6[,$
$[24.6, 28.7[, [28.7, 32.8[, [32.8, 36.9[, [36.9, 41[, [41, 45.1[.$

We use their center points to construct a framework space U consists of 11 controlling points,

$$
\begin{aligned}
U &= \{u_1, u_2, \cdots, u_{11}\} \\
&= \{2.05, 6.15, 10.25, 14.35, 18.45, 22.55, 26.65, 30.75, 34.85, 38.95, 43.05\}.
\end{aligned}
$$

Its step length is $\Delta = h = 4.1$.

Employing formula (3.25) with $h = 4.1$ and X in (4.19) to estimate density $p(x)$, we obtain

$$\begin{aligned}
\widehat{P}_3 &= \{\hat{p}(u_1), \hat{p}(u_2), \cdots, \hat{p}(u_{11})\} \\
&= \{0, 0, 0.0128, 0.0642, 0.0513, 0.0513, 0.0128, 0.0385, 0, 0, 0.0128\}.
\end{aligned}$$

Employing formula (4.9) with $\Delta = 4.1$ and the same X, we Obtain SHE

$$\begin{aligned}
\widetilde{P}_3 &= \{\tilde{p}(u_1), \tilde{p}(u_2), \cdots, \tilde{p}(u_{11})\} \\
&= \{0, 0.0029, 0.0139, 0.0588, 0.0593, 0.0395, 0.0240, 0.0278, 0.0050, \\
&\quad\; 0, 0.0127\},
\end{aligned}$$

The true probability distribution calculated by formula (4.18) is

$$\begin{aligned}
P_3 &= \{p(u_1), p(u_2), \cdots, p(u_{11})\} \\
&= \{0, 0.0020, 0.0237, 0.0488, 0.0529, 0.0424, 0.0291, 0.0184, 0.0111, \\
&\quad\; 0.0065, 0.0038\}.
\end{aligned}$$

The corresponding non-log-divergences are $\hat{\rho} = 0.000229$ and $\tilde{\rho} = 0.000150$. Fig. 4.5 shows HE and SHE with respect to a lognormal distribution.

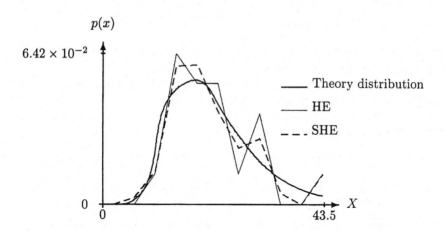

Fig. 4.5   Probability density estimators by HE and SHE
with respect to a lognormal distribution.

$\forall n \in \{10, 12, \cdots, 30\}$, we run 90 lognormal experiments and obtain the corresponding average non-log-divergence ρ and the relative error e shown in Table 4-9.

Table 4-9 Average non-log-divergence ρ and the relative error e for $p_3(x)$

n	10	12	14	16	18	20
$\hat{\rho}$	6.023E-4	5.699E-4	4.926E-4	4.434E-4	4.008E-4	3.491E-4
$\tilde{\rho}$	4.951E-4	4.588E-4	3.925E-4	3.464E-4	3.185E-4	2.778E-4
e	0.178	0.195	0.203	0.219	0.205	0.204
n	22	24	26	28	30	
$\hat{\rho}$	3.253E-4	3.049E-4	2.794E-4	2.621E-4	2.496E-4	
$\tilde{\rho}$	2.658E-4	2.476E-4	2.276E-4	2.130E-4	2.000E-4	
e	0.183	0.188	0.185	0.187	0.199	

It proves that SHE is also better than HE with respect to a lognormal distribution. Note, the reduction of the error is 19.5%, which is the same as one in the normal experiment.

From Table 4-9 we obtain

$$\hat{\rho}_{16} = 4.434E - 4 \approx 4.588E - 4 = \tilde{\rho}_{12}, \quad t_1 = \frac{16 - 12}{16} = 0.25,$$

$$\hat{\rho}_{20} = 3.491E - 4 \approx 3.464E - 4 = \tilde{\rho}_{16}, \quad t_2 = \frac{20 - 16}{20} = 0.2,$$

$$\hat{\rho}_{26} = 2.794E - 4 \approx 2.658E - 4 = \tilde{\rho}_{22}, \quad t_3 = \frac{26 - 22}{26} = 0.15,$$

$$T_3 = \frac{1}{3} \sum_{i=1}^{3} t_i = \frac{0.25 + 0.2 + 0.15}{3} = 0.2.$$

Therefore, the work efficiency of method of information distribution is 20% higher than histogram method is.

To verify that the result, SHE is better than HE, is independent of the parameters μ and σ, we simulate other two experiments, one with $\mu = 2$ and $\sigma = 2$, another With $\mu = 5$ and $\sigma = 0.1$. The simulation results are shown in Table 4-10 and Table 4-11, respectively.

Table 4-10 Average non-log-divergence ρ and the relative error e to estimate $p(x) = 1/(2x\sqrt{2\pi}) \exp[-(\log x - 2)^2/8]$

n	10	12	14	16	18	20
$\hat{\rho}$	1.894E-4	1.872E-4	1.121E-4	7.225E-5	5.495E-5	4.019E-5
$\tilde{\rho}$	1.58E-4	1.529E-4	8.702E-5	5.306E-5	4.163E-5	2.966E-5
e	0.166	0.183	0.224	0.266	0.242	0.262
n	22	24	26	28	30	
$\hat{\rho}$	2.891E-5	2.599E-5	2.352E-5	2.110E-5	1.871E-5	
$\tilde{\rho}$	2.091E-5	1.925E-5	1.690E-5	1.526E-5	1.356E-5	
e	0.277	0.259	0.281	0.277	0.275	

Table 4-11 Average non-log-divergence ρ and the relative error e to estimate
$$p(x) = 1/(0.1x\sqrt{2\pi})\exp[-(\log x - 5)^2/0.02]$$

n	10	12	14	16	18	20
$\hat{\rho}$	3.526E-5	3.503E-5	3.268E-5	2.994E-5	2.791E-5	2.633E-5
$\tilde{\rho}$	2.739E-5	2.652E-5	2.460E-5	2.266E-5	2.079E-5	1.942E-5
e	0.223	0.243	0.247	0.243	0.255	0.262

n	22	24	26	28	30
$\hat{\rho}$	2.539E-5	2.382E-5	2.278E-5	2.093E-5	1.984E-5
$\tilde{\rho}$	1.885E-5	1.836E-5	1.740E-5	1.638E-5	1.546E-5
e	0.258	0.229	0.236	0.217	0.221

Because

$$T = \frac{T_1 + T_2 + T_3}{3} = \frac{0.32 + 0.17 + 0.20}{3} = 0.23,$$

it may be argued that the work efficiency of the method of information distribution is about 23% higher than histogram method.

4.4.5 Comparison with maximum likelihood estimate

Based on a correct assumption for X in terms of distribution function, as normal or exponential function, we can obtain a more accuracy estimator of the population density $p(x)$ by maximum likelihood estimate (MLE) (see subsection 3.4.2). However, if the assumption is wrong, the estimator must be far away from the true distribution $p(x)$. For example, when we assume that X in (4.14) obeys normal distribution, with

$$f(x) = \frac{1}{s\sqrt{2\pi}}\exp[-\frac{(\bar{x} - x_i)^2}{2s^2}],$$

where sample mean

$$\bar{x} = \frac{1}{19}\sum_{i=1}^{19} x_i = 6.86,$$

and estimator of the standard deviation

$$s = \sqrt{\frac{1}{18}\sum_{i=1}^{19}(\bar{x} - x_i)^2} = 0.34,$$

we can obtain the following estimator,

$$\begin{aligned}F = {} & \{f(u_1), f(u_2), \cdots, f(u_{11})\} \\ = {} & \{0.01, 0.07, 0.23, 0.56, 0.95, 1.16, 1.01, 0.62, 0.27, 0.09, 0.02\},\end{aligned}$$

and its non-log-divergence, $\rho_{normal} = 0.022$.

However, for same X, if the assumption is an exponential function, with

$$g(x) = \hat{\lambda}e^{-\hat{\lambda}x}, \quad x \geq 0,$$

where

$$\hat{\lambda} = \frac{1}{\bar{x}} = \frac{1}{6.86} = 0.1458,$$

we obtain another estimator,

$$
\begin{aligned}
G = \ & \{g(u_1), g(u_2), \cdots, g(u_{11})\} \\
= \ & \{0.0622, 0.0604, 0.0587, 0.0570, 0.0554, 0.0538, \\
& \ 0.0522, 0.0507, 0.0493, 0.0479, 0.0465\},
\end{aligned}
$$

and another non-log-divergence, $\rho_{exponential} = 0.321$, which is much larger than $\hat{\rho}(= 0.073)$ and $\tilde{\rho}(= 0.041)$.

Generally speaking, the MLE would meet more risk than the histogram methods in the case of that it is impossible to assume a reasonable function for probability estimation. It is easy to see as

$$\bar{\rho} = \frac{\rho_{normal} + \rho_{exponential}}{2} = \frac{0.022 + 0.321}{2} = 0.172 > \hat{\rho} > \tilde{\rho}.$$

Same thing happens to other populations.

4.4.6 Results

Study of the above simulation experiments reveals that the superiority of SHE is largely dependent on whether we are blind to population and the size of a given sample is small. In the case, the given sample would be considered as fuzzy and some benefit can be obtained by SHE. The work efficiency of SHE is about 23% higher than HE. That is, if both of a statistician and a fuzzy engineer haven't any knowledge about the population from which the given sample is drawn, and if the sample size is small (less 30), then, the statistician has to obtain more observations, adding about 23%, to guarantee that his estimation is as better as one given by the fuzzy engineer.

If we have a lot of knowledge about the population to confirm an assumption, it implies that the statistical object with respect to a given sample is more clear. In the case, MLE is better than SHE.

If the size of a given sample is large, it implies that there is plentiful statistical information in the sample. In this case, it is unnecessary to replace HE by SHE although little benefit can still be obtained by using SHE.

4.5 Non-Linear Distribution

From the linear distribution, by modifying operations on fuzzy sets, we can obtain some non-linear distributions.

Let $\mu(x, u)$ be an information distribution of X on U. The class of all information distributions of X on U will be denoted $\mathcal{F}(X \times U)$.

Definition 4.5 Concentration is defined as a unary operation on $\mathcal{F}(X \times U)$:

$$\text{con} : \mathcal{F}(X \times U) \quad \to \quad \mathcal{F}(X \times U)$$
$$\mu \quad \mapsto \quad \mu^2, \forall \mu \in \mathcal{F}(X \times U)$$

i.e., $\text{con}(\mu(x, u)) = \mu^2(x, u)$.

Definition 4.6 Dilatation is defined as a unary operation on $\mathcal{F}(X \times U)$.

$$\text{dil} : \mathcal{F}(X \times U) \quad \to \quad \mathcal{F}(X \times U)$$
$$\mu \quad \mapsto \quad \mu^{\frac{1}{2}}, \forall \mu \in \mathcal{F}(X \times U)$$

i.e., $\text{dil}(\mu(x, u)) = \mu^{\frac{1}{2}}(x, u)$.

When $\mu(x, u)$ is the linear distribution, $\mu^2(x, u)$ and $\mu^{\frac{1}{2}}(x, u)$ are non-linear distributions.

Concentrating or dilating the last term of the linear distribution, i.e., $\frac{|x - u|}{\Delta}$, we obtain the simplest non-linear distributions:

$$\mu^{(2)} = 1 - \frac{(x - u)^2}{\Delta^2}, \tag{4.20}$$

$$\mu^{(\frac{1}{2})} = 1 - \frac{|x - u|^{(1/2)}}{\Delta^{1/2}}. \tag{4.21}$$

$\mu^{(2)}$ is called the *quadratic distribution* (qd), $\mu^{(\frac{1}{2})}$ is called the *quadratic-root distribution* (qtd). One easily verifies that both of them are reflexive and decreasing.

For a fixed u_j, and $x \in \mathbb{R}$, due to the decrease, $\mu(x, u)$ can be regarded as a membership function in terms of the fuzzy concept "around u_j". The fuzzy set can be written as $\mu_{u_j}(x)$, i.e.,

$$\underset{\sim}{u_j} = \int_{\mathbb{R}} \mu(x, u_j)/x$$

The linear distribution, quadratic distribution and quadratic-root distribution are shown in Fig. 4.6.

As we known, from subsection 4.4.2 (normal experiment), SHE is better than SH. If we replace the linear distribution in the experiment with the quadratic distribution or quadratic-root distribution, what will happen?

Let $S = Q_1 + Q_2 + \cdots + Q_m$. For the linear distribution, due to the information conserved, we know that S is equal to the size of the given sample, n. However, for qd and qtd, in general, $S \neq n$. To estimate $p(x)$, we must modify formula (4.9) into

$$\check{p}(x) = \frac{1}{S\Delta}(\text{total information to controlling point in same interval as } x). \tag{4.22}$$

which is called a *pseudo histogram estimate* (PHE) of $p(x)$.

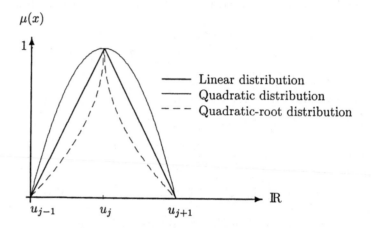

Fig. 4.6   The linear distribution, quadratic distribution and
quadratic-root distribution as fuzzy sets.

With respect to $p_1(x)$ in (1.13), the corresponding PHE to the quadratic distribution or quadratic-root distribution is easily calculated. Running 90 simulation experiments with different seed numbers, we obtain the average non-log-divergences shown in Table 4-12. Plotting (n, ρ) in Table 4-3 and Table 4-12, and linking them, we obtain Fig. 4.7 to show the errors. From the figure, we know that the quadratic distribution is the best one to estimate a normal distribution.

Table 4-12 Average non-log-divergence ρ for $p_1(x)$

n	10	12	14	16	18	20	22	24	26	28	30
ρ_{qd}	0.146	0.139	0.131	0.121	0.115	0.107	0.102	0.097	0.092	0.085	0.080
ρ_{qtd}	0.169	0.157	0.149	0.136	0.129	0.122	0.116	0.110	0.103	0.098	0.094

We also replace the linear distribution in the exponential and lognormal experiments with the quadratic distribution and quadratic-root distribution, and obtain the errors curves with respect to $p_2(x)$ and $p_3(x)$, respectively. They are shown in Fig. 4.8 and Fig. 4.9, respectively. They also show that the quadratic distribution is the best one.

Strictly speaking, $\mu^{(2)}$ is not an information distribution because it does not satisfy the information conserved. However, for probability distribution estimation, it is better than the linear distribution. The benefit may come because the quadric distribution dilates the information carried by an observation.

$\forall x \in X$, no loss in generality, we suppose $u_j \leq x_i \leq u_{j+1}$. If $x_i = u_j$ or $x_i = u_{j+1}$, one easily verifies that $\mu^{(2)}(x, u_j) + \mu^{(2)}(x, u_{j+1}) = 1$. Else, i.e., $u_j < x_i < u_{j+1}$, we have

$$\mu^{(2)}(x, u_j) + \mu^{(2)}(x, u_{j+1})$$

$$= 1 - \frac{(x - u_j)^2}{\Delta^2} + 1 - \frac{(x - u_{j+1})^2}{\Delta^2}$$

$$= 1 - \frac{(x - u_j)^2}{\Delta^2} + \frac{\Delta^2}{\Delta^2} - \frac{(x - u_{j+1})^2}{\Delta^2}$$

$$= 1 - \frac{(x - u_j)^2}{\Delta^2} + \frac{(u_j - u_{j+1})^2}{\Delta^2} - \frac{(x - u_{j+1})^2}{\Delta^2}$$

$$= 1 - \frac{2[x^2 + u_j u_{j+1} - (u_j + u_{j+1})x]}{\Delta^2}.$$

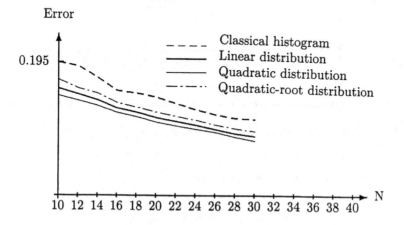

Fig. 4.7 Average non-log-divergence curves with respect to estimating $p_1(x)$

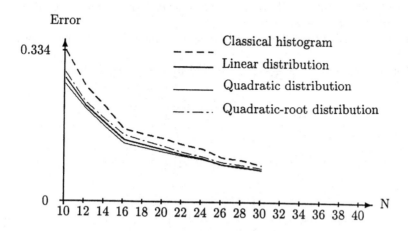

Fig. 4.8 Average non-log-divergence curves with respect to estimating $p_2(x)$

Error

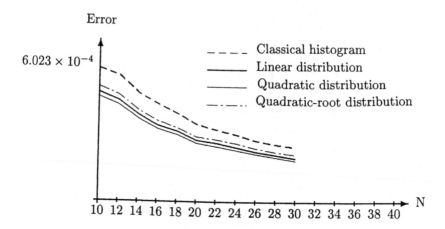

Fig. 4.9 Average non-log-divergence curves with respect to estimating $p_3(x)$

Because $u_j < x_i < u_{j+1}$, therefore $x < u_{j+1}$ and $x - u_j > 0$. Then

$$(x - u_j)x < (x - u_j)u_{j+1}$$
$$\Rightarrow x^2 - u_j x < u_{j+1}x - u_i u_{j+1}$$
$$\Rightarrow x^2 + u_j u_{j+1} - (u_j + u_{j+1})x < 0$$
$$\Rightarrow -\frac{2[x^2 + u_j u_{j+1} - (u_j + u_{j+1})x]}{\Delta^2} > 0.$$

Therefore,

$$\mu^{(2)}(x, u_j) + \mu^{(2)}(x, u_{j+1}) > 1.$$

The quadric-root distribution, on the contrary, concentrates the information carried by an observation.

Does it implies that the more dilating, the more better? The answer is Positive.

Let $\mu^{(s)}(x, u) = 1 - \frac{|x - u_j|^s}{\Delta^s}$, $s > 0$, and $u_j < x_i < u_{j+1}$, then $\frac{|x - u_j|}{\Delta} < 1$, so

$$\lim_{s \to \infty} \mu^{(s)}(x, u) = 1 - \frac{|x - u_j|^s}{\Delta^s} = 1$$

In this case, we have

$$\mu^{(\infty)}(x, u_j) + \mu^{(\infty)}(x, u_{j+1}) = 1 + 1 = 2$$

where

$$\mu^{(\infty)}(x, u_j) = \begin{cases} 1 & \text{for } x = u_j, \\ 1 & \text{for } 0 < |x - u_j| < \Delta, \\ 0, & \text{for} \Delta < |x - u_j|. \end{cases} \qquad (4.23)$$

A PHE based on $\mu^{(\infty)}$ is called an *artificial histogram estimate* (AHE). We use $\mu^{(\infty)}(x, u_j)$ to replace the linear distribution in the experiments of $p_1(x)$, $p_2(x)$ and $p_3(x)$. Fig. 4.10 shows the comparison of the errors.

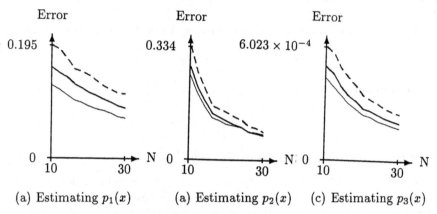

(a) Estimating $p_1(x)$ (a) Estimating $p_2(x)$ (c) Estimating $p_3(x)$

Fig. 4.10   Average non-log-divergence curves of HS (dash line), SHE (thick line) and AHE (thin line)

In spite of these simulation studies, it is practically impossible to use the artificial mathematical model $\mu^{(\infty)}$ in probability distribution estimation because of two main reasons. First, the PDF is strictly defined by probability in terms of events. It is difficult to explain the effect of $S > n$. Second, In $\mu^{(\infty)}$ model, all observations in the interval (u_j, u_{j+1}) play a same role because each observation will give information, gain at 1, to controlling point u_j. That is, the model neglects the difference between the observations near u_j. This model does add artificial information and lose real information.

4.6 *r*-Dimension Distribution

Let $X = \{x_1, x_2, \cdots, x_n\}$ be a r-dimension random sample. Then $\forall x_i \in X$, we know that it is a r-dimension vector, i.e.,

$$x_i = (x_{1i}, x_{2i}, \cdots, x_{ri}).$$

Denote $K = \{1, 2, \cdots, r\}$ and let

$$X_k = \{x_{ki}|i = 1, 2, \cdots, n\}, \quad k \in K \tag{4.24}$$

$\forall k \in K$, we suppose that the chosen framework space of X_k is

$$U_k = \{u_{kj}|j = 1, 2, \cdots, m_k\}, \quad k \in K. \tag{4.25}$$

Then,

$$U = U_1 \times U_2 \times \cdots \times U_r = \prod_{k \in K} U_k$$

is the chosen framework space of X. There are $m = m_1 m_2 \cdots m_r$ elements in U.

Let $\mu_{(k)}$ be an information distribution of X_k on U_k, it can be written as

$$\mu_{(k)}(x_i, u_j) = \mu_{(k)}(x_{ki}, u_{kj}).$$

Definition 4.7 $\forall x_i \in X$, $\forall u_j \in U$,

$$\mu(x_i, u_j) = \prod_{k \in K} \mu_{(k)}(x_{ki}, u_{kj}) \tag{4.26}$$

is called a r-*dimension distribution* of X on U.

Definition 4.8 Let $X = \{x_1, x_2, \cdots, x_n\}$ be a r-dimension random sample, and $U = \{u_1, u_2, \cdots, u_m\}$ with step lengths $u_{kj} - u_{kj-1} \equiv \Delta_k$, $j = 2, 3, \cdots, m_k$, $k \in K$, be the chosen framework space. For any $x_i \in X$, and $u_j \in U$, the following formula is called r-*dimension linear-information-distribution*:

$$\mu(x_i, u_j) =$$
$$\begin{cases} \prod_{k=1}^{r}(1 - \frac{|x_{ki} - u_{kj}|}{\Delta_k}), & (|x_{1i} - u_{1j}| \leq \Delta_1) \wedge \cdots \wedge (|x_{ri} - u_{rj}| \leq \Delta_r), \\ 0, & \text{otherwise.} \end{cases}$$
$$\tag{4.27}$$

Where, Δ_k is called k-th step length. In short, this μ is called the r-dimension linear distribution.

Theorem 4.2 A r-dimension linear distribution μ is reflexive, decreasing and conserved.
Proof: One easily verifies that μ is reflexive and decreasing.

$\forall x_i \in X$, if there exists $u_j \in U$ such that $x_i = u_j$, one easily verifies that

$$\sum_{u \in U} \mu(x_i, u) = \mu(x_i, u_j) = 1.$$

Otherwise, there is no loss in generality when we suppose

$$u_{kj} < x_{ki} < u_{kj+1}, k \in K.$$

(Note, for different k, in general, j is different.) That is

$$\left. \begin{array}{l} \mu_{(k)}(x_{ki}, u_{kj}) > 0, \quad \mu_{(k)}(x_{ki}, u_{kj+1}) > 0 \\ \mu_{(k)}(x_{ki}, u_{kj}) + \mu_{(k)}(x_{ki}, u_{kj+1}) = 1 \end{array} \right\} \tag{4.28}$$

We write

$$\delta_{k1} = \mu_{(k)}(x_{ki}, u_{kj}), \quad \delta_{k2} = \mu_{(k)}(x_{ki}, u_{kj+1}).$$

With the help from the knowledge in the combination, we know that there are 2^r points, in U, which satisfy the condition represented by (4.28). The total distributed information from x_i to these controlling points is

$$q_r = \sum_{(i_1,i_2,\cdots,i_r)\in\{1,2\}^N} \delta_{1i_1}\delta_{1i_2}\cdots\delta_{1i_r}.$$

We employ the mathematical induction to prove $q_r = 1$.

$$\begin{aligned}
r = 1: \quad q_1 &= \delta_{11} + \delta_{12} \\
&= 1 \\
r = 2: \quad q_2 &= \delta_{21}(\delta_{11} + \delta_{12}) + \delta_{22}(\delta_{11} + \delta_{12}) \\
&= \delta_{21} + \delta_{22} \\
&= 1 \\
r = 3: \quad q_3 &= \delta_{31}(\delta_{21}(\delta_{11} + \delta_{12}) + \delta_{22}(\delta_{11} + \delta_{12})) \\
&\quad + \delta_{32}(\delta_{21}(\delta_{11} + \delta_{12}) + \delta_{22}(\delta_{11} + \delta_{12})) \\
&= \delta_{31} + \delta_{32} \\
&= 1
\end{aligned}$$

Obviously, $q_r = \delta_{r1}q_{r-1} + \delta_{r2}q_{r-1}$. We suppose that, when $r = l$, $q_r = 1$ holds. Then, if $r = l + 1$, we have

$$q_r = q_{l+1} = \delta_{l1}q_l + \delta_{l2}q_l = \delta_{l1} + \delta_{l2} = 1.$$

Therefore, for an arbitrary positive integer r, we have $q_r = 1$.

$\square$

Example 4.3 The data in Table 4-13 were obtained from the pamphlet published by a company which manufactures prefabricated houses[10].

Table 4-13 Data related to prefabricated houses

Observation	Area of the first floor (m^2)	Sale price (10,000 Yen)
x_1	38.09	606
x_2	62.10	710
x_3	63.76	808
x_4	74.52	826
x_5	75.38	865
x_6	52.99	852
x_7	62.93	917
x_8	72.04	1031
x_9	76.12	1092
x_{10}	90.26	1203
x_{11}	85.70	1394
x_{12}	95.27	1420
x_{13}	105.98	1601
x_{14}	79.25	1632
x_{15}	120.50	1699

From Table 4-13, we obtain

$$
\begin{aligned}
X_1 &= \{x_{1i}|i = 1, 2, \cdots, 15\} \\
&= \{38.09, 62.10, 63.76, 74.52, 75.38, 52.99, 62.93, 72.04, \\
&\quad 76.12, 90.26, 85.70, 95.27, 105.98, 79.25, 120.50\}, \\
X_2 &= \{x_{2i}|i = 1, 2, \cdots, 15\} \\
&= \{606, 710, 808, 826, 865, 852, 917, 1031, 1092, \\
&\quad 1203, 1394, 1420, 1601, 1632, 1699\}.
\end{aligned}
$$

Let

$$
a_k = \max_{1 \le i \le n} \{x_{ki}\}, \quad b_k = \min_{1 \le i \le n} \{x_{ki}\}.
$$

From X_1 and X_2 we obtain

$$
a_1 = 120.50, \quad b_1 = 38.09, \quad a_2 = 1699, \quad b_2 = 606.
$$

We let $\Delta_1 = (a_1 - b_1)/2 = 41.21$, $\Delta_2 = (a_2 - b_2)/2 = 546.5$ and take

$$
U_1 = \{u_{1j}|j = 1, 2, 3\} = \{38.09, 79.29, 120.50\},
$$

$$
U_2 = \{u_{2k}|k = 1, 2, 3\} = \{606, 1152.5, 1699\}
$$

for being the chosen framework spaces of X_1 and X_2, respectively. Then

$$
U = U_1 \times U_2 = \{v_{jk}|v_{jk} = (u_{1j}, u_{2k}), j = 1, 2, 3; k = 1, 2, 3\}
$$

is the chosen framework space of X.

Using the 2-dimension linear distribution, we obtain all distributed information q_{ijk} shown in Table 4-14.

For example, for $x_3 = (x_{13}, x_{23}) = (63.76, 808)$, because

$$
u_{11} = 38.09 < x_{13} < 79.29 = u_{12}, \quad u_{21} = 606 < x_{23} < 1152.5 = u_{22},
$$

we have

$$
\begin{aligned}
q_{311} &= \mu(x_3, v_{11}) \\
&= (1 - \frac{|x_{13} - u_{11}|}{\Delta_1})(1 - \frac{|x_{23} - u_{21}|}{\Delta_2}) \\
&= (1 - \frac{|63.76 - 38.09|}{41.21})(1 - \frac{|808 - 606|}{546.5}) \\
&= (1 - 0.62)(1 - 0.37) \\
&= 0.38 \times 0.63 \\
&= 0.24,
\end{aligned}
$$

Table 4-14 Distributed information on v_{jk} from x_i

x_i	v_{11}	v_{12}	v_{13}	v_{21}	v_{22}	v_{23}	v_{31}	v_{32}	v_{33}
x_1	1	0	0	0	0	0	0	0	0
x_2	0.34	0.08	0	0.47	0.11	0	0	0	0
x_3	0.24	0.14	0	0.39	0.23	0	0	0	0
x_4	0.07	0.05	0	0.53	0.36	0	0	0	0
x_5	0.05	0.05	0	0.48	0.43	0	0	0	0
x_6	0.35	0.29	0	0.20	0.16	0	0	0	0
x_7	0.17	0.23	0	0.26	0.34	0	0	0	0
x_8	0.04	0.14	0	0.18	0.64	0	0	0	0
x_9	0.01	0.07	0	0.10	0.82	0	0	0	0
x_{10}	0	0	0	0	0.67	0.07	0	0.24	0.02
x_{11}	0	0	0	0	0.47	0.37	0	0.09	0.07
x_{12}	0	0	0	0	0.31	0.30	0	0.20	0.19
x_{13}	0	0	0	0	0.06	0.29	0	0.12	0.53
x_{14}	0	0	0	0	0.12	0.88	0	0	0
x_{15}	0	0	0	0	0	0	0	0	1
$\sum_{i=1}^{15} q_{ijk}$	2.26	2.61	0	1.03	4.73	0.64	0	1.91	1.81

$$q_{312} = \mu(x_3, v_{12})$$
$$= (1 - \frac{|x_{13} - u_{11}|}{\Delta_1})(1 - \frac{|x_{23} - u_{22}|}{\Delta_2})$$
$$= (1 - \frac{|63.76 - 38.09|}{41.21})(1 - \frac{|808 - 1152.5|}{546.5})$$
$$= (1 - 0.62)(1 - 0.63)$$
$$= 0.38 \times 0.37$$
$$= 0.14,$$

$$q_{321} = \mu(x_3, v_{21})$$
$$= (1 - \frac{|x_{13} - u_{12}|}{\Delta_1})(1 - \frac{|x_{23} - u_{21}|}{\Delta_2})$$
$$= (1 - \frac{|63.76 - 79.29|}{41.21})(1 - \frac{|808 - 606|}{546.5})$$
$$= (1 - 0.38)(1 - 0.37)$$
$$= 0.62 \times 0.63$$
$$= 0.39,$$

$$q_{322} = \mu(x_3, v_{22})$$

$$= (1 - \frac{|x_{13} - u_{12}|}{\Delta_1})(1 - \frac{|x_{23} - u_{22}|}{\Delta_2})$$

$$= (1 - \frac{|63.76 - 79.29|}{41.21})(1 - \frac{|808 - 1152.5|}{546.5})$$

$$= (1 - 0.38)(1 - 0.63)$$

$$= 0.62 \times 0.37$$

$$= 0.23.$$

For other v_{jk}, $q_{3jk} = 0$. The values all are shown in 4-th row of Table 4-14.

Then, the primary information distribution of X on U is

$$Q = (Q_{11}, Q_{12}, Q_{13}, Q_{21}, Q_{22}, Q_{23}, Q_{31}, Q_{32}, Q_{33})$$

$$= (2.26, 2.61, 0, 1.03, 4.73, 0.64, 0, 1.91, 1.81).$$

It can be represented as an information matrix

$$Q = \begin{array}{c} \\ u_{11}(38.09) \\ u_{12}(79.29) \\ u_{13}(120.50) \end{array} \begin{array}{ccc} u_{21}(606) & u_{22}(1152.5) & u_{23}(1699) \\ \left(\begin{array}{ccc} 2.26 & 2.61 & 0 \\ 1.03 & 4.73 & 0.64 \\ 0 & 1.91 & 1.81 \end{array} \right) \end{array}$$

Appendix 4.A gives a Fortran program to do this example.

4.7 Fuzzy Relation Matrix from Information Distribution

Let $X = \{w_1, w_2, \cdots, w_n\}$ be a r-dimension random sample. There is no loss in generality when we suppose $\forall w_i \in X$, $x_i = (w_{1i}, w_{2i}, \cdots, w_{r-1i})$ and $y_i = w_{ri}$ are drawn from input space and output space, respectively. Hence, a given sample can be written as

$$X = \{(x_1, y_2), (x_2, y_2) \cdots, (x_n, y_n)\}. \tag{4.29}$$

Correspondingly, we suppose that the chosen framework space of X in input space is U, and the chosen framework space of X in output space is V such as

$$U = \{u_1, u_2, \cdots, u_m\}, \quad V = \{v_1, v_2, \cdots, v_t\}. \tag{4.30}$$

Let Q_{jk} be the total distributed information to controlling point (u_j, v_k). We use an information matrix to represent the primary information distribution of X on U:

$$Q = \begin{array}{c} \\ u_1 \\ u_2 \\ \\ u_m \end{array} \begin{array}{cccc} v_1 & v_2 & \cdots & v_t \\ \begin{pmatrix} Q_{11} & Q_{12} & \cdots & Q_{1l} \\ Q_{21} & Q_{22} & \cdots & Q_{2l} \\ \cdots & \cdots & \cdots & \cdots \\ Q_{t1} & Q_{t2} & \cdots & Q_{mt} \end{pmatrix} \end{array} \tag{4.31}$$

It seems to be same as one in (2.6), but the elements are calculated in different approaches.

According to the character of Q, there are three types of fuzzy relation matrices can be obtained.

4.7.1 R_f based on fuzzy concepts

When $U \times V$ is a factor space, the elements of V may be fuzzy concepts.

The theory of *factor space* was proposed by Wang in 1990 (see [14]). Peng, Kandel and Wang (1991, see [7]) studied the representation of knowledge based on factor space. Then, Peng, Wang and Kandel (1996, see [8]) develop the theory and use it in knowledge acquisition by random sets. In the theory, factors are used to capture the characteristics of a concept (intension) and fuzzy sets are used to describe the membership degree of objects to a concept (extension). Various forms of facts and rules are presented in the factor space structure.

For representation of a fuzzy concept in a fuzzy set, the most important and interesting problem is to select the universe. For example, the concept **Young** may be represented as a fuzzy set of a given group of persons P (each with a membership degree). If P is very large (e.g., $|P| > 1000$), it is not practical to list the membership degree for every one. But from another point of view, the concept **Young** can be represented by a fuzzy set of the "age" universe $[0,150]$. This type of representation captures the most important factors that describe the concept and is more highly *integrated* than the first. Also, if there are enough names to describe the shapes of a " face", then **Young** can also be represented by a fuzzy set of the universe of "faces." The describing variables such as "age" and "face" are called *factors*.

Sometimes a simple factor is not enough for the representation of a concept. For example, **Young Men** should be described by the compound factor "age and sex". It is a fuzzy set of the universe of "age and sex," which is the

Cartesian product of [0, 150] and {male, female}. Factors such as "age-and-sex" are called *compound factors*.

Every factor α corresponds to a universe X_α in which α takes possible values depending on various objects. X_α is called the *state space* of α. For example, if α="age," then X_α may be an interval [0,150] or a set of linguistic lables {*young, very young, old, middle age,* $\cdots$}. If β="intensity of earthquake," then X_β may be an interval [0,8.5] or a set of MMI (Modified Mercalli Intensity) Leve[19] lables {$I, II, III, IV, V, VI, VII, VIII, IX, X, XI, XII$}.

The most fundamental factors are simple ones, and every compound factor is a union of simple factors. A factor space is a collection of state spaces indexed by all related factors. All factors relating a given problem compose a Boolean algebra.

Definition 4.9 (Wang and Sugeno[17]) A factor space is a family of sets $\{X_\alpha\}(\alpha \in L)$, denoted as $(X_\alpha, \alpha \in L)$, where L is a Boolean algebra $L = (L, \vee, \wedge, c)$ satisfying

(1) $X_0 = \{\emptyset\}$

(2) If $T \subseteq L$ is an independent set, i.e., for any $\alpha, \beta \in T$, $\alpha \wedge \beta = 0$ whenever $\alpha \neq \beta$, then

$$X_\gamma = \prod_{\alpha \in T} X_\alpha \qquad (4.32)$$

where $\gamma = \vee\{\alpha | \alpha \in T\}$, 0 and 1 are the smallest and largest elements of L, respectively, $\prod$ is the Cartesian product operator, and $X \times \{\emptyset\} = X$.

For a factor space $(X_\alpha, \alpha \in L)$, $\alpha \in L$, α is called a factor, X_α is called the state space of α, and X_1 is called the whole state space. From this definition, for two factors α and β, if $\alpha > \beta$, then the state space X_α can be projected onto the state space X_β, and the natural projective mapping $X_\alpha \to X_\beta$ is denoted by $i_{\alpha\beta}$. If $\alpha \leq \beta \leq \gamma$, the $i_{\beta\gamma} \circ i_{\alpha\beta} = i_{\alpha\gamma}$ and the representation mapping can be described by the family of mapping $\{j_\alpha\}(\alpha \in A)$, where A is a complete set of union-generating factors, i.e., every factor may be represented as a union of a subset of A.

A factor space is like the coordinate system in mathematics. It provides a describing environment for things. In a factor space, every thing can be described, it must have its own manifestation on each related factor. Each concept can be viewed as a fuzzy set in the factor space, it must have its own manifestation on each related factor, if we can obtain all these factors' manifestations, then we can get a whole understanding of the concept. Each factor manifestation can be described by a fuzzy set on the state space of the factor, the fuzzy set can be obtained by projecting the concept set onto the factor axis.

For a given problem, there certainly exists a corresponding factor space, but different persons may build different factor spaces.

By operations $\vee$, $\wedge$, and c, we can define subfactor, independent factors, difference factor, complementary factor and atom factor:

- Factor α and factor β are independent if and only if $\alpha \wedge \beta = 0$.
- If factor γ satisfies:$(\alpha \wedge \beta) \vee \gamma = \alpha, \gamma \wedge \beta = 0$, then γ is called difference factor of α by β.
- A factor β is called a proper subfactor of factor α, denoted by $\alpha > \beta$, if there exists a set $Y : Y \neq \emptyset$ and $Y \neq \{\theta\}$, satisfying $X_\alpha = X_\beta \times Y$; A factor β is called a subfactor of factor α, denoted by $\alpha \geq \beta$, if $\alpha > \beta$ or $\alpha = \beta$.
- The complementary factor of factor α, denoted by α^c, is defined as $\alpha^c \triangleq 1 - \alpha$.
- A factor α is called an atom factor if α has no proper subfactor except zero factor.

Usually we can build a factor space corresponding to a given problem by listing all the atom factors. Factors can be parted into four kinds according to their state spaces:

- Switch-Kind(KS) the state space of each factor of this kind consists of two opposite states, for instance, the state space of factor $\alpha=$"sex", is $X_\alpha = \{male, female\}$.
- Degree-Kind(KD) the state space of each factor of this kind is $[0,1]$, each state means a degree given from this factor. For instance, factor $\alpha=$"satisfaction degree", and factor $\beta=$"difficulty degree" are factors of this kind.
- Measure-Kind(KM) the state space of each factor of this kind consists of measurable values by some scale. For example, factor $\alpha=$"body temperature" and factor $\beta=$"body height" are factors of this kind, their states are $[36,44)(°C)$ and $(0,260)(cm)$ respectively.
- Enumeration-Kind(KE) the state space of each factor of this kind consists of different kind-values which can be enumerating. For instance, factor $\alpha = occupation$ is a KE-factor whose state space is

$$X_\alpha = \{teacher, worker, peasant, soldier, \cdots\}.$$

In fact, a KD-factor may be regarded as a KE-factor, KD-factors are particular cases of KE-factors.

Particularly, the states of a KD-factor may be fuzzy sets of the universe regarded as another factor space. For example, KD-factor

$$\alpha = \text{``age''}, \tag{4.33}$$

whose state space is

$$X_\alpha = \{young, \; very \; young, \; old, \; middle \; age, \; \cdots\}. \tag{4.34}$$

If there are enough names to describe the shapes of a " face", **Young** can be represented by a fuzzy set of the universe of "faces." In this case, in fact, each element of X_α also can be represented by a fuzzy set of the universe of "faces."

Another example is more engineering. In seismology a scale of seismic intensity is a way of measuring or rating the effects of an earthquake at different sites. The Modified Mercalli Intensity Scale is commonly used in the United States by seismologists seeking information on the severity of earthquake effects. Intensity ratings are expressed as Roman numerals between I at the low end and XII at the high end. Let

$$\alpha = \text{``seismic intensity''}, \tag{4.35}$$

whose state space is

$$X_\alpha = \{I, II, III, IV, V, VI, VII, VIII, IX, X, XI, XII\}. \tag{4.36}$$

They are defined by the table of Modified Mercalli Intensity Scale (see Appendix 4.B). Each element of X_α, in fact, is a fuzzy concept because there does not exist any crisp bound between two neighboring levels.

A very interesting phenomenon occurs when an earthquake engineer want to use the seismic intensity from an isoseismal map for designing an earthquake-resistant structure. Then, he has to consider the intensity as some scale being seismic ground motion such as acceleration. In other words, an element of X_α may be regarded as a fuzzy set of the universe of ground motion acceleration.

Now, let us return to discuss how to construct a fuzzy relation if $U \times V$ is a factor space and the elements of V are fuzzy concepts. In this case, there is a factor α such that $V = X_\alpha$, and, naturally, U will act as the universe for the concepts based on the primary information distribution $Q = \{Q_{jk}\}_{m \times t}$.

First, we let

$$s_k = \max_{1 \le j \le m} Q_{jk}, \quad k = 1, 2, \cdots, t, \tag{4.37}$$

and

$$\mu_k(u_j) = \frac{Q_{jk}}{s_k}, \quad j = 1, 2, \cdots, m. \tag{4.38}$$

Then, based on Q, each element v_k of V can be represented by a fuzzy set μ_k of the universe U.

Second, we let

$$r_{jk} = \mu_k(u_j), \quad j = 1, 2, \cdots, m; k = 1, 2, \cdots, t, \tag{4.39}$$

and

$$R_f = \{r_{jk}\}_{m \times t}. \tag{4.40}$$

Then, based on Q, we obtain a fuzzy relation R_f. Where we use the first letter "f" of *factor space* to indicate this kind of fuzzy relation.

4.7.2 R_m based on fuzzy implication theory

Let $Q = \{Q_{jk}\}_{m \times t}$ be the primary information distribution from data X on $U \times V$. If the most of elements of Q are nonzero, with common sense, we know that, data points of X must be scattering strongly. We have to filter more disturbance in the given Q. In this case, we suggest the well known Mamdani implication[6] for constructing a fuzzy relation.

First, as same as doing in (4.37) and (4.38), we can produce m and t fuzzy sets on U and V, respectively, from Q by

$$\begin{cases} \mu_{A_k}(u_j) = Q_{jk}/a_k, & j = 1, 2, \cdots, m; k = 1, 2, \cdots, t, \\ \mu_{B_j}(v_k) = Q_{jk}/b_j, & k = 1, 2, \cdots, t; j = 1, 2, \cdots, m. \end{cases} \tag{4.41}$$

Where

$$a_k = \max_{1 \le j \le m} Q_{jk}, \quad k = 1, 2, \cdots, t,$$

$$b_j = \max_{1 \le k \le t} Q_{jk}, \quad j = 1, 2, \cdots, m.$$

Notice that there is not assumption if V consists of fuzzy concepts.

A_k is the fuzzy set as an input that will cause behavior v_k, and B_j is the fuzzy set as an output that is caused by input u_j.

Let

$$\begin{cases} \mathcal{A} = \{A_k | k = 1, 2, \cdots, t\}, \\ \mathcal{B} = \{B_j | j = 1, 2, \cdots, m\}. \end{cases} \tag{4.42}$$

Second, we use $\mathcal{A} \times \mathcal{B}$ to learn data X and find a set of fuzzy rules[1],[11], such as "if input u is A, then output v is B", where A is a fuzzy set on U, and B is a fuzzy set on V.

$\forall (x_i, y_i) \in X$, we distribute x_i on U by the $r - 1$-dimension formula derived from one which is used to produce Q. And, we distribute y_i on V with the relating 1-dimension formula. For example, in Example 4.3, we use the 2-dimension linear distribution to obtain Q. Both of the $r-1$-dimension formula and the relating 1-dimension formula are the linear distribution shown in (4.43) and (4.44).

$$\mu^{(r-1)}(x_i, u) = \begin{cases} 1 - \frac{|u - x_i|}{h_u}, & \text{if } |u - x_i| \le h_u, \\ 0, & \text{otherwise.} \end{cases} \tag{4.43}$$

$$\mu^{(1)}(y_i, v) = \begin{cases} 1 - \frac{|v - y_i|}{h_v}, & \text{if } |v - y_i| \le h_v, \\ 0, & \text{otherwise.} \end{cases} \tag{4.44}$$

Where h_u and h_v are the step lengths with respect to U and V, respectively.

Hence, we can represent x_i and y_i by fuzzy set C_{x_i} and D_{y_i}, respectively, as

$$
\begin{cases}
C_{x_i} = \frac{\mu^{(r-1)}(x_i,u_1)}{u_1} + \frac{\mu^{(r-1)}(x_i,u_2)}{u_2} + \cdots + \frac{\mu^{(r-1)}(x_i,u_m)}{u_m}, \\
D_{y_i} = \frac{\mu^{(1)}(y_i,v_1)}{v_1} + \frac{\mu^{(1)}(y_i,v_2)}{v_2} + \cdots + \frac{\mu^{(1)}(y_i,v_t)}{v_t}
\end{cases}
\tag{4.45}
$$

C_{x_i} and D_{y_i} are called the *derived fuzzy sets* of x_i and y_i, respectively, in terms of the information distribution.

Given a universe of discourse U. The set of all fuzzy subsets of U is called the fuzzy power set of U and is denoted by $\mathcal{F}(U)$.

Definition 4.10 $\forall A, B \in \mathcal{F}(U)$,

$$
g(A, B) = \sup_{u \in U} \{\mu_A(u) \wedge \mu_B(u)\}
\tag{4.46}
$$

is called the degree of compatibility of A and B.

Definition 4.11 Let $A \in \mathcal{A}$, $B \in \mathcal{B}$ and $(x_i, y_i) \in X$. Let the derived fuzzy set of x_i be C_{x_i} and the derived fuzzy set of y_i be D_{y_i}. If $\forall A_k \in \mathcal{A}$, $\forall B_j \in \mathcal{B}$, we have

$$
g(A, C_{x_i}) > 0, \ g(A, C_{x_i}) \geq g(A_k, C_{x_i}),
\tag{4.47a}
$$

$$
g(B, D_{y_i}) > 0, \ g(B, D_{y_i}) \geq g(B_j, D_{y_i}),
\tag{4.47b}
$$

we say that (x_i, y_i) makes A and B to be compatible.

That A is compatible with B based on (x_i, y_i) means that the sample point fires the fuzzy implication $A \to B$. That is, if A is true, then B is almost true. In fact, when we obtain sample point (x_i, y_i), it is regarded a fact that x_i causes y_i. That is, if x_i is the input, then the output is almost y_i.

$\forall (x_i, y_i) \in X$, $\forall (A_k, B_j) \in \mathcal{A} \times \mathcal{B}$. If (x_i, y_i) makes A_k and B_j to be compatible, we say that (x_i, y_i) provides information, in value 1, to Cartesian space point (A_k, B_j). Else, we say that (x_i, y_i) provides information, in value 0, to the Cartesian space point. For k, j, accumulating all information at (A_k, B_j) from X, we obtain total information $\mathcal{Q}_{kj}$. Therefore, a new information matrix $\mathcal{Q} = \{\mathcal{Q}_{kj}\}_{m \times t}$ is produced.

For k, let $i \in \{1, 2, \cdots, t\}$, if $\mathcal{Q}_{ij} > 0$ and $\forall k \in \{1, 2, \cdots, t\}$, $\mathcal{Q}_{ij} \geq \mathcal{Q}_{kj}$, we say that $A_i \to B_j$ is true in the largest strength with respective to the j-column.

For k, let $i \in \{1, 2, \cdots, m\}$, if $\mathcal{Q}_{ki} > 0$ and $\forall t \in \{1, 2, \cdots, m\}$, $\mathcal{Q}_{ki} \geq \mathcal{Q}_{kt}$, we say that $A_k \to B_i$ is true in the largest strength with respective to the k-row.

In other words, if there are nonzero elements in every column and row of $\mathcal{Q}$, we can obtain $t \times m$ fuzzy implication rules.

Collecting all fuzzy implication rules being in the largest strength from all columns and rows, we obtain $t \times m$ fuzzy implication rules, denote the set of these rules as $\mathcal{F}(\mathcal{A} \to \mathcal{B})$. That is, $\forall r \in \mathcal{F}(\mathcal{A} \to \mathcal{B})$, there must exist $A \in \mathcal{A}$ and $B \in \mathcal{B}$, such that

$$r \stackrel{\triangle}{=} A \to B. \tag{4.48}$$

We denote A and B in the formula (4.48) as A_r and B_r, respectively.

$\mathcal{F}(\mathcal{A} \to \mathcal{B})$ can construct a fuzzy relation by the Mamdani model:

$$R_r(u,v) = \mu_{A_r \to B_r}(u,v) = \mu_{A_r}(u) \wedge \mu_{B_r}(v), \tag{4.49}$$

and

$$R_m = \bigcup_{r \in \mathcal{F}(\mathcal{A} \to \mathcal{B})} R_r(u,v) = \bigvee_{r \in \mathcal{F}(\mathcal{A} \to \mathcal{B})} \{\mu_{R_r}(u,v)\}. \tag{4.50}$$

Then, using Q and X to construct a new information matrix $\mathcal{Q}$, we obtain a fuzzy relation R_m. Where we use the first letter "m" of *Mamdani model* to indicate this kind of fuzzy relation.

4.7.3 R_c based on conditional falling shadow

If the size n of X is very large, we suggest the falling shadow theory for constructing a fuzzy relation.

The falling shadow theory is also called the theory of the set-valued statistics. This theory was proposed by Wang, in 1985,[13]. He has detected a link between the (theoretical) random sets and (practical) fuzzy sets. Because of this relation, we can import many formulae from probability to fuzzy sets. Wang, Liu and Sanchez use it in earthquake engineering (1985, see [16]). Wang, Huang and Zhang use it to show the essential differences between fuzziness and randomness (1992, see [15]).

In set-valued statistics, each trial gets an ordinary subset of the phase space. This general statistics gives us a possible way to acquire concepts by statistics. For example, for a concept C on a universe U, we may ask n experts to give their estimations $C_1, C_2, \cdots, C_n$ (all are subsets of U) of C on U under the basic rules of psychological measurement. Then we can get the membership function for C as

$$C(u) = \frac{1}{n} \sum_{i=1}^{n} C_i(u), \tag{4.51}$$

where C_i is the identity function on C_i. Of course, set-valued statistics can be considered as general statistics on the power set of the phase space. For the sake of completeness of the method, we cite some definitions and theorems without proof.

Definition 4.12 (Wang[13]) Let X be the universe, $\mathcal{B}$ is a σ-algebra of which X is an element, $(\mathcal{P}(X), \mathcal{B})$ is a measurable space, a *random set* ξ

on X is a measurable mapping from some field of probability $(\Omega, \mathcal{A}, P)$ to $(\mathcal{P}(X), \mathcal{B})$:

$$\xi : \Omega \to \mathcal{P}(X)$$

$$\xi^{-1}(C) = \{\omega \in \Omega | \xi(\omega) \in C\} \in \mathcal{A}, \forall C \in \mathcal{B}$$

Let $\Xi(\mathcal{A}, \mathcal{B})$ be the set of all random sets from some field of probability $(\Omega, \mathcal{A}, p)$ to $(\mathcal{P}(X), \mathcal{B})$.

Definition 4.13 (Wang[13]) Suppose $\xi \in \Xi(\mathcal{A}, \mathcal{B})$, the *falling shadow* of ξ at $x \in X$ is defined as

$$\mu_\xi(x) = P\{\omega \in \Omega | \xi(\omega) \ni x\}$$

$\mu_\xi(x)$ is called *falling shadow function of random set ξ.*

Theorem 4.3 (Wang[13]) (law of large numbers of the falling shadow of random sets) If $\xi_i \in \Xi(\mathcal{A}, \mathcal{B})(i = 1, 2, \cdots, n, \cdots)$ are random sets which are independent and of same distributions, $\mu_{\xi_i}(x) = \mu(x)$, then

$$\bar{\xi}_n(x) = \frac{1}{n} \sum_{j=1}^{n} \chi_{\xi_i}(x) \to \mu(x) \ (a.e.)(n \to \infty)$$

where "a.e." stands for "almost everywhere", and $\chi_{\xi_i}(x)$ is the characteristic function of set ξ_i, i.e., if $x \in \xi_i$, then $\chi_{\xi_i}(x) = 1$, else $\chi_{\xi_i}(x) = 0$.

Further generating each trial to get a fuzzy set, we have the following conclusions.

Definition 4.14 (Luo[5]) Suppose $(\Omega, \mathcal{A}, P)$ is a field of probability, let $X^* = \{x^*_\lambda | x \in X, \lambda \in [0, 1]\}$, $x^*_\lambda = \{A \in \mathcal{F}(X) | x \in A_\lambda\}$, $\mathcal{B} = \sigma(X^*)$, $(\mathcal{F}(X), \mathcal{B})$ is a fuzzy measurable space on X. A random fuzzy set ξ on X is a mapping defined below

$$\xi : \Omega \to \mathcal{F}(X), \omega \to \xi(\omega),$$

$$\xi^{-1}(C) = \{\omega \in \Omega | \xi(\omega) \in C\} \in \mathcal{A}, \forall C \in \mathcal{B}$$

Let $\Xi^*(\mathcal{A}, \mathcal{B})$ be the set of all random fuzzy sets from some field of probability $(\Omega, \mathcal{A}, P)$ to $(\mathcal{F}(X), \mathcal{B})$.

Definition 4.15 (Luo[5]) Suppose $\xi \in \Xi^*(\mathcal{A}, \mathcal{B})$, the falling shadow of ξ at $x \in X$ is defined as

$$\mu_\xi(x) = P_\xi(x^*) = \int x^*(A)P_\xi$$

$\mu_\xi(x)$ is called *falling shadow function of random fuzzy set ξ* , where $\forall C \in \mathcal{B}$, $P_\xi(C) \stackrel{\triangle}{=} P(\xi^{-1}(C))$, $\forall C^* \in \mathcal{B}^* = \{C \in \mathcal{F}(\mathcal{F}(X)) | C_\lambda \in \mathcal{B}, \lambda \in (0, 1]\}$, $P_\xi(C^*) \stackrel{\triangle}{=} \int C^*(A)P_\xi$.

Theorem 4.4 (Luo[5]) (Law of large numbers of the falling shadow of random fuzzy sets) If $\xi_i \in \Xi^*(\mathcal{A}, \mathcal{B})(i = 1, 2, \cdots, n, \cdots)$ are random fuzzy sets which are independent and of same distributions, $\mu_{\xi_i}(x) = \mu(x)$, then

$$\bar{\xi}_n(x) = \frac{1}{n} \sum_{i=1}^{n} \chi_{\xi_i}(x) \to \mu(x) \ (a.e.)(n \to \infty)$$

where $\chi_{\xi_i}(x)$ is the membership function of fuzzy set ξ_i.

In the falling shadow theory, $(\mathcal{F}(X), \mathcal{B})$ is called a *hyper measurable structure* on X.

Definition 4.16 Suppose that $(\mathcal{F}(X), \mathcal{B}_1)$ and $(\mathcal{F}(Y), \mathcal{B}_2)$ are hyper measurable structures on X and Y, respectively. Let $\xi \in \Xi^*(\mathcal{A}, \mathcal{B}_1)$ and $\eta \in \Xi^*(\mathcal{A}, \mathcal{B}_2)$, then

$$\mu_{(\xi,\eta)}(x, y) = P\{\omega \in \Omega | \xi(\omega) \ni x, \eta(\omega) \ni y\}, \tag{4.52}$$

is called the *joint falling shadow* of ξ and η.

Definition 4.17 Suppose that $\mu_\xi(x) > 0$, then

$$\mu_{(\eta|\xi)}(y|x) = P\{\eta(\omega) \ni y | \xi(\omega) \ni x\}, \tag{4.53}$$

is called the *conditional falling shadow* of η given $\xi(\omega) \ni x$.

According to the falling shadow theory, we have

$$\mu_{(\eta|\xi)}(y|x) = \frac{\mu_{(\xi,\eta)}(x, y)}{\mu_\xi(x)}, \tag{4.54}$$

and the formula for the *marginal falling shadow* in terms of X is

$$\mu_\xi(x) = \frac{\int_Y \mu_{(\xi,\eta)}(x, y) dy}{\int_X \int_Y \mu_{(\xi,\eta)}(x, y) dx dy}. \tag{4.55}$$

Now, let us return to discuss how to construct a fuzzy relation if n is large.

As we know, by the information distribution, we change the crisp observations of a given sample into fuzzy sets. Because we regard the given sample X as a random sample, then we obtain n random fuzzy sets. Hence, the primary information distribution $Q = \{Q_{jk}\}_{m \times t}$ plays the role of a joint falling shadow except a coefficient. Therefore, from Q we can obtain the joint falling shadow of X on $U \times V$

$$\mu_{(\xi,\eta)}(u_j, v_k) = \frac{Q_{jk}}{\sum_{j=1}^{m} \sum_{k=1}^{t} Q_{jk}}, \tag{4.56}$$

and the marginal falling shadow in terms of U

$$\mu_\xi(u) = \frac{\sum_{k=1}^t Q_{jk}}{\sum_{j=1}^m \sum_{k=1}^t Q_{jk}}. \qquad (4.57)$$

Hence, we obtain the conditional falling shadow of η given $\xi(\omega) \ni u_j$.

$$\mu_{(\eta|\xi)}(v_k|u_j) = \frac{\mu_{(\xi,\eta)}(u_j, v_k)}{\mu_\xi(u_j)} = \frac{Q_{jk}}{\sum_{k=1}^t Q_{jk}}. \qquad (4.58)$$

We let

$$r_{jk} = \mu_{(\eta|\xi)}(v_k|u_j), \quad j = 1, 2, \cdots, m; k = 1, 2, \cdots, t, \qquad (4.59)$$

and

$$R_c = \{r_{jk}\}_{m \times t}. \qquad (4.60)$$

Then, based on Q, we obtain a fuzzy relation R_c. Where we use the first letter "c" of *conditional falling shadow* to indicate this kind of fuzzy relation.

4.8 Approximate Inference Based on Information Distribution

Suppose we are given an input fuzzy set A of independent universe U as

$$\mu_A(u) = \frac{\mu_A(u_1)}{u_1} + \frac{\mu_A(u_2)}{u_2} + \cdots + \frac{\mu_A(u_m)}{u_m}, \qquad (4.61)$$

and a fuzzy relation matrix $R = \{r_{jk}\}_{m \times t}$ on $U \times V$ (U and V are shown in (4.30)). Now, suppose we want to answer the question:

What is the output fuzzy set of dependent universe V?

It is the problem of fuzzy inference[2],[3],[18]. We denote B as the output fuzzy set. In general, we use the following formula to calculate B:

$$B = A \circ R, \qquad (4.62)$$

where operator "$\circ$" is chosen according to the character of R. When R is produced from information distribution, we suggest the corresponding operators described in next subsections.

4.8.1 Max-min inference for R_f

When R_f seems more uncertain (due to n is too small or other reasons), it is impossible to make a more accurate inference. In this case, we suggest the well known max-min fuzzy composition rule to be the operator "$\circ$", i.e.,

$$\mu_B(v_k) = \max_{u_j \in U}\{\min\{\mu_A(u_j), r_{jk}\}\}, \quad v_k \in V. \qquad (4.63)$$

It can be written as

$$\mu_B(v_k) = \sup_{u_j \in U}\{\mu_A(u_j) \wedge r_{jk}\}, \quad v_k \in V.$$

4.8.2 Similarity inference for R_f

When R_f is regular with respective to each column, we suggest the so-called lattice-similarity composition rule to be the operator "∘". For example, for k, if $r_{jk} = \mu_k(u_j)$ seems a normal distribution as

$$\mu_k(u_j) = \exp[-\frac{(u_j - a_k)^2}{b_k^2}],$$

we say that R_f is regular appearing a normal form. The composition rule is defined by other two concepts, internal product and external product[12].

Definition 4.18 Let $A, B \in \mathcal{F}(U)$. The *internal product* of A and B is defined by

$$A \bullet B = \bigvee_{u \in U} (\mu_A(u) \wedge \mu_B(u)). \tag{4.64}$$

Definition 4.19 Let $A, B \in \mathcal{F}(U)$. The *external product* of A and B is defined by

$$A \odot B = \bigwedge_{u \in U} (\mu_A(u) \vee \mu_B(u)). \tag{4.65}$$

Definition 4.20 Let $A, B \in \mathcal{F}(U)$. The *lattice-similarity* of A and B is defined by

$$(A, B) = \frac{1}{2}[A \bullet B + (1 - A \odot B)]. \tag{4.66}$$

Employing the lattice-similarity composition rule to be the operator "∘", we obtain

$$\mu_B(v_k) = \frac{1}{2}[\bigvee_{u_j \in U} (\mu_A(u_j) \wedge r_{jk}) + (1 - \bigwedge_{u_j \in U} (\mu_A(u_j) \vee r_{jk}))], \quad v_k \in V. \tag{4.67}$$

4.8.3 Max-min inference for R_m

Because we only choose the largest strength implication rules for constructing R_m, the fuzzy relation matrix must be rather stable, i.e., without significant local waving. Hence, for R_m, we also suggest the max-min composition rule to infer the output fuzzy set, which has been described in(4.63).

4.8.4 Total-falling-shadow inference for R_c

The inference with respect to R_c will match with the conditional falling shadow. We suggest the total-falling-shadow formula shown in (4.68).

$$\mu_B(v_k) = \frac{\sum_{u_j \in U} \mu_A(u_j) r_{jk}}{\sum_{u_j \in U} \mu_A(u_j)}, \quad v_k \in V. \tag{4.68}$$

4.2 Conclusion and Discussion

In many cases, the core part of an expert system is a fuzzy relation matrix as the integration of some fuzzy rules. If we can construct the fuzzy relation matrix from a given sample directly, it may become easier to improve an expert system. The method of information distribution is considered as the approach to do it. The method has the advantage of identifying the fuzzy relation when the given sample is small or scattering strongly.

By three typical simulation experiments for the estimation of probability distribution, we confirm that 1-dimension linear-information-distribution can improve a classical histogram estimate in raising 23% work efficiency.

Employing the method of information distribution, it is easy to construct an information matrix to illustrate the given sample. In fact, the matrix from the information distribution is a fuzzy-interval information matrix being complete (all observations can be illustrated) and sufficient (the position difference between the observations can be shown). The matrix is also called the *primary information distribution Q*.

According to the character of Q, we suggest three models to transform Q into fuzzy relation matrix, based on factor space theory, fuzzy implication theory, and falling shadow theory.

And then, we suggest the fuzzy inference operators to infer the output fuzzy set B with an given input fuzzy set A through a fuzzy relation matrix R.

In part II, we will give some applications to show how to use these models and operators.

Choosing controlling points is the so-called "bottle neck" problem of the information distribution. That is, we know how many controlling points is good for constructing a framework space. We do that only according to engineering experience. In the next chapter we will discuss in detail the problem and propose the principle of information diffusion to solve it.

References

1. Berkan, R. C. and Trubatch, S. L. (1997]), Fuzzy Systems Design Principles: Building Fuzzy If-Then Rule Bases. IEEE, New York
2. Gupta, M. M. and Qi, J. (1991), Theory of T-norms and fuzzy inference methods. Fuzzy Sets and Systems, Vol.40, pp.431-450
3. Jang, J. -S. R. (1993), ANFIS: Adaptive-network-based fuzzy inference system. IEEE Transactions on Systems, Man, and Cybernetics, Vol.23, No., pp.665-685
4. Kullback, S. (1959), Information Theory and Statistics. Wiley, New York
5. Luo, C. Z. (1993), The Fundamental Theory of Fuzzy Sets(II). Beijing Normal University Press, Beijing (in Chinese)
6. Mamdani, E. H. (1977), Application of fuzzy logic to approximate reasoning using linguistic synthesis. IEEE Trans. Computers, Vol.26, No.12, pp. 1182-1191
7. Peng, X. T., Kandel, A., Wang, P. Z. (1991), Concepts, rules, and fuzzy reasoning: a factor space approach. IEEE Trans. Systems, Man, and Cybernetics, Vol.21, No.2, pp.194-205
8. Peng, X. T., Wang, P. Z., and Kandel, A. (1996), Knowledge acquisition by random sets. International Journal of Intelligent Systems, Vol.11, No.3, pp.113-147
9. Pérez-Neira, A., Sueiro, J. C., Rota, J., and Lagunas, M. A. (1998), A dynamic non-singleton fuzzy logic system for DSKDMA communications. Proceedings of FUZZ-IEEE'98, Anchorage, USA, pp.1494-1499
10. Tanaka, H. (1996), Possibility model and its applications. in: Ruan, D. Ed., Fuzzy Logic Foundations and Industrial Applications, Kluwer Academic Publishers, Boston, pp.93-110
11. Wang, L. X. and Mendel, J. M. (1992), Generating fuzzy rules by learning through examples. IEEE Trans. on Systems, Man, and Cybernetics, Vol.22, No.6, pp. 1414-1427
12. Wang, P. Z. (1983), Fuzzy Sets Theory and Applications. Shanghai Publishing House of Science and Technology, Shanghai (in Chinese)
13. Wang, P. Z. (1985), Fuzzy Sets and Falling Shadows of Random Sets. Beijing Normal University Press, Beijing (in Chinese)
14. Wang, P. Z. (1990), A factor space approach to knowledge representation. Fuzzy Sets and Systems, Vol.36, pp.113-124
15. Wang, P. Z, Huang, M. and Zhang, D. Z. (1992), Reeaxmining fuzziness and randomness using falling shadow theory. Proceedings of the Tenth International Conference on Multiple Criteria Decision Making, Taipei, pp.101-110
16. Wang, P. Z., Liu, X. H. and Sanchez E. (1986), Set-valued statistics and its application to earthquake engineering. Fuzzy sets and Systems, Vol.18, pp.347-356
17. Wang, P. Z. and Sugeno, M. (1982), The factor fields and background structure for fuzzy subsets. Fuzzy Mathematics, Vol.2, pp.45-54
18. Yager, R. R. and Filev, D. P. (1994), Essentials of Fuzzy Modeling and Control, John Wiley, New York
19. Zsutty, T. C. (1985), A deterministic and fuzzy method or relating intensity, building damage, and earthquake ground motion. Fen Deyi and Liu Xihui (eds): Fuzzy Mathematics in Earthquake Researches. Seismological Press, Beijing , pp.79-94

Appendix 4.A: Linear Distribution Program

Fortran program to do Example 4.3 (2-dimension linear-information-distribution
Program)

```
      PROGRAM MAIN
      INTEGER N,M1,M2,P
      REAL X1(15),X2(15),U1(4),U2(4)
      CHARACTER*26 DAT
      DATA X1/38.09,62.10,63.76,74.52,75.38,52.99,62.93,72.04,
    1 76.12,90.26,85.70,95.27,105.98,79.25,120.50/
      DATA X2/606,710,808,826,865,852,917,1031,1092,
    2 1203,1394,1420,1601,1632,1699/
      N=15
      M1=3
      M2=3
      DAT='PROGRM4A.RST'
      OPEN(1,FILE=DAT,STATUS='NEW')
      WRITE(1,*)'Table 4-13 Data related to prefabricated houses
      WRITE(1,*)'————————————————————'
      WRITE(1,*)'Observation Area of the first floor Sale price'
      DO 10 I=1,N
      P=X2(I)
      WRITE(1,20)I,X1(I),P
10    CONTINUE
20    FORMAT(1X,' x',I2, F12.2,' ',I12)
      WRITE(1,*)'————————————————————'
      WRITE(1,*)
      WRITE(1,*)'For X1, we have'
      CALL DISRETE(N,X1,M1,U1)
      WRITE(1,30)1,1,(U1(J),J=1,M1)
      WRITE(1,*)
      WRITE(1,*)'For X2, we have'
      CALL DISRETE(N,X2,M2,U2)
      WRITE(1,30)2,2,(U2(J),J=1,M1)
30    FORMAT(1X,'The chosen framework space of X',I1 ' is U',I1,
    3 '=',2(F8.2,','),F8.2,'')
      CALL DISTRIBUTION(N,X1,X2,M1,U1,M2,U2)
      CLOSE(1)
      STOP
      END

      SUBROUTINE DISRETE(N,X,M,U)
      INTEGER N,M
```

```
        REAL X(15),U(4),DELTA
        CALL MAXMIN(N,X,A,B)
        AM=M-1
        DELTA=(A-B)/AM
        WRITE(1,10)A,B,DELTA
10      FORMAT(1X,'a=',F8.2,' b=',F8.2,' Delta=',F8.2)
        DO 20 I=1,M
        U(I)=B+(I-1)*DELTA
20      CONTINUE
        RETURN
        END

        SUBROUTINE DISTRIBUTION(N,X1,X2,M1,U1,M2,U2)
        INTEGER N,M1,M2,P(4)
        REAL X1(15),X2(15),U1(4),U2(4),H1,H2,Q(15,4,4),R(4,4)
        H1=U1(2)-U1(1)
        H2=U2(2)-U2(1)
        DO 10 K=1,M2
        P(K)=U2(K)
10      CONTINUE
        DO 40 I=1,N
        DO 30 J=1,M1
        DO 20 K=1,M2
        A=ABS(X1(I)-U1(J))
        B=ABS(X2(I)-U2(K))
        IF(A.GE.H1) GOTO 20
        IF(B.GE.H2) GOTO 20
        Q(I,J,K)=(1-A/H1)*(1-B/H2)
20      CONTINUE
30      CONTINUE
40      CONTINUE
        WRITE(1,*)' '
        WRITE(1,*)'Table 4-13 Data related to prefabricated houses'
        WRITE(1,*)'————————————————————————'
        WRITE(1,*)'xi v11 v12 v13 v21 v22 v23 v31 v32 v33'
        WRITE(1,*)'————————————————————————'
        DO 50 I=1,N
        WRITE(1,60)I,((Q(I,J,K),K=1,M2),J=1,M1)
50      CONTINUE
60      FORMAT(1X,'x',I2,' ',9F5.2)
        WRITE(1,*)'————————————————————————'
        DO 80 J=1,M1
        DO 80 K=1,M2
        A=0
```

```
      DO 70 I=1,N
      A=A+Q(I,J,K)
70    CONTINUE
      R(K,J)=A
80    CONTINUE
      WRITE(1,90)((R(J,K),K=1,M2),J=1,M1)
90    FORMAT(1X,'SUM ',9F5.2)
      WRITE(1,*)'————————————————————'
      WRITE(1,*)' '
      WRITE(1,*)' Information matrix Q'
      WRITE(1,*)' ————————————————'
      WRITE(1,100)(P(K),K=1,M2)
      WRITE(1,*)' ————————————————'
100   FORMAT(1X,' ',3I8)
      DO 110 J=1,M1
      WRITE(1,120)U1(J),(R(J,K),K=1,M2)
110   CONTINUE
120   FORMAT(1X,' ',F7.2,'—',3F8.2)
      RETURN
      END

      SUBROUTINE MAXMIN(N,X,A,B)
      INTEGER N
      REAL X(15)
      A=0
      B=1E+10
      DO 10 I=1,N
      IF(A.LT.X(I)) A=X(I)
      IF(B.GT.X(I)) B=X(I)
10    CONTINUE
      RETURN
      END
```

Appendix 4.B: Intensity Scale

Modified Mercalli Intensity Scale

I. People do not feel any Earth movement.

II. A few people might notice movement if they are at rest and/or on the upper floors of tall buildings.

III. Many people indoors feel movement. Hanging objects swing back and forth. People outdoors might not realize that an earthquake is occurring.

IV. Most people indoors feel movement. Hanging objects swing. Dishes, windows, and doors rattle. The earthquake feels like a heavy truck hitting the walls. A few people outdoors may feel movement. Parked cars rock.

V. Almost everyone feels movement. Sleeping people are awakened. Doors swing open or close. Dishes are broken. Pictures on the wall move. Small objects move or are turned over. Trees might shake. Liquids might spill out of open containers.

VI. Everyone feels movement. People have trouble walking. Objects fall from shelves. Pictures fall off walls. Furniture moves. Plaster in walls might crack. Trees and bushes shake. Damage is slight in poorly built buildings. No structural damage.

VII. People have difficulty standing. Drivers feel their cars shaking. Some furniture breaks. Loose bricks fall from buildings. Damage is slight to moderate in well-built buildings; considerable in poorly built buildings.

VIII. Drivers have trouble steering. Houses that are not bolted down might shift on their foundations. Tall structures such as towers and chimneys might twist and fall. Well-built buildings suffer slight damage. Poorly built structures suffer severe damage. Tree branches break. Hillsides might crack if the ground is wet. Water levels in wells might change.

IX. Well-built buildings suffer considerable damage. Houses that are not bolted down move off their foundations. Some underground pipes are broken. The ground cracks. Reservoirs suffer serious damage.

X. Most buildings and their foundations are destroyed. Some bridges are destroyed. Dams are seriously damaged. Large landslides occur. Water is thrown on the banks of canals, rivers, lakes. The ground cracks in large areas. Railroad tracks are bent slightly.

XI. Most buildings collapse. Some bridges are destroyed. Large cracks appear in the ground. Underground pipelines are destroyed. Railroad tracks are badly bent.

XII. Almost everything is destroyed. Objects are thrown into the air. The ground moves in waves or ripples. Large amounts of rock may move.

5. Information Diffusion

This chapter describes the principle of information diffusion. First we introduce the concept of incomplete-data set and study "fuzziness" of a given sample. Some mathematical proofs are given to show that the principle holds, at least, in the case of probability density distribution (PDF) estimation if we take information diffusion functions which satisfy the conditions as same as that in the kernel estimator, and use the formulae which are employed in the kernel estimator. The chapter is organized as follows: in section 5.1, we discuss the "bottle neck" problems of the information distribution. In section 5.2, we give the definition of incomplete-data set. In section 5.3, we discuss the incompleteness and fuzziness of a given sample. In section 5.4, we mathematically give the definition of information diffusion. In section 5.5, we review some properties of random sets. In section 5.6, the principle of information diffusion as an assertion is described with respect to function approximation. Section 5.7 proves that the principle of information diffusion holds, at least, in the case of estimating a PDF.

5.1 Problems in Information Distribution

Information distribution, a method by which a given sample can be completely and sufficiently illustrated by an information matrix, is a new approach to fuzzy function approximation. The main advantage of the method is that we can construct a fuzzy relation matrix from a given sample without any assumptions. Particularly, when we can't guarantee that an assumption is correct, the new method is useful to avoid the artificial destruction to the data structure. However, it is difficult to choose both framework space $U = \{u_1, u_2, \cdots, u_m\}$ and distribution function μ when we face a practical problem. In general, the controlling points are experientially chosen such that the primary information distribution Q is neither strongly fluctuating (there are some conspicuous wave troughs) nor too dull (many neighboring elements have same value larger than zero). For filling in the wave troughs, we would choose a larger step length Δ in U. However, if the length is too large, Q will become dull, and we have to let it be little. The adjustment work must be done repeatedly. It is a hard work and requires some engineering experience.

The same problem appears in the classical histogram. Where researchers have to choose the number m of intervals according to the size n of the given sample $X = \{x_1, x_2, \cdots, x_n\}$. If X is drawn from a population with a normal density function, Otness and Encysin (1972, see [27]) proved that the asymptotic optimum estimate of m is

$$m = 1.87(n-1)^{2/5}. \tag{5.1}$$

It seems that we can also use the formula to choose the number m for the framework space. However the proof of Otness and Encysin is done for a large n and based on the normal assumption. It means that we cannot use the formula to fix the framework space U.

Another problem in the new method is to choose the distribution function μ. Although we have demonstrated, in simulation technique, that the 1-dimension linear-information-distribution can give a better probability estimate, it is almost impossible to discuss its mathematical properties. The information distribution would be regarded as an engineering method rather than a scientific method.

Lu, Shang, Xu and Chen (1999, see [21],[35]) proposed an improving information distribution method (IDM) to choose a non-linear distribution function independent of the condition $\sum_{j=1}^{m} \mu(x_i, u_j) = 1$, $i = 1, 2, \cdots, n$, as follows:

$$\mu(x_i, u_j) = \begin{cases} 1 - \frac{|x_i - u_j|^\lambda}{\Delta^\lambda}, & \text{for } |x_i - u_j| \leq \Delta, \\ 0, & \text{otherwise.} \end{cases}$$

They use Matlab[1] and apply the linear search algorithm to optimize λ and Δ. The improving IDM is applied to determine the illness fuzzy set for the groups with respect to the prevalence rate of diseases such as hypertension, coronary heart diseases, and diabetes mellitus. The research involves risk analysis. They claim that the result from the improving IDM is quite well.

The problem lies in that we have to construct an object function before we employ any optimizing technique. In paper [21], researchers take the result from a large sample to be the real one and compare the result from IDM with it to adjust λ and Δ to obtain the least error. However, in many fields, as geology, astronomy, archaeology, and economics, we never obtain any large sample with respect to some special phenomena. This means that an improving IDM is efficient only for some special fields. Certainly, in this field, when we chosen λ and Δ, for a similar case, with the improving IDM a small sample will act as a large sample. This suggests that significant monetary saving can occur. For example, suppose the researchers have been given a large sample in Shanghai with respect to the prevalence rate of diseases and chosen λ and Δ with an optimizing technique for a small n, then when they want to obtain the fuzzy set of groups in terms of the same illness in Beijing, it is enough to sample n observations.

---

[1] Matlab is registered trademarks of The MathWorks Inc.

5.2 Definition of Incomplete-Data Set

The information distribution approach provides an important advance for developing fuzzy set theory for small samples without any other information. The successes in engineering and the demonstration in simulation show that changing observations in a given small-sample into fuzzy sets may bring some benefit for fuzzy function approximation.

It is well known that a small sample implies insufficient knowledge. Of cause, a large sample does not guarantee sufficient knowledge. In fact, for estimating a continuous probability distribution, at any size n the sample is always insufficient. For example, suppose we use a classical histogram to estimate a PDF. For any n, it is easy to choose a small bin width h such that the histogram is strongly fluctuating. Therefore errors exist between the real function and the histogram. To estimate the real function by a histogram, we have to let $n \to \infty$ and $h \to 0$. If we use other traditional methods, as maximum likelihood method, to estimate it, the situation is same. In other words, if the population has a continuous distribution function such as a normal or exponential distribution, a given sample must be incomplete for representing the population, from which the sample is drawn. In a sense, the main task of exploring in sciences is to approximate real world systems with incomplete information and knowledge by observation, experiment, learning and inference. In other words, it is difficult to avoid this so-called incompleteness.

In Chapter 1, we simply deal with the concept of incompleteness with respect to a given sample X. X is called an incomplete-data set if it cannot provide sufficient evidence for discovering the relations we want to know precisely. In fact, the intension of the concept is too plentiful to describe in common terms.

5.2.1 Incompleteness

The notion of incompleteness obsesses scientists. Kurt Gödel's incompleteness theorem[4] seemed to pull the rug out from under mathematical certainty, objectivity, and rigor. Chaitin (1990, see [5]) developed the incompleteness concept as it relates to randomness.

Gödel's theorem is directed against David Hilbert's sixth problem which had to do with establishing the fundamental universal truths, or axioms, of physics. Gödel proved that it is impossible to obtain both a consistent and complete axiomatic theory of mathematics and a mechanical procedure for deciding whether an arbitrary mathematical assertion is true or false, or is provable or not.

Chaitin's ideas are aimed at randomness in series of binary digits. Chaitin shows that the halting (stop) probability of a computer program is algorithmically random. It cannot be compressed into a shorter program. To get N bits of the number out of a computer, you need to put in a program at least

N bits long. His result that the halting probability is random corresponds to Alan Turing's assertion that the halting problem is undecidable.

Little did Gödel and Chaitin imagine that incompleteness and fuzziness are subtly related. This was because they had known that simple mathematical questions do not always have clear answers and some questions can give answers that are completely random and look grey, rather than black or white.

In fact, the incompleteness which intends to break the law of excluded middle is fuzziness, and is relevant to the lack of causality is randomness.

An incomplete-data set is both random and fuzzy. For example, consider these two incomplete series of binary digits:

$$01010101010101010101$$

$$01101100110111100101$$

The first is obviously constructed according to a simple rule; it consists of the number 01 repeated ten times. If one were asked to speculate on how the series might continue, one could predict with considerable confidence that the next two digits would be 0 and 1. Inspection of the second series of digits yields no such comprehensive pattern. There is no obvious rule governing the formation of the number, and there is no rational way to guess the succeeding digits. The arrangement seems haphazard; in other words, the sequence appears to be a random assortment of 0's and 1's.

The second series of binary digits was generated by flipping a coin 20 times and writing a 1 if the outcome was heads and a 0 if it was tails. Tossing a coin is a classical procedure for producing a random number. Tossing a coin 20 times can produce any one of 2^{20} binary series, and each of them has exactly the same probability which is 2^{-20}.

Obviously, it needs a very long series of binary digits as the second to precisely calculate the probability of head occurring to toss a coin in one time. The series of binary digits in our example is too short, we cannot obtain a precise result by it; in other words, the sample only provides incomplete information for calculating the probability we want to know. The result from the series of binary digits must be imprecise, vague, and ambiguous. In a word, fuzzy.

If we are interested in whether an event occurs, we study randomness of a sample. If we study an incomplete sample, we cannot avoid its fuzziness. When we talk fuzziness of an incomplete-data set, we focus on the data themselves rather than guessing the succeeding digits.

For example, when Muriel, Nicotra and Lambas (1995,see[24]) used the data from the APM Bright Galaxy Survey to estimate the shape of the Luminosity Function, by the Maximum Likelihood method for the determination of the best fitting parameters, they applied Monte Carlo techniques to deal with incompleteness effects in the data. They are interested in whether an event occurred.

In recent years, the terminology "incomplete information" frequently appears in many literatures. In order to extend the ability to handle incomplete information in a definite deductive database, a Horn clause-based system representing incomplete information as incomplete constants is proposed[19]. By using the notion of incomplete constants the deductive database system handles incomplete information in the form of sets of possible values, thereby giving more information than null values. The resulting system extends Horn logic to express a restricted form of indefiniteness. Although a deductive database with this kind of incomplete information is, in fact, a subset of an indefinite deductive database system, it represents indefiniteness in terms of value incompleteness, and therefore it can make use of the existing Horn logic computation rules. The inference rules for such a system are presented, its model theory discussed, and a model theory of indefiniteness proposed. The theory is consistent with minimal model theory and extends its expressive power. The researchers met the fuzziness with respect to indefiniteness.

Taking into account that incompleteness is concerned with both of randomness and fuzziness, obviously, we know that the term "incompleteness" cannot be directly defined by randomness or fuzziness. The incompleteness also differs from the uncertainty.

"Incomplete" is not complete; not filled up; not finished; not having all its parts, or not having them all adjusted; imperfect; defective.

"Uncertain" is not certain; not having certain knowledge; not assured in mind; distrustful.

Obviously, incompleteness is more related to knowledge representation and the intension of a concept. However, uncertainty is more related to event prediction and the extension of a concept.

In the framework of information diffusion, the concept of incompleteness is narrow sense, which is defined only for a given sample. It is necessary to give the definition of correct-data set before we study incompleteness with respect to a given sample.

5.2.2 Correct-data set

Let $X = \{x_1, x_2, \cdots, x_n\}$ be a given sample, called a data set, drawn from a population Ω meaning that every element of the sample has been assigned a value or a vector whose components are valuations. We assume that X will be employed to estimate a relation R in Ω.

Definition 5.1 Give a sample X. Let R be the real relation we want to know. X is called a correct-data set to R if and only if there exists a model γ in which we can obtain an estimator R_X^γ such that $R_X^\gamma = R$.

Example 5.1 In a box, there are 320 balls. Among them, 20 balls are black balls, 80 brown, 120 red, 60 orange and 40 yellow. The color set is

$$A = \{black, brown, red, orange, yellow\}.$$

Let
$$V = \{v_1, v_2, \cdots, v_5\} = \{1, 2, 3, 4, 5\},$$
we can quantify A by a mapping f

$$
\begin{array}{rcl}
f: & A & \to & V \\
& black & \mapsto & 1 \\
& brown & \mapsto & 2 \\
& red & \mapsto & 3 \\
& orange & \mapsto & 4 \\
& yellow & \mapsto & 5
\end{array}
$$

The real relation R from U to probability is

$$p(v_1) = p(1) = \frac{20}{320} = 0.0625, \quad p(v_2) = p(2) = \frac{80}{320} = 0.25,$$

$$p(v_3) = p(3) = \frac{120}{320} = 0.375, \quad p(v_4) = p(4) = \frac{60}{320} = 0.1875,$$

$$p(v_5) = p(5) = \frac{40}{320} = 0.125.$$

The following matrix can represent this relation:

$$
R_{A \to P} =
\begin{array}{c}
\\
black \\
brown \\
red \\
orange \\
yellow
\end{array}
\begin{array}{c}
\begin{array}{ccccc}
0.0625 & 0.25 & 0.375 & 0.1875 & 0.125
\end{array} \\
\left(
\begin{array}{ccccc}
1 & 0 & 0 & 0 & 0 \\
0 & 1 & 0 & 0 & 0 \\
0 & 0 & 1 & 0 & 0 \\
0 & 0 & 0 & 1 & 0 \\
0 & 0 & 0 & 0 & 1
\end{array}
\right)
\end{array}.
$$

The R can be estimated by the following model with two steps.

(1) For a ball x drawn from the box, its characteristic function is

$$\chi_v(x) = \begin{cases} 1, & \text{for } x = v, \\ 0, & \text{otherwise.} \end{cases} \quad \forall v \in V.$$

(2) Let $X = \{x_1, x_2, \cdots, x_n\}$ be a sample drawn from the box. Let

$$\hat{p}(v) = \frac{1}{n} \sum_{i=1}^{n} \chi_v(x), \quad \forall v \in V.$$

Obviously, the sample of all balls from the box is a correct-data set to R.

It is interesting to note that, using a sample of only some balls from the box, sometimes we may obtain the real relation R. For example, if the sample consists of 2 black balls, 8 brown balls, 12 red balls, 6 orange balls and 4 yellow balls, by above model, we obtain

$$\hat{p}(v_1) = \frac{2}{32} = 0.0625, \quad \hat{p}(v_2) = \frac{8}{32} = 0.25,$$

$$\hat{p}(v_3) = \frac{12}{32} = 0.375, \quad \hat{p}(v_4) = \frac{6}{32} = 0.1875,$$

$$\hat{p}(v_5) = \frac{4}{32} = 0.125.$$

i.e., in this case, $\hat{p}(v_i) = p(v_i), i = 1, 2, 3, 4, 5$.

5.2.3 Incomplete-data set

We use Ω to denote the population we will study. Let X be a sample drawn from Ω, its size is n. Let U be the universe of the relation R described by Ω.

In Example 5.1, the set of all balls from the box is Ω. Because the universe of probability is interval $[0,1]$, the universe of the relation R is $U = V \times [0, 1]$. The relation can be written as $R = \{r(v,p)\}_{V \times [0,1]}$, where

$$r(v,p) = \begin{cases} 1, & \text{for } p = \frac{1}{n} \sum_{i=1}^{n} \chi_v(x), \\ 0, & \text{otherwise.} \end{cases} \quad v \in V, p \in [0, 1].$$

The set of all random samples with size n drawn from Ω is called the *n-sample space* of Ω, denoted by $\mathcal{X}_n$. The set of all models by which we can estimate R with a given sample is called the *operator space*, denoted by Γ. $\forall X \in \mathcal{X}_n, \forall \gamma \in \Gamma$, we use $r_X^\gamma(u)$ to denote the estimate of R at a point $u \in U$ with X by γ.

Definition 5.2 Let $X \in \mathcal{X}_n$. X is called a *handicapped sample* if and only if $\forall \gamma \in \Gamma, \exists u \in U$, such that

$$|r_X^\gamma(u) - r(u)| > 0. \tag{5.2}$$

For example, $X = \{0,1,1,0,1,1,0,0,1,1,0,1,1,1, 1,0,0,1,0,1\}$ is a handicapped sample to calculate the probability of "head" occurring to toss a coin in one time. There is only one operator that can be used to deal with this X. The result is $r_X^\gamma(head) = 12/20 = 0.6$. However, real one is $r(head) = 0.5$. That is, $|r_X^\gamma(head) - r(head)| = 0.6 - 0.5 = 0.1 > 0$.

Definition 5.3 $\mathcal{X}_n$ is called an *incomplete sample space* of Ω if and only if $\exists X \in \mathcal{X}_n$ that is handicapped.

Definition 5.4 $X \in \mathcal{X}_n$ is called an *incomplete-data set* if and only if $\mathcal{X}_n$ is incomplete.

When X is an incomplete-data set and no confusion can arise, we also say that X is an incomplete sample.

Corollary 5.1 If $\mathcal{X}_n$ is incomplete, each element of $\mathcal{X}_n$ is an incomplete-data set.

An incomplete-data set may be a correct-data set. In Example 5.1, $\mathcal{X}_{32}$ is an incomplete sample space. According to Corollary 5.1, any sample with size 32 must be an incomplete-data set. Therefore, the sample of 2 black balls, 8 brown balls, 12 red balls, 6 orange balls and 4 yellow balls is an incomplete-data set. However, it is a correct-data set because we can find a model by which the estimated relation is just same as one described by the population.

Example 5.2 Tossing a coin $n(< \infty)$ times, we obtain a sample with size n. The set of all samples with size n is an incomplete sample space because there must exist elements in the sample space which are handicapped. Therefore, any sample with size n is an incomplete-data set to estimate the relation between events (head and tail) and probabilities of events occurrence of the coin.

Example 5.3 The data set of middle and strong earthquake given by Appendix 2.A is an incomplete-data set because it is impossible to know, precisely, the relationship between magnitude, M, and epicentral intensity, I_0, with the data set by any model.

Corollary 5.2 For a nonlinear relation, any sample with size less than the size of its population is an incomplete-data set.

5.3 Fuzziness of a Given Sample

In general, people are more interested in randomness with respect to an incomplete-data set. They focus on whether an observed value is recorded (i.e., an event occurs) rather than the value itself (suppose that the value is crisp). A few people consider the fuzziness of an incomplete-data set. Even fuzzy engineers consider the fuzziness of an incomplete-data set, in general, they attribute the fuzziness to mistaking, losing, missing, and roughing. In the process of data analysis, Meier, Weber and Zimmermann (1994, see[23]) stated that, if either features or classes of objects studied are fuzzy, the use of fuzzy approaches is desirable. (When we consider an element of a given sample as an object, one of its features is representation, in crisp number or natural language). Most fuzzy engineers believe that no fuzziness exists in a random sample if each element can be recorded correctly and precisely in a number or a vector. By the viewpoint, there does not exist any fuzziness in a sample from tossing a coin.

The traditional viewpoint is grounded in a concept: the fuzziness must be related either by simplifying in the terms of a macroscopic grad or by employing non-precise data. In other words, fuzziness is related to a macroscopic grad or non-precise data.

Therefore, for most scientists and engineers, it is a natural action to introduce fuzzy set theory to study natural language systems. For example, "tall" and "young" can be expressed in terms of fuzzy sets and employed to classify a group of students. "Tall" is regarded as a macroscopic measure for height and "young" for age. Hence, scientists in "soft" sciences, such as management science, can use or develop a variety of fuzzy approaches to deal with natural language. Some scientists in "hard" sciences, such as controlling, employ fuzzy rules represented by natural language to construct the relationship between input and output.

Researchers are also interested in studying the so-called non-precise data with fuzzy sets. The results of measurements are often not precise real numbers or vectors but more or less *non-precise* numbers or vectors. This uncertainty, different from measurement errors and stochastic uncertainty, is called *imprecision*. Imprecision is a feature of single observations. Errors are described by statistical models and should not be confused with imprecision. In general imprecision and errors are superimposed. A special case of non-precise data is *interval data*. Viertl (1996, see[37]) suggested fuzzy subsets to study non-precise data. Here, imprecision almost equals fuzziness.

In fact, except for macroscale and imprecision, incompleteness would be regarded a general resource of fuzziness. Obviously, macroscale and imprecision in a given sample must cause the sample to be incomplete. As such, a given sample including macroscale or imprecision recorders can be described by fuzzy sets. Our interest here is to find the answer for the question "Is there any fuzziness in an incomplete-data set if there isn't any macroscale or imprecision?" In other words, "If all observations of an incomplete sample are crisp numbers or vectors, is there any fuzziness in the sample?"

The answer is positive. To explain it, let us first review fuzziness.

5.3.1 Fuzziness in terms of fuzzy sets

There are probably as many definitions of fuzziness (or fuzzy uncertainty) as there are books on the subject.

In fuzzy set literature, fuzziness is defined as an index to measure the grade of difficulty of determining the membership or nonmembership of the elements of a universe with respect to a given fuzzy set. In other words an index of fuzziness tries to indicate the degree of fuzziness. Concrete constructions of such indexes are based on the concept of distance between fuzzy sets. In detail, fuzziness is manifested in three different forms:
- degrees of membership in fuzzy set theory
- degrees of truth in fuzzy logic
- degrees of possibility in possibility theory

These three manifestations are strongly interconnected, and each has meaningful interpretations in the other two domains. The three areas — fuzzy set theory, fuzzy logic, and possibility theory — are sometimes referred to as fuzzy theory.

The question of how to measure the degree of vagueness or fuzziness of a fuzzy set has been one of the basic issues of fuzzy set theory. Various measures of vagueness, more often called *measures of fuzziness*, have been proposed[18]. One way of measuring fuzziness, suggested by Yager (1979, see[41]), and further investigated by Higashi and Klir (1982, see[13]), is to view it as the lack of distinction between the fuzzy set and its complement. Clearly, the less a fuzzy set differs from its complement, the fuzzier it is. Using this approach, the measure of fuzziness depends on the complementation operator employed (which is not unique, see [13]) and on the distance function by which the distinction between the set and its complementation is expressed.

In terms of sets, probabilistic and fuzzy uncertainties are different facets of uncertainty. Fuzziness deals with vagueness between the overlapping sets[1][16], meanwhile probability concerns the likelihood of randomness of a phenomenon[20]. Fuzziness does not depend on the occurrence of the event; whereas probability does. Fuzziness lies in the subsets defined by the linguistic variables, like tall, big, whereas indiscernibility is a property of the referential itself, as perceived by some observers, not of its subsets[34].

5.3.2 Fuzziness in terms of philosophy

Many fuzziness definitions are relatively experiential or abstract.

Fuzziness began as vagueness in the late nineteenth century. Pragmatist philosopher Charles Sanders Peirce was the first logician to have dealt with Vagueness[29]: "Vagueness is not more to be done away with in the world of logic than friction in mechanics." A concept is vague just it has blurred boundaries. The concept "mountain" is vague because we do not know where a mountain ends and a hill begins.

In China, most accept the following definition "Fuzziness intends to break the law of excluded middle. It is not caused by the lack of causality but of hard division." This is derived from Russell's statement[33] (first identified vagueness at the level of symbolic logic): concept A is vague if and only if it break Aristotle's "law" of excluded middle — if and only if A or not-A fails to hold.

The need to bridge the gap between a mathematical model and experience is well characterized in a penetrating study by the American philosopher Max Black (1937, see[3]): *It is a paradox, whose important familiarity fails to diminish, that the most highly developed and useful scientific theories are ostensibly expressed in terms of objects never encountered in experience. The line traced by a draftsman, no matter how accurate, is seen beneath the microscope as a kind of corrugated trench, far removed from the ideal line of pure geometry. And the "point-planet" of astronomy, the "perfect gas" or thermodynamics, or the "pure species" of genetics are equally remote from exact realization. Indeed the unintelligibility at the atomic or subatomic level of the notion of a rigidly demarcated boundary shows that such objects not merely are not but could not be encountered. While the mathematician constructs a theory in terms of "perfect" objects, the experimental scientist observes objects of which the properties demanded by theory are and can, in the very nature of measurement, be only approximately true. Mathematical deduction is not useful to the physicist if interpreted rigorously. It is necessary to know that its validity is unaltered when the premise and conclusion are only "approximately true". But the indeterminacy thus introduced, it is necessary to add in criticism, will invalidate the deduction unless the permissible limits of variation are specified. To do so, however, replaces the original mathematical deduction by a more complicated mathematical theory in respect of whose interpretation the same problem arises, and whose exact nature is in any case unknown. This lack of exact correlation between a scientific theory and its empirical interpretation can be blamed either upon the world or upon the theory. We can regard the shape of an orange or a tennis ball as imperfect copies of an ideal form of which perfect knowledge is to be had in pure geometry, or we can regard the geometry of spheres as a simplified and imperfect version of the spatial relations between the members of a certain class of physical objects. On either view there remains a gap between scientific theory and its application which ought to be, but is not, bridged. To say that all language (symbolism, or thought) is vague is a favorite method for evading the problems involved and lack of analysis has the disadvantage of tempting even the most eminent thinkers into the appearance of absurdity. We shall not assume that "laws" of logic or mathematics prescribe modes of existence to which intelligible discourse must necessarily conform. It will be argued, on the contrary, that deviations from the logical or mathematical standards of precision are all pervasive in symbolism; that to label them as subjective aberrations sets an impassable gulf between formal laws and experience and leaves the usefulness of the formal sciences an insoluble mystery.*

It is easy to see that the concepts of vagueness and precision mentioned in this quote are closely related to the concepts of fuzziness and crispness. According to his viewpoint, vagueness and precision are not inherent in the real world, but emerge from language (symbolism, or natural language) in which we describe our constructions (classes, systems, etc.).

A general conclusion reached by Black in the quoted paper[3] is that dealing with vague propositions requires a logic that does not obey some of the laws of classical logic, notably the law of excluded middle and the law of contradiction. The same conclusion was arrived at by Bertrand Russell in 1923 (see [33]), as shown above.

5.3.3 Fuzziness of an incomplete sample

In the following when we study an incomplete sample, unless stated otherwise, it is assumed that all elements of X are crisp observations and the sample does not involve mistaking, losing, missing, or roughing.

Most may feel uncomfortable when we say that there exists fuzziness in a sample from tossing a coin, where every observation is crisp. This should not be surprising, because the sample can be employed to estimate an objective probability: based on the calculation of absolute and relative frequencies in order to describe the value of a random variable in a repetitive experiment. However, when the sample is incomplete, the estimated probability must be imprecise.

That the sample X is incomplete implies that the observations in X are the representatives of all possible observations. The observations as representatives, in some sense, are similar to the electoral votes of the states in America's presidential election, and all possible observations can be regarded as all electors of in the United States. In general, the behavior of a representative should show, in some degree, the thought of a group which consists of some bodies who are around the representative. Because the boundaries of the groups are fuzzy, it is impossible to inverse every elector's opinion from the sample of the electoral votes.

Incompleteness is not inherent in the sample, but emerges from our desire to know more. If our task is not to describe a natural phenomenon (or relationship) controlled by the population, but to show something controlled by the sample itself, then any sample of crisp observations is not incomplete.

Science aims at explaining natural phenomena, while engineering aims at solving real problems using knowledge from science. In some sense, science is, based on partial facts, to discover "what is,". In this case, the facts we collect play spokesmen for some groups with vague boundaries. Combining all groups, we obtain the universe of all possible facts.

It may be argued that for a given sample, if it is incomplete, any element of it would be regarded a spokesman of a fuzzy group when we want to use the sample to describe a natural phenomenon.

In the model of the histogram, all observations falling into the same interval are spokesmen in the same power for the interval. However, in the model of the information distribution, the intervals are fuzzified and the power of the spokesmen for a same fuzzy interval is different. The nearer the center of the fuzzy interval, the bigger power the spokesman has. Fuzziness of a given sample is due to the fact that we have to use partial knowledge to estimate a PDF. And, when want to know more than we known, the fuzziness is inherent in the incomplete sample.

In the model of the information distribution, q_{ij} (distributed information on u_j from x_i) would be regarded as the power of x_i as a spokesman for the fuzzy interval whose center is controlling point u_j. The nearer x_i is u_j, the bigger power x_i has for the corresponding interval. The largest power is 1 when x_i is just located at u_j. Firstly we have to choose the controlling points in which to form some fuzzy intervals, then the power of an observation as a spokesman is measured according to the distance between the observation and a controlling point. As stated in the last section, it is difficult to choose the controlling points when we face a practical problem. In other words, we meet a "bottle neck" problem when we want to divide an universe into several fuzzy groups before we discuss the calculation of powers. Figuratively speaking, it is difficult to divide all electors into some fuzzy groups before we discuss the powers of the votes as spokesmen for the groups, respectively. To overcome this obstacle, an approach is to study which fuzzy group empowers a given observation to be its chief spokesman. That is, we would use a given observation x_i to look for a reasonable fuzzy group which empowers x_i to be chief spokesman, or study the area affected by x_i. Obviously, the area hasn't any crisp boundary and center of the area is just located at x_i. We use a mathematical framework called information diffusion to form the procedure.

5.4 Information Diffusion

Let $X = \{x_1, x_2, \cdots, x_n\}$ be a random sample. We let U stand for the set of all possible values of the random variable X and call it the universe of discourse of X. In short, X and U are called sample and universe, respectively.

Example 5.4 Tossing a coin 8 times, we obtain a random sample

$$X = \{1, 0, 1, 1, 1, 0, 1, 0\},$$

its universe is $U = \{0, 1\}$.

Example 5.5 Collecting all middle and strong earthquake data observed in China from 1900 to 1975, we obtain Appendix 2.A (see Chapter 2), the given sample is

$$
\begin{aligned}
X &= \{x_1, x_2, x_3, \cdots, x_{134}\} \\
&= \{(M_i, I_i)|i = 1, 2, \cdots, 134\} \\
&= \{(5\tfrac{3}{4}, VII), (5\tfrac{3}{4}, VII), (6.4, VIII), \cdots, (7.3, IX)\},
\end{aligned}
$$

its universe is $U = [4.2, 8.6] \times \{V, VI, \cdots, XII\}$.

Definition 5.5 Let X be a sample and V be a subset of U. A mapping from $X \times V$ to [0,1]

$$
\begin{aligned}
\mu: \quad & X \times V && \to [0, 1] \\
& (x, v) && \mapsto \mu(x, v), \forall (x, v) \in X \times V
\end{aligned}
$$

is called an *information diffusion* of X on V, if it is decreasing: $\forall x \in X$, $\forall v', v'' \in V$, if $||v' - x|| \leq ||v'' - x||$ then $\mu(x, v') \geq \mu(x, v'')$. μ is called a *diffusion function* and V is called a *monitoring space*. In short, μ is called a diffusion.

Example 5.6 There are 30 students in a class located at a small town. One day their teacher asked them to inform their fathers or other elder members in families attending a meeting. In this case, the set of 30 students is a sample, denoted by X, and the set of members of the students' families is a monitoring space, denoted by V. Furthermore, we can define a distance between a student and an elder member of his family. The distance between a student and his father may be the least. To his mother, the distance is little large, and to his elder brother and sisters it is more large, and so on. Then, from the students, the fathers know when and where the meeting will be held and what the theme of the meeting is. For other members of the families, the students may provide little information about the meeting. The procedure can be formed by an information diffusion of X on V. Obviously, it seems that each family empowers its child in the class to be its chief spokesman collecting the information before the meeting. In general, the boundary of a family is fuzzy, because some families have a blood relationship.

Corollary 5.3 A distribution function is a diffusion function.

Definition 5.6 $\mu(x,v)$ is sufficient if and only if $V = U$, denoted by $\mu(x,u)$.

Corollary 5.4 A distribution function is not sufficient.

It is enough to note that, in the framework of the information distribution, U is employed to denote the discrete universe of X, but, in the framework of the information diffusion, U is employed to denote the whole universe of a random variable. Certainly, the discrete universe is a subset of the whole universe. Therefore, a distribution function is not sufficient.

Definition 5.7 $\mu(x,u)$ is conservative if and only if $\forall x \in X$,

$$\int_U \mu(x,u)du = 1.$$

When the whole universe U is discrete, suppose $U = \{u_1, u_2, \cdots, u_m\}$, the conservative condition is

$$\sum_{j=1}^{m} \mu(x,u_j) = 1, \quad \forall x \in X.$$

For example, according to the property 3 of an information diffusion, we know that a distribution function is a conservative diffusion function.

Definition 5.8 For U, given an origin u_0 and a bin width h, we can construct the intervals $I_j = [u_0 + (j-1)h, u_0 + jh[, \; j = 1, 2, \cdots, m$. An *interval diffusion* is defined by

$$\mu(x,u) = \begin{cases} \frac{1}{h}, & \text{if } x \text{ is in the same interval as } u; \\ 0, & \text{otherwise.} \end{cases} \tag{5.3}$$

where "interval as u" is one of $I_j, j = 1, 2, \cdots, m$.

For example, let $X = \{1, 3, 6, 2, 7\}$, $u_0 = 0$, and $h = 2$. The interval diffusion of X on U is defined by Table 5.1.

Table 5.1   The interval diffusion of $X = \{1, 3, 6, 2, 7\}$
with $u_0 = 0$, and $h = 2$

$\mu(x,u)$	$u \in [0,2[$	$u \in [2,4[$	$u \in [4,6[$	$u \in [6,8[$
$x_1 = 1$	1/2	0	0	0
$x_2 = 3$	0	1/2	0	0
$x_3 = 6$	0	0	0	1/2
$x_4 = 2$	0	1/2	0	0
$x_5 = 7$	0	0	0	1/2
$\sum$	1/2	1	0	1

One easily verifies an interval diffusion is conservative. An interval diffusion is sufficient if and only if

$$\bigcup_{j=1}^{m} I_j = U.$$

We can use a sufficient interval diffusion to obtain a histogram estimate of PDF, that is

$$\hat{p}(x) = \frac{1}{n} \sum_{1=1}^{n} \mu(x_i, x), \quad x \in \mathbb{R}.$$

Definition 5.9 The *trivial diffusion* is defined by

$$\mu(x, u) = \begin{cases} 1, & \text{if } u = x; \\ 0, & \text{otherwise.} \end{cases} \tag{5.4}$$

For example, let x be an observation from tossing a coin. If "tail" appears, $x = 0$; else, $x = 1$. The characteristic function of x based on universe $U = \{0,1\}$

$$\chi_u(x) = \begin{cases} 1, & \text{for } x = u, \\ 0, & \text{otherwise.} \end{cases} \forall u \in \{0,1\}$$

is a trivial diffusion.

One easily verifies the trivial diffusion is sufficient. However, the trivial diffusion with a continuous universe is not conservative.

With the trivial diffusion, we express observations by fuzzy singletons. Because a singleton is just another form of a crisp observation, we say that the trivial diffusion do nothing for an observation. In other words, the trivial diffusion would be regarded as a non-diffusion operator.

Definition 5.10 $\mathcal{D}(X) = \{\mu(x, u) | x \in X, u \in U\}$ is called the *sample of fuzzy sets* derived from X on U by information diffusion. In short, $\mathcal{D}(X)$ is called the *fuzzy sample* (FS).

When each element of $\mathcal{D}(X)$ is a classical subset of U, the elements are just so-called random sets. Before we give the principle of information diffusion, it is necessary to study some properties of random sets.

5.5 Random Sets and Covering Theory

Random sets were first studied in 1940s in the Euclidean space[32],[38]. Later, they were studied in a more topological setting. After all, random sets are random elements taking subsets as values, and as such, they can be formulated as abstract settings involving measurability concepts. Wang in 1985 gave a definition of a random set based on a probability space (see Definition 4.12). Then, Wang set up a mathematical link between random sets and a fuzzy set.

5.5.1 Fuzzy logic and possibility theory

Consider an arbitrary proposition in the canonical form "u is C," where u is an object from some universal set U and C is a predicate relevant to the object. To qualify for a treatment by classical logic, the proposition must be devoid of any uncertainty. That is, it must be possible to determine whether the proposition is true or false. Any proposition that does not satisfy this requirement, due to some inherent uncertainty in it, is thus not admissible to classical logic.

An example is the fuzzy proposition "Inflation rate is very low." In this case, the predicate is fuzzy and, at the same time, information regarding the object (the actual value of the inflation rate) is deficient. The proposition is true to some specific degree, but this degree cannot be determined without the knowledge of the actual inflation rate. However, the proposition is a source of information about the actual inflation rate, under which the set of possible values of inflation rate becomes appropriately constrained. In this sense, it is meaningful to interpret the degree of possibility that the actual inflation rate is equal to r (rate). This interpretation may then serve as a basis for developing a measure-theoretic counterpart of fuzzy set theory.

Let $\mathcal{F}(U)$ be the set of all fuzzy sets of U and $\mathcal{P}(U)$ be the set of all crisp sets of U. In the sequel, we do not distinguish a subset and its indicator function, so we have $\mathcal{P}(U) \subseteq \mathcal{F}(U)$.

There are many interpretation for the meaning of a fuzzy set $A \in \mathcal{F}(U)$. The most important and widely accepted one is the possibility theory due to Zadeh (1978, see [9][42]). Suppose A is related to some variable u, then A may be considered as a restriction on the values of u. The membership degree $A(u_0)$ may be translated to be the possibility that $u = u_0$. Thus, we get a so-called possibility distribution $\pi(u)$ on U associated with variable u from A:

$$\pi(u) = A(u), \quad u \in U. \tag{5.5}$$

From this distribution, we can easily get the possibility that $u \in G$ for any subset G of U:

$$\Pi(G) = \bigvee_{u \in G} \pi(u), \tag{5.6}$$

Π is called a *possibility measure*.

Definition 5.11 A possibility measure is a function $\Pi: \mathcal{P}(U) \to [0,1]$ such that

(1) $\Pi(\emptyset) = 0$;
(2) $\Pi(\bigcup_{i \in I} A_i) = \bigvee_{i \in I} \Pi(A_i)$.

Since u is in the universe U, and $\Pi(G)$ describes the possibility of that $u \in G$, we may require the normal condition:

$$\Pi(U) = 1. \tag{5.7}$$

5.5.2 Random sets

First, we recall several basic definitions from measurement and probability theory[2]. A class of subsets of U is called a σ-algebra over U if it contains U itself and is closed under the formation of complement and countable unions. If $\mathcal{A}$ is a σ-algebra, then the pair $(U, \mathcal{A})$ is called a *measure space*, and the subsets in $\mathcal{A}$ are called *measurable sets*. For two measurable space $(U, \mathcal{A})$ and $(V, \mathcal{B})$, a mapping $f : U - V$ is $\mathcal{A} \to \mathcal{B}$ *measurable* if for every $B \in \mathcal{B}$, $f^{-1}(B) \in \mathcal{A}$. The *$\sigma$-algebra generated* by $\mathcal{A}[\subseteq \mathcal{P}(U)]$, denoted by $\sigma(\mathcal{A})$, is the intersection of all σ-algebras containing $\mathcal{A}$. A *measure* on a measurable space $(U, \mathcal{A})$ is a function $m : \mathcal{A} \to [0, \infty]$ such that $m(\emptyset) = 0$ and m is countably additive (i.e., $m(\bigcup_{i=1}^{\infty} A_i) = \sum_{i=1}^{\infty} m(A_i)$ if A_i 's are disjoint, measurable sets). For example, the extension of the length in $\mathbb{R}$ to Borel sets is a measure. In general, for R^n, there is one and only one measure on its Borel extending the ordinary volume:

$$m(\{x = (x_1, x_2, \cdots, x_n) | a_i \leq x_i \leq b_i, i = 1, 2, \cdots, n\}) = \prod_{i=1}^{n} (b_i - a_i).$$

This unique measure is called *Lebesque measure* on Borel sets[2]. If m is a measure on $(U, \mathcal{A})$, then the triple $(U, \mathcal{A}, m)$ is called a *measure space*.

We let Ω stand for a sample space, also called a foundation space. For example Ω can be all possible outcomes or results of an experiment, game (Ω can be one dimensional or multidimensional; Ω can be finite or infinite). Let $\mathcal{A}$ be a σ-algebra over Ω and P be a measure on $(\Omega, \mathcal{A})$. A *probability space* is a measure space $(\Omega, \mathcal{A}, P)$ satisfying $P(\Omega) = 1$.

Suppose $(U, \mathcal{G})$ is a locally compact, Hausdorff space[17]. Let $\mathcal{F}$ and $\mathcal{K}$ be the set of all closed and compact subsets, respectively. Matheron[22] defined three classes of random sets by three topological structures on subsets of $\mathcal{P}(U)$. For any subset $\mathcal{A}$ and element P of $\mathcal{P}(U)$, define

$$\mathcal{A}_P = \{Q | Q \in \mathcal{A}, P \cap Q = \emptyset\},$$
$$\mathcal{A}^P = \{Q | Q \in \mathcal{A}, P \cap Q \neq \emptyset\},$$
$$_P\mathcal{A} = \{Q | Q \in \mathcal{A}, P \subseteq Q\},$$
$$^P\mathcal{A} = \{Q | Q \in \mathcal{A}, P \not\subseteq Q\}.$$

With these definitions, we have the following three hypertopologic spaces: $(\mathcal{F}, \mathcal{T}_f), (\mathcal{G}, \mathcal{T}_g)$ and $(\mathcal{K}, \mathcal{T}_k)$ generated respectively by sub-spaces $\{\mathcal{F}_A, \mathcal{F}^B | a \in \mathcal{G}, B \in \mathcal{K}\}$, $\{^A\mathcal{G}, _B\mathcal{G} | a \in \mathcal{G}, B \in \mathcal{K}\}$ and $\{\mathcal{K}_A, \mathcal{K}^B | a \in \mathcal{G}, B \in \mathcal{F}\}$.

Let $\mathcal{B}_f$, $\mathcal{B}_g$, and $\mathcal{B}_k$ denote the σ-algebras generated by these three topologies on $\mathcal{F}$, $\mathcal{G}$, and $\mathcal{K}$, respectively (i.e., Borel algebras). Suppose $(\Omega, \mathcal{A}, P)$ is a probability space. Matheron defined a *random closed (open, compact, respectively) set* as an $\mathcal{A}$ - $\mathcal{B}_f$ ($\mathcal{A}$ - $\mathcal{B}_g$, $\mathcal{A}$ - $\mathcal{B}_k$, respectively) measurable mapping from Ω to $\mathcal{F}$ ($\mathcal{G}$, $\mathcal{K}$, respectively).

A closed random set ξ defines a probability measure on $\mathcal{B}_f$ by

$$P_\xi(\mathcal{D}) = P\xi^{-1}(\mathcal{D}) = P(\omega|\xi(\omega) \in \mathcal{D}).$$

the probability measure $Q = P_\xi$ is defined for every Borel set of $\mathcal{F}$. The topology on $\mathcal{F}$ is already complicated and its Borel algebra is a set of Sets of subsets. In 1953, Choquet[7] Proved that Q is completely determined by a capacity defined on $\mathcal{K}$.

Definition 5.12 A $\mathcal{K}$-capacity or Choquet capacity is a set function $T : \mathcal{P}(U) \to \mathbb{R}$ such that
(1) T is nondecreasing: $A \subseteq B \Rightarrow T(A) \leq T(B)$;
(2) T is upper continuous: $A_n \uparrow A$ in $\mathcal{P}(U) \Rightarrow T(A_n) \uparrow T(A)$;
(3) T is continuous on $\mathcal{K}$: $A_n \downarrow A$ in $\mathcal{K} \Rightarrow T(A_n) \downarrow T(A)$.

For $K, K_1, \cdots, K_n \in \mathcal{K}$, define

$$\Delta_1(K; K_1) = T(K \cup K_1) - T(K),$$

$$\Delta_2(K; K_1, K_2) = \Delta_1(K; K_1) - \Delta_1(K \cup K_2; K_1),$$

and generally,

$$\Delta_n(K; K_1, \cdots, K_n) = \Delta_{n-1}(K; K_1, \cdots, K_{n-1}) - \Delta_{n-1}(K \cup K_n; K_1, \cdots, K_{n-1})$$

A Choquet capacity is said to be alternating of infinite order is $\Delta_n \geq 0$ for every $n \geq 1$.

Theorem 5.1 (Choquet[7]) Let U be locally compact, Hausdorff space, and $T : \mathcal{K} \to \mathbb{R}$. Then there exists a unique probability measure Q on $\mathcal{B}_f$ satisfying

$$T(K) = Q(\mathcal{F}_K), \ \forall K \in \mathcal{K}$$

if and only if T is a Choquet capacity, alternating of infinite order, such that $T(\emptyset) = 0$ and $0 \leq T \leq 1$.

For the study of fuzzy sets, we only need a measurable structure on a subset $\mathcal{B}$ of $\mathcal{P}(U)$. A σ-algebra $\check{\mathrm{C}}$ cover $\mathcal{B}$ is called a *super-σ algebra*. Another approach is to define some hypertopology first on $\mathcal{B}$ and then study the Borel algebra. Before we define covering function, we need to assume that

$$_u\mathcal{B} =_{\{u\}} \mathcal{B} \in \check{\mathrm{C}}, \forall u \in U, \tag{5.8}$$

Definition 5.13[11][12] Suppose $(\Omega, \mathcal{A}, P)$ is a probability space, $\mathcal{B} \subseteq \mathcal{P}(U)$. A $\mathcal{B}$-valued primitive random set over U is a mapping $\xi : \Omega \to \mathcal{A}$ such that for every $u \in U$, $\xi^{-1}(_u\mathcal{B}) \in \mathcal{A}$. If $\check{\mathrm{C}}$ satisfies Eq.(5.8), then a random set over U is an $\mathcal{A}$-$\check{\mathrm{C}}$ measurable mapping. The set of all random sets over U is denoted by $\Xi(\Omega, \mathcal{A}, P; U, \mathcal{B}, \check{\mathrm{C}})$, or simply by $\Xi(\Omega, U)$ if there is no confusion.

5.5.3 Covering function

Definition 5.14[12][39] For $\xi \in \Xi(\Omega, U)$, its covering function is defined on U by

$$C_\xi(u) = P\{\omega | u \in \xi(\omega)\} = P[\xi^{-1}(_u\mathcal{B})] = P(u \in \xi). \qquad (5.9)$$

Although it was used by Robbins[32] in 1944, the function definition defined by Eq.(5.9) has not been formally defined until it was rediscovered and related to the membership function of fuzzy sets[26].

Theorem 5.2(Nguyen[26]**, Pratt**[31] **and Robbin**[32]**)** (1) Let U be a subset of R^n, m be the Lebesgue measure, and ξ be a random set over U. If $m(\xi)$ is well defined (i.e., $\xi(\omega)$ is Lebesque measurable for all ω) and $C_\xi(u) = P(u \in \xi)$ is defined for almost all $u \in U$, then the expected volume of ξ is the integral of $P(u \in \xi)$ over U. That is,

$$E[m(\xi)] = \int_U P(u \in \xi) dm(u) = \int_U C_\xi(u) dm(u).$$

(2) Let U be a finite set, and $\xi \in \Xi[\Omega; U, \mathcal{P}(U)]$. Then

$$E[card(\xi)] = \sum_{u \in U} C_\xi(u).$$

That is, when U is finite, the expected cardinality of ξ is the $\sum$-count cardinality of C_ξ, an extension of cardinality to fuzzy sets[43].

For a general universe U, if m is a measure over U, $\xi \in \Xi(\Omega, U)$, and C_ξ is measurable over measurable space $(U, \mathcal{B})$. By generalizing Robbins' theorem, Peng, Wang and Kandel (1996, see [30]) defined the *volume* of ξ as:

$$\overline{m}(\xi) = \int_U C_\xi(u) dm(u). \qquad (5.10)$$

A random set defines a fuzzy set by Eq. (5.9). A more important theorem is the existence of random sets for every fuzzy set, which ensures that all concepts can be approximated by random sets.

Since $\sup\{\pi(x) | x \in \cup_i A_i\} = \sup_i\{\sup\{\pi(x) | x \in A_i\}\}$, then The restriction of the possibility measure Π on J is a space law. So by Matheron's theorem, we know Π is approximatable by a random set. That is the following theorem:

Theorem 5.3 (Nguyen[25]**)** Let $\pi : U \to [0, 1]$ be a fuzzy set and Π be the possibility measure defined by Eq.(5.6). There exists a random set ξ such that for any $A \subseteq U$, we have

$$\Pi(A) = \sup\{\Pi(I) | I \in J, I \subseteq A\}$$

with

$$\Pi(I) = P\{\omega | \xi(\omega) \cap I \neq \emptyset\} = P\xi^{-1}(\mathcal{P}(U)_I).$$

Goodman and Nguyen[12] presented several very interesting theorem on existence of the random sets. They also studied the operations on fuzzy sets isomorphic to that for random sets.

Because of the need for a practical method to set up membership functions for fuzzy sets, Wang[39] proposed the set-valued statistics based on the random set theory. Set-valued statistics serves a link between the (theoretical) random sets and (practical) fuzzy sets.

Definition 5.15 A probability $(\Omega, \mathcal{A}, P)$ is sufficient for a given measurable space $(V, \mathcal{B})$ if for any probability measure m defined on $\mathcal{B}$, there is a Bore-measurable mapping $\xi : \Omega \to V$ such that m is the induced probability measured by ξ, i.e.,

$$m(B) = P(\xi^{-1}(B)), \quad \forall B \in \mathcal{B}. \tag{5.11}$$

Theorem 5.4 (Wang[39]) Suppose that $(\Omega, \mathcal{A}, P)$ is sufficient for $(\mathbb{R}, \mathcal{B})$, where $\mathcal{B}$ is the Borel field on $\mathbb{R}$. For any fuzzy set A on $\mathbb{R}$, if the membership function μ_A is Borel measurable, then there is a random set $\xi \in \Xi(\Omega, \mathbb{R})$ such that μ_A is the falling shadow function of random set ξ, i.e.,

$$\mu_A(x) = \mu_\xi(x), \quad x \in \mathbb{R}. \tag{5.12}$$

Corollary 5.5 Suppose that $(\Omega, \mathcal{A}, P)$ is sufficient for $(U \times V, \mathcal{B})$, where $\mathcal{B}$ is the Borel field on $U \times V$. For any fuzzy relationship R on $U \times V$, if the membership function $\mu_R(u, v)$ is Borel measurable, then there is a random set $\xi \in \Xi(\Omega, U \times V)$ such that μ_R is the falling shadow function of random set ξ, i.e.,

$$\mu_R(u, v) = \mu_\xi(u, v), \quad (u, v) \in U \times V. \tag{5.13}$$

The proof is trivial.

5.5.4 Set-valuedization of observation

In many situations, the observations (data) are not set values. Peng, Wang and Kandel (1996, see [30]) suggested the so-called set-valuedization of observation to set up membership functions for fuzzy sets with a give sample. Suppose $x_1, x_2, \cdots, x_n$ are single-valued data drawn from random set sample $A_1, A_2, \cdots, A_n$ of random set ξ; that is, $x_i \in A_i$. They introduced kernel methods and the nearest neighbor principle to estimate the covering function from x_i's. The main idea is to estimate the random set sample A_i's first and the apply histograms or parametric methods. A *set-valuedization* of x_i is an estimation of A_i from x_i. To set-valuedize the observations is to estimate the random set sample A_i's from the observations. On the linear line, a very general class of estimators is the so-called kernel estimators:

Definition 5.16 A kernel estimator of the covering function from x_i' is defined by

$$\widehat{C}_K(x) = \frac{1}{n}\sum_{i=1}^{n} K(\frac{x - x_i}{h_i}), \quad x \in \mathbb{R}, \tag{5.14}$$

where h_i is a constant and $K(x)$, the kernel function, satisfies $0 \le K(x) \le 1$ and $K(0) = 1$.

If the kernel function $K(x)$ takes only 0 and 1, then the Kernel Estimator (5.14) is the naive histogram of the set-valuedization A_i's with $A_i(x) = K((x - x_i)/h_i)$. There are many definition of kernels. Following are some simple forms.

Neighbor

$$K(x) = I_{[-1,1]}, \tag{5.15}$$

Right Neighbor

$$K(x) = I_{[0,1]}, \tag{5.16}$$

Application of this kernels is equivalent to the intervalization of every Observation, i.e., estimate A_i by an interval.

$$\text{Neighbor: } x_i \rightarrow A_i = [x_i - h_i, x_i + h_i].$$

Triangular

$$K(x) = \begin{cases} 1 - |x|, & \text{if } |x| \le 1; \\ 0, & \text{otherwise.} \end{cases} \tag{5.17}$$

Gaussian

$$K(x) = \exp(-x^2), \tag{5.18}$$

Quadratic

$$K(x) = \begin{cases} 1 - x^2, & \text{if } |x| \le 1; \\ 0, & \text{otherwise.} \end{cases} \tag{5.19}$$

The last three kernels are equivalent to fuzzy-valuedization for every observation.

The kernel function $K(x)$ decides the shape of the covering function, and the width parameter h_i decides the influence of the point x_i.

In paper [30], the nearest neighbor principle is employed to estimate the parameter h_i's. The principle was proposed by Fix and Hodges[10][14] and later refined by Cover and Hart[8].

Definition 5.17 Among $x_1, x_2, \cdots, x_n$, x_j is called a *nearest neighbor* of x_i, a specified one is denoted by nn_i, if

$$d(x_i, x_j) = \min_{k \ne i} d(x_i, x_k), \tag{5.20}$$

where d is a distance for x's.

In pattern recognition, roughly speaking, the nearest neighbor principle says that an object may be classified to the class which contains objects that are its nearest neighbor.

Based on the principle, we may assume that x_i comes from the same random observation as its nearest neighbor nn_i. That is, $x_i, nn_i \in A_i$. Thus, $m(A_i) \geq |x_i - nn_i|$ on the real line. that is, on $\mathbb{R}$, we may take

$$h_i = |nn_i - x_i|, \tag{5.21}$$

as an estimation of $m(A_i)$. If we estimate A_i itself by a kernel function, the estimator is a kernel estimation with the Length (5.21).

To intervalize the observations with fixed-length random intervals, the only value we have to estimate is the length l. We may estimate l by the average distances of nearest neighbors:

$$2l = \frac{1}{n} \sum_{i=1}^{n} |nn_i - x_i|. \tag{5.22}$$

5.6 Principle of Information Diffusion

Obviously, the set-valuedization of observation acts as some information diffusion. That is, there are some researchers who have considered changing a given sample X into the fuzzy sample $\mathcal{D}(X)$ (see Definition 5.10). However, they didn't consider the benefit from the changing. In fact, when X is an incomplete sample, the changing many be useful to fill up gaps caused by a scarcity of data for estimating a relationship as function approximation. In this section, we give the principle of information diffusion to describe the benefit. The principle is an assertion. In the next section, we will prove that the principle holds in the case of estimating a PDF. In next two chapters, we will discuss the properties of two diffusion functions.

Firstly, let us study how to construct relationships by using the so-called associated characteristic function[15].

5.6.1 Associated characteristic function and relationships

Given a sample $X = \{x_i | i = 1, 2, \cdots, n\}$ in a domain of deterministic crisp measurements which is denoted by U.

Definition 5.18 The *associated characteristic function of X on U* is defined as

$$\chi_{X,U}(x_i, u) = \begin{cases} 1 & \text{if } u = x_i, \\ 0 & \text{if } u \neq x_i. \end{cases} \tag{5.23}$$

where $x_i \in X$, and $u \in U$. $\chi_{X,U}(x_i, u)$ is denoted simply by $\chi(x_i, u)$ if there is no confusion.

Suppose that U is separated to be m subsets A_j, $j = 1, 2, \cdots, m$, such that

$$U = \bigcup_{j=1}^{m} A_j, \quad \text{and} \quad A_i \bigcap A_j = \emptyset \ (i \neq j). \tag{5.24}$$

Taking an element of A_j such as u_j to stands for j-th subset A_j, we obtain a discrete universe $U_m = \{u_1, u_2, \cdots, u_m\}$. In fact, U_m is a hard m-partitions about U.

Definition 5.19 The *associated characteristic function of X on U_m* is defined as

$$\chi_m(x_i, u_j) = \begin{cases} 1 & \text{if } x_i \in A_j \\ 0 & \text{if } x_i \notin A_j \end{cases} \tag{5.25}$$

Using X and employing a model (operator) γ, we can obtain an estimator of the relationship among the factors in the domain U. The estimated relationship $\widehat{R}$ depends on the given sample X and the employed operator γ, denoted as

$$\widehat{R}(\gamma, X) = \{\gamma(\chi(x_i, u)) | x_i \in X, u \in U\}. \tag{5.26}$$

Example 5.7 We introduce the *frequency operator* γ_F as

$$\gamma_F(\chi_m(x_i, u_j)) = \frac{1}{n} \sum_{i=1}^{n} \chi_m(x_i, u_j), \tag{5.27}$$

then the frequency distribution about event A_j can be obtained by

$$f(A_j) = \frac{1}{n} \sum_{i=1}^{n} \chi_m(x_i, u_j). \tag{5.28}$$

This is the probability estimation about A_j by the sample X, and the estimation is as same as the histogram when U is in the real number axis. In other words, we can employ the frequency operator and the associated characteristic function to construct a relationship between events and the probability of occurrence.

Example 5.8 In mechanics of materials, the relationship $\sigma = f(\epsilon)$ between the stress, σ, and strain, ϵ, is a most fundamental concept. It can be used to measure such diverse physical parameters as force, pressure, torque and acceleration. In general, the relationship can be obtained by tests on the stress-strain experiment. Here the domain is a 2-dimension Cartesian space, and x, y values represent the strain of the elastic element and the corresponding stress, respectively. When we get a sample $X = \{(x_1, y_2), (x_2, y_2), \cdots, (x_n, y_n)\}$, where $(x_i, y_i) = (\epsilon_i, \sigma_i)$, from an experiment on a linear elastic element, we can take the *linear least squares operator* γ_{LS} to study the stress-strain relationship about the element. That is

$$\widehat{R}(\gamma_{LS}, X) = \begin{cases} \sigma = a + b\epsilon, \\ \min \sum_{i=1}^{n} (a + b\epsilon_i - \sigma_i)^2. \end{cases} \tag{5.29}$$

where $\min \sum_{i=1}^{n} (a + b\epsilon_i - \sigma_i)^2$ can be regarded as the core part of the operator γ_{LS}, and $\sigma = a + b\epsilon$ as the last result represented by a linear function, which, in fact, is a relationship on the stress-strain product space $X \times Y$.

Eq. (5.29) is a regression result. From subsection 2.6.2 we note that the regression estimates a conditional expectation $E(Y|x)$ versus x. That is, the stress-strain relationship can be calculated from a probability distribution. On the other hand, a probability distribution can be estimated by using the associated characteristic function. Therefore, the stress-strain relationship can also be calculated from the associated characteristic function.

It may be argued that, it is always possible to estimate the relationship among the factors in the domain U by using the associated characteristic function of X on U or U_m.

5.6.2 Allocation function

Naturally, as Zadeh generalized a characteristic function in terms of a classical set into a membership function in terms of a fuzzy set, we can also generalize the associated characteristic function to take more values.

Suppose that U_m is a hard m-partitions about U, i.e., $U_m = \{A_1, A_2, \cdots, A_m\}$, A_j, $j = 1, 2, \cdots, m$, are classical subsets of U, such that $U = \bigcup_{j=1}^{m} A_j$, and $A_i \cap A_j = \emptyset$ $(i \neq j)$. Let u_j $(j = 1, 2, \cdots, m)$ be the center of the subset A_j, and h_j be a distance from u_j to the boundary of A_j.

Definition 5.20 The *allocation function of X on U_m* is defined as

$$\mu_{h_j}(x_i, u_j) = \begin{cases} 1 - ||x_i - u_j||/h_j, & \text{if } ||x_i - u_j|| \leq h_j, \\ 0, & \text{if } ||x_i - u_j|| > h_j. \end{cases} \tag{5.30}$$

$$i = 1, 2, \cdots, n; j = 1, 2, \cdots, m.$$

In fact, $\mu_{h_j}(x_i, u_j)$ is a membership function which indicates that x_i can belong to more than one subset A.

Example 5.9 The linear distribution function is an allocation function.

Obviously, using an allocation function, we can obtain a soft estimator of the relationship we want to know. For the probability distribution, the soft estimator is

$$\tilde{f}(A_j) = \frac{1}{n} \sum_{i=1}^{n} \mu_{h_j}(x_i, u_j) \tag{5.31}$$

In the last chapter we learned that, the linear distribution function would likely lead to 23% higher in the work efficiency than the histogram method for the estimation of probability distribution.

5.6.3 Diffusion estimate

Using a diffusion function, we can change a given sample X into a FS $\mathcal{D}(X)$ (see Definition 5.10, FS stands for "fuzzy sample"). If we can find a model to deal with the FS for estimating the relationship we want to know, the corresponding result is called a diffusion estimate. Formally, we give the following definition.

Definition 5.21 Let X be a given sample which can be used to estimate a relationship R by the operator γ. If the estimate is calculated by using the FS $\mathcal{D}(X)$, the estimate is called the *diffusion estimate* of R, denoted by

$$\widetilde{R}(\gamma, \mathcal{D}(X)) = \{\gamma(\mu(x_i, u)) | x_i \in X, u \in U\}. \tag{5.32}$$

where $\mu(x_i, u)$ is a diffusion function of X on U.

Correspondingly, the estimated relationship $\widehat{R}(\gamma, X)$ in Eq.(5.26) is called the *non-diffusion estimate* of R. Because a distribution function is a diffusion function, the soft estimator in Eq.(5.32) is the diffusion estimate.

Corollary 5.6 The trivial diffusion estimate is the non-diffusion estimate.

It reflects the fact that the trivial diffusion function is just the associated characteristic function.

Example 5.10 Let $x_i \in \mathbb{R}$, $U_m = \{u_1, u_2, \cdots, u_m\}$ be the discrete universe of X, and $\mu(x_i, u_j)$ be a diffusion function of X on U. Then, the diffusion estimate

$$\gamma(\mu(x_i, u_j)) = \frac{1}{n} \sum_{i=1}^{n} \frac{\mu(x_i, u_j)}{\sum_j \mu(x_i, u_j)},$$

is an estimation of the frequency distribution about event u_j, which is also an estimation of the relationship between events and probabilities of occurrence.

Theorem 5.5 A kernel estimator of the covering function from x_i' defined by Eq.(5.14) is the diffusion estimate.
Proof: The kernel function $K(x)$ is a mapping from $X \times \mathbb{R}$ to $[0, 1]$ because it satisfies $0 \leq K(x) \leq 1$ and $K(0) = 1$. Hence, $K(\frac{x-x_i}{h_i})$ is a diffusion function. Obviously, $\frac{1}{n} \sum_{i=1}^{n}$ can be regarded an operator (model). Therefore, a kernel estimator is the diffusion estimate.

$\square$

Corollary 5.7 A kernel estimator of the covering function is the sufficient diffusion estimate.
One easily verifies the kernel function $K(x)$ is a sufficient diffusion function.

The following are some kernel functions[36] satisfying $\int K(x)dx = 1$ which can produce conservative diffusion estimates.

Epanechnikov

$$K_E(x) = \begin{cases} \frac{3}{4}(1 - \frac{1}{5}x^2)/\sqrt{5}, & \text{for } |x| < \sqrt{5}, \\ 0, & \text{otherwise.} \end{cases} \qquad (5.33)$$

Biweight

$$K_B(x) = \begin{cases} \frac{15}{16}(1 - x^2)^2, & \text{for } |x| < 1, \\ 0, & \text{otherwise.} \end{cases} \qquad (5.34)$$

Triangular

$$K_T(x) = \begin{cases} 1 - |x|, & \text{for } |x| < 1, \\ 0, & \text{otherwise.} \end{cases} \qquad (5.35)$$

Gaussian

$$K_G(x) = \frac{1}{\sqrt{2\pi}}e^{-(1/2)x^2}. \qquad (5.36)$$

Rectangular

$$K_R(x) = \begin{cases} \frac{1}{2}, & \text{for } |x| < 1, \\ 0, & \text{otherwise.} \end{cases} \qquad (5.37)$$

5.6.4 Principle of Information Diffusion

When we studied the information matrix which illustrates a given small-sample for showing its information structure, we known that the so-called fuzzy-interval matrix is a more intelligent architecture. Fuzzifying intervals would likely be of benefit to show the position difference between the observations and to obtain the extra information for pattern recognition.

The fuzzifying intervals means that an observation will belong to more than one interval (fuzzy interval). In terms of information distribution, we say that the information carried by an observation has been distributed over two intervals. An information distribution acts as a singleton fuzzifier[23] mapping the crisp measurement x_i into a fuzzy set $\underset{\sim}{x_i}$ whose membership function μ_{x_i} is described by a distribution function. The simplest distribution function is the 1-dimension linear-information-distribution. Computer simulation shows the work efficiency of the distribution method is about 23% higher than histogram method for the estimation of probability distribution. In other words, the extra information is of remarkable benefit.

Interval estimation is the first tool to study the small-sample problem. Empirical Bayes methods provide a way by which such historical data can be used in the assessment of current results. The set-valuedization of observation acts as some information diffusion to cover functions. The kernel functions can be employed to set-valuedize the observations.

A common result with all approaches is that their estimates can be represented as the diffusion estimates.

In fact, according to the connection between the fuzzy intervals and distribution function and Corollary 5.3, we know that the fuzzy-interval matrix is a diffusion estimate with respect to the information structure. The interval estimation is a set-valuedization estimate of parameters. It is a special diffusion estimate. In a broad sense, an Empirical Bayes estimate is also a diffusion estimate if we regard the prior distribution as tool to set-valuedize current data for smoothing the result.

It may be argued that, in some cases, the diffusion estimate is better than the non-diffusion estimate. Else, it is difficult to explain why so many researchers are full of enthusiasm for exploring all the above methods. At least, in small-sample problems, non-diffusion estimating is considered to be rough and some diffusion estimate is needed to improve it.

Therefore, as an assertion, we propose the following principle.

The principle of information diffusion   Let $X = \{x_1, x_2, \cdots, x_n\}$ be a given sample which can be used to estimate a relationship R on universe U. Assuming that γ is a reasonable operator, $\chi(x_i, u)$ is the associated characteristic function, and by them the estimate about R is

$$\widehat{R}(\gamma, X) = \{\gamma(\chi(x_i, u)) | x_i \in X, u \in U\}.$$

Then, if and only if X is incomplete, there must exist a reasonable diffusion function $\mu(x_i, u)$ and a corresponding operator γ' to replace $\chi(x_i, u)$ and adjust γ, respectively, which leads to a diffusion estimate

$$\widetilde{R}(\gamma', \mathcal{D}(X)) = \{\gamma'(\mu(x_i, u)) | x_i \in X, u \in U\}$$

such that

$$||R - \widetilde{R}|| < ||R - \widehat{R}|| \tag{5.38}$$

where $||.||$ denotes the error between the real relationship and the estimated one.

The principle of information diffusion guarantees the existence of reasonable diffusion functions to improve the non-diffusion estimates when the given samples are incomplete. In other words, when X is incomplete, there must exist some approach to pick up fuzzy information of X for more accurately estimating a relationship as function approximation. However, the principle does not provide any indication on how to find the diffusion functions.

Fig. 5.1 gives an illustration of the information diffusion.

Corollary 5.8 Let Q be a simple information matrix of the given sample X. If X is incomplete, there must exist a fuzzy-interval information matrix $\mathcal{Q}$ of X, such that $\mathcal{Q}$ is better than Q to hang X for showing its information structure.

Reviewing the study in Chapter 2, we know that the corollary is true.

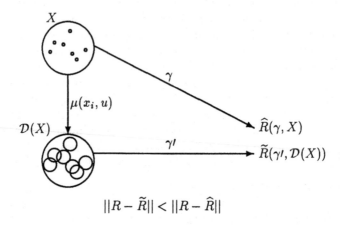

$$||R - \tilde{R}|| < ||R - \hat{R}||$$

Fig. 5.1   Illustration of the principle of information
            diffusion. X is a given sample. There must
            exist a diffusion function $\mu(x_i, u)$ that
            changes X to be a fuzzy sample $\mathcal{D}(X)$.
            Diffusion estimate $\tilde{R}$ is nearer to
            the real relationship than non-diffusion
            estimate $\hat{R}$.

Corollary 5.9 Let X be a given sample drawn from a population with
PDF $p(x)$, and $\hat{p}(x)$ be a non-diffusion estimate about $p(x)$ based on X. If
X is incomplete, there must exist a diffusion estimate $\tilde{p}(x)$ based on X, such
that $\tilde{p}(x)$ is more accurate than $\hat{p}(x)$.

The next section confirms that Corollary 5.9 is also true.

5.7 Estimating Probability by Information Diffusion

Definition 5.22 Let $x_i (i = 1, 2, ..., n)$ be $i.i.d$ sample drawn from a population with density $p(x)$, $x \in \mathbb{R}$. Suppose $\mu(y)$ is a Borel measurable function
in $(-\infty, \infty)$, $\Delta_n > 0$ is a constant, called the *diffusion coefficient*. Then,

$$\tilde{p}_n(x) = \frac{1}{n\Delta_n} \sum_{i=1}^{n} \mu\left(\frac{x - x_i}{\Delta_n}\right) \tag{5.39}$$

is called the *diffusion estimate about* $p(x)$.

$\tilde{p}_n(x)$ is a more accurate estimate if it satisfies:
1. Asymptotically unbiased property, i.e.,

$$\lim_{n\to\infty} E\tilde{p}_n(x) = p(x); \qquad (5.40)$$

2. Mean squared consistent property, i.e.,

$$\lim_{n\to\infty} E[\tilde{p}_n(x) - p(x)]^2 = 0; \qquad (5.41)$$

3. $\tilde{p}_n(x) \to p(x)(n \to \infty)$ with faster convergence rate.

In this section we will prove property 1 and 2. For property 3, we prove that $\tilde{p}_n(x)$ converges much faster than the histogram estimate. For that, firstly, we give the following lemma.

Lemma 5.1 Let $\mu(y)$, $p(x)$ be Borel measurable functions in $(-\infty, \infty)$, if they satisfy:

(1) μ is a bounded function in $(-\infty, \infty)$;

(2) $\displaystyle\int_{-\infty}^{\infty} |\mu(y)|\, dy < \infty$;

(3) $\displaystyle\lim_{|y|\to\infty} y\mu(y) = 0$, or p is a bounded function in $(-\infty, \infty)$;

(4) $\displaystyle\int_{-\infty}^{\infty} |p(x)|\, dx < \infty$.

Let:

$$p_n(x) = \frac{1}{\Delta_n} \int_{-\infty}^{\infty} \mu\Big(\frac{y}{\Delta_n}\Big) p(x - y)\, dy, \qquad (5.42)$$

where $\Delta_n > 0$ is a constant, and $\Delta_n \to 0(n \to \infty)$. When x is a continuous point, then:

$$\lim_{n\to\infty} p_n(x) = p(x) \int_{-\infty}^{\infty} \mu(y)\, dy. \qquad (5.43)$$

If p is a bounded and uniformly continuous function in$(-\infty, \infty)$, then:

$$\lim_{n\to\infty} \Big\{ \sup_x \Big| p_n(x) - p(x) \int_{-\infty}^{\infty} \mu(y)\, dy \Big| \Big\} = 0. \qquad (5.44)$$

Proof: Suppose the first condition in (3) is true. Take a constant $\delta > 0$ and fix it, then:

$$\Big| p_n(x) - p(x) \int_{-\infty}^{\infty} \mu(y)\, dy \Big|$$

$$= \Big| \int_{-\infty}^{\infty} [p(x - y) - p(x)] \frac{1}{\Delta_n} \mu\Big(\frac{y}{\Delta_n}\Big) dy \Big|$$

$$\leq \Big| \int_{|y|\leq\delta} [p(x-y)-p(x)] \frac{1}{\Delta_n} \mu\Big(\frac{y}{\Delta_n}\Big) dy \Big| + \Big| \int_{|y|\geq\delta} [p(x-y)-p(x)] \frac{1}{\Delta_n} \mu\Big(\frac{y}{\Delta_n}\Big) dy \Big|$$

$$\leq \sup_{|y|\leq\delta} |p_n(x-y) - p(x)| \int_{-\infty}^{\infty} |\mu(y)| dy \tag{5.45}$$

$$+ \left| \int_{|y|\geq\delta} p(x-y)\frac{1}{\Delta_n}\mu\left(\frac{y}{\Delta_n}\right) dy \right| + \left| p(x) \int_{|y|\geq\delta} \frac{1}{\Delta_n}\mu\left(\frac{y}{\Delta_n}\right) dy \right|$$

$$\leq \sup_{|y|\leq\delta} |p(x-y) - p(x)| \int_{-\infty}^{\infty} |\mu(y)| dy$$

$$+ \sup_{|y|\geq\delta} \left| \frac{y}{\Delta_n}\mu\left(\frac{y}{\Delta_n}\right) \right| \int_{-\infty}^{\infty} |p(y)| dy + |p(x)| \int_{|y|\geq\delta} \left| \frac{1}{\Delta_n}\mu\left(\frac{y}{\Delta_n}\right) \right| dy$$

$$= J_{1n} + J_{2n} + J_{3n}.$$

Because (2) is true and p is continuous at x point, when $\delta > 0$ is sufficient small, then J_{1n} is arbitrary small. Fix this δ. When $n \to \infty$, in the range $|y| \geq \delta$, $|\frac{y}{\Delta_n}| \to \infty$. According to (4) and the first condition in (3), when $n \to \infty$, $J_{2n} \to 0$. Because:

$$\int_{|y|\geq\delta} \left| \frac{1}{\Delta_n}\mu\left(\frac{y}{\Delta_n}\right) \right| dy = \int_{|y|\geq\frac{\delta}{\Delta_n}} |\mu(y)| \, dy, \tag{5.46}$$

and

$$\lim_{n\to\infty} \delta/\Delta_n \to \infty, \tag{5.47}$$

by the condition (2), we know that when $n \to \infty$, $J_{3n} \to 0$. It means that (5.43) is true. If the second condition in (3) is true, denote $C = \sup_x |p(x)|$, replace $J_{2n} + J_{3n}$ in (5.45) by

$$2C \int_{|y|\geq\delta} \left| \frac{1}{\Delta_n}\mu\left(\frac{y}{\Delta_n}\right) \right| dy = 2C \int_{|y|<\frac{\delta}{\Delta_n}} |\mu(y)| \, dy \tag{5.48}$$

the condition (2) will support the same result (now, the condition (4) is unnecessary).

Let us go to prove (5.44) is true. In fact:

$$\sup_x \left| p_n(x) - p(x) \int_{-\infty}^{\infty} \mu(y)dy \right| \leq \sup_x J_{1n} + \sup_x J_{2n} + \sup_x J_{3n}. \tag{5.49}$$

Because p is uniformly continuous, $\forall \varepsilon > 0$, $\exists \delta = \delta(\varepsilon) > 0$, for any x: when $|y| < \delta$, $|p(x-y) - p(x)| < \varepsilon$. So, if $\delta > 0$ is sufficient small, $\sup_x |p(x-y) - p(x)|$ is sufficient small too, and $\sup_x J_{1n}$ is arbitrary small. Fix the δ, analyse J_{2n} and J_{3n}. Because J_{2n} is uncorrelation to x, according to fore proof, we know $\lim_{n\to\infty} \sup_x J_{2n} = 0$. But p is a bounded function in $(-\infty, \infty)$, let $C = \sup_x |p(x)|$, then

$$\sup_x J_{3n} = C \int_{|y|\geq\delta/\Delta_n} |\mu(y)| \, dy \to 0, (n \to \infty). \tag{5.50}$$

$\square$

5.7.1 Asymptotically unbiased property

Theorem 5.5 Let diffusion function μ be satisfying the conditions in Lemma 5.1, and

$$\int_{-\infty}^{\infty} \mu(y)dy = 1, \tag{5.51}$$

if $\lim_{n\to\infty} \Delta_n = 0$, then

$$\lim_{n\to\infty} E\tilde{p}_n(x) = p(x), \tag{5.52}$$

where x is a continuous point of $p(x)$. If $p(x)$ is an uniformly continuous function in $(-\infty, \infty)$, then

$$\lim_{n\to\infty} \left\{ \sup_x | E\tilde{p}_n(x) - p(x) | \right\} = 0. \tag{5.53}$$

Proof:

$$
\begin{aligned}
E\tilde{p}_n(x) &= E\frac{1}{n\Delta_n} \sum_{i=1}^{n} \mu\left(\frac{x - x_i}{\Delta_n}\right) \\
&= \frac{1}{n\Delta_n} \sum_{i=1}^{n} E\mu\left(\frac{x - x_i}{\Delta_n}\right) \\
&= \frac{1}{n\Delta_n} \sum_{i=1}^{n} \int_{-\infty}^{\infty} \mu\left(\frac{x - y}{\Delta_n}\right) p(y)dy \\
&= \frac{1}{\Delta_n} \int_{-\infty}^{\infty} \mu\left(\frac{y}{\Delta_n}\right) p(x - y)dy.
\end{aligned}
\tag{5.54}
$$

Let $p_n(x) = E\tilde{p}_n(x)$, which is the same as equality in (5.42). Since μ and p are satisfying the conditions in Lemma 5.1, according to the Lemma 5.1:

$$\lim_{n\to\infty} E\tilde{p}_n(x) = \lim_{n\to\infty} p_n(x) = p(x). \tag{5.55}$$

If $p(x)$ is uniformly continuous, $p(x)$ must be a bounded function. If that is not true, in other words, suppose $p(x)$ is unbounded, since $p(x)$ is a density function, we know that $p(x) \geq 0$, $| p(x) | = p(x)$. Therefore, there exists subsequence t_m s.t. $| t_m | \to \infty$, and $\lim_{n\to\infty} p(t_m) = \infty$. That p is uniformly continuous means that for $\varepsilon = 1$, there must exists $\delta_1 > 0$, when $| x - y | < \delta_1$, $| p(x) - p(y) | < 1$. Let $I_m = [t_m - \delta_1, t_m + \delta_1]$, then $\forall x \in I_m, | p(x) - p(t_m) | < 1$, hence $p(t_m) - p(x) \leq | p(x) - p(t_m) | < 1 \Rightarrow p(x) \geq p(t_m) - 1$. That is $\min_{x \in I_m} p(x) \geq p(t_m) - 1$. So:

$$\int_{I_m} p(x)dx \geq 2\delta_1(p(t_m) - 1) \to \infty, (n \to \infty). \tag{5.56}$$

It contradicts to that $p(x)$ is a density function and $\int_{-\infty}^{\infty} p(x)dx = 1$, which is the condition we known. The contradiction means that $p(x)$ must be bounded.

Because $p(x)$ is satisfying the conditions in Lemma 5.1, according to the Lemma 5.1, we obtain that:

$$\lim_{n\to\infty}\left\{\sup_x |\, E\tilde{p}_n(x) - p(x)\,|\,\right\} = \lim_{n\to\infty}\left\{\sup_x |\, p_n(x) - p(x)\,|\,\right\} = 0. \quad (5.57)$$

$\square$

5.7.2 Mean squared consistent property

Theorem 5.6 Let diffusion function μ be satisfying the conditions in Lemma 5.1, Eq.(5.51) and

$$\lim_{n\to\infty} \Delta_n = 0, \ \lim_{n\to\infty} n\Delta_n = \infty \quad (5.58)$$

then, for any continuous point x of $p(x)$, Eq.(5.41) must be true.
Proof: Obviously, the variance of $\tilde{p}_n(x)$ is:

$$Var(\tilde{p}_n(x)) = \frac{1}{n^2\Delta_n^2}\sum_{i=1}^{n} Var\mu\left(\frac{x - x_i}{\Delta_n}\right)$$

$$= \frac{1}{n^2\Delta_n^2}\sum_{i=1}^{n} Var\mu\left(\frac{x - x_1}{\Delta_n}\right)$$

$$= \frac{1}{n\Delta_n^2} Var\mu\left(\frac{x - x_1}{\Delta_n}\right)$$

$$= \frac{1}{n\Delta_n^2} E\left(\mu\left(\frac{x - x_1}{\Delta_n}\right) - E\mu\left(\frac{x - x_1}{\Delta_n}\right)\right)^2$$

$$= \frac{1}{n\Delta_n^2}\left[E\mu^2\left(\frac{x - x_1}{\Delta_n}\right) - \left(E\mu\left(\frac{x - x_1}{\Delta_n}\right)\right)^2\right] \quad (5.59)$$

$$\le \frac{1}{n\Delta_n^2} E\mu^2\left(\frac{x - x_1}{\Delta_n}\right)$$

$$= \frac{1}{n\Delta_n}\cdot\frac{1}{\Delta_n}\int_{-\infty}^{\infty}\mu^2\left(\frac{x - y}{\Delta_n}\right)p(y)dy$$

$$= \frac{1}{n\Delta_n}\cdot\frac{1}{\Delta_n}\int_{-\infty}^{\infty}\mu_1\left(\frac{y}{\Delta_n}\right)p(x - y)dy$$

$$(\mu_1 = \mu^2).$$

It is simple to verify that if μ is satisfying the conditions in Lemma 5.1, μ_1 will be too. Because p is continuous at x, according Lemma 5.1, we obtain that:

$$\lim_{n\to\infty}\frac{1}{\Delta_n}\int_{-\infty}^{\infty}\mu_1\left(\frac{y}{\Delta_n}\right)p(x - y)dy = p(x)\int_{-\infty}^{\infty}\mu_1(y)dy < \infty. \quad (5.60)$$

According to (5.59), (5.60) and $n\Delta_n \to \infty (n \to \infty)$, we can know:

$$\lim_{n\to\infty} Var(\tilde{p}_n(x)) = 0.$$

Moreover, according to Theorem 5.5, $\lim_{n\to\infty} E\tilde{p}_n(x) = p(x)$. Therefore, when $n \to \infty$, we obtain that:

$$E[\tilde{p}_n(x) - p(x)]^2$$
$$= E\tilde{p}_n^2(x) - 2p(x)E\tilde{p}_n(x) + p^2(x)$$
$$= E\tilde{p}_n^2(x) - [E\tilde{p}_n(x)]^2 + [E\tilde{p}_n(x)]^2 - 2p(x)E\tilde{p}_n(x) + p^2(x)(5.61)$$
$$= Var(\tilde{p}_n(x)) + [E\tilde{p}_n(x) - p(x)]^2 \to 0.$$

$\square$

5.7.3 Asymptotically property of mean square error

We have proved that the *mean square error* (MSE) tend towards zero when sample size $n \to \infty$ (see (5.61)). Now, we start to calculate the main part of mean square error from which we can obtain some knowledge about convergence rate of $\tilde{p}_n(x)$, and get some information about Δ_n and μ.

Suppose that population density function satisfies :

$$p''(x) \text{ in } (-\infty, \infty) \text{ is bounded and continuous everywhere.} \qquad (5.62)$$

Furthermore, let diffusion function μ be satisfying the conditions in Lemma 5.1, and

$$K_1 = \int_{-\infty}^{\infty} y\mu(y)dy = 0, \quad K_2 = \int_{-\infty}^{\infty} y^2\mu(y)dy < \infty. \qquad (5.63)$$

We define $b_n(x) = E\tilde{p}_n(x) - p(x)$, then:

$$b_n(x)/\Delta_n^2 = \frac{1}{\Delta_n^2}\Big[\frac{1}{\Delta_n}\int_{-\infty}^{\infty}\mu\Big(\frac{y}{\Delta_n}\Big)p(x-y)dy - p(x)\int_{-\infty}^{\infty}\mu(y)dy\Big]$$
$$= \frac{1}{\Delta_n^2}\Big[\int_{-\infty}^{\infty}\mu(y)p(x-y\Delta_n)dy - p(x)\int_{-\infty}^{\infty}\mu(y)dy\Big]$$
$$= \frac{1}{\Delta_n^2}\Big\{\int_{-\infty}^{\infty}\mu(y)\Big[p(x-y\Delta_n)-p(y)\Big]dy + \frac{1}{\Delta_n}p'(y)\int_{-\infty}^{\infty}y\mu(y)dy\Big\}$$
$$= \int_{-\infty}^{\infty}\mu(y)\frac{p(x-y\Delta_n)+\Delta_n yp'(x)-p(x)}{\Delta_n^2}dy$$
$$= \int\mu(y)y^2p''(x-\theta y\Delta_n^2)dy/2 \qquad (5.64)$$

(We utilized Cauchy mean-value theorem and Lagrange mean-value theorem) where $|\theta| \le 1$ (θ is in connection with x, y, n). According to (5.62), (5.63) and control convergent theorem, we obtain that if $p''(x) \ne 0$, then:

$$b_n(x) = \frac{1}{2}p''(x)K_2\Delta_n^2 + o(\Delta_n^2). \qquad (5.65)$$

According (5.59) and (5.60), we obtain that if $p(x) \neq 0$, then:

$$Var(\tilde{p}_n(x)) = \frac{1}{n\Delta_n}p(x)\int_{-\infty}^{\infty}\mu^2(y)dy + o\Big(\frac{1}{n\Delta_n}\Big). \qquad (5.66)$$

Finally, according to (5.65) and (5.66), we have mean square error:

$$
\begin{aligned}
MSE&(\tilde{p}_n(x))\\
&= E[\tilde{p}_n(x) - p(x)]^2\\
&= Var(\tilde{p}_n(x)) + [E\tilde{p}_n(x) - p(x)]^2 \qquad (5.67)\\
&= (n\Delta_n)^{-1}p(x)\int_{-\infty}^{\infty}\mu^2(y)dy + \frac{1}{4}(p''(x))^2 K_2^2 \Delta_n^4\\
&\quad + \text{(infinitesimal of higher order)} .
\end{aligned}
$$

Therefore, the main part of mean square error is the sum of first and second term at the right side of (5.67). Denote the sum of them in $g(\Delta_n)$, then:

$$g'(\Delta_n) = -n^{-1}\Delta_n^{-2}p(x)\int_{-\infty}^{\infty}\mu^2(y)dy + (p''(x))^2 K_2^2 \Delta_n^3. \qquad (5.68)$$

Let $g'(\Delta_n) = 0$, then:

$$\Delta_n = \Big(\frac{p(x)\int_{-\infty}^{\infty}\mu^2(y)dy}{n(p''(x))^2 K_2^2}\Big)^{1/5}. \qquad (5.69)$$

Δ_n in (5.69) is the diffusion coefficient which make the main part of mean square error be going to be this least. Using this Δ_n for (5.67), we have:

$$
\begin{aligned}
MSE&(\tilde{p}_n(x))\\
&= n^{-1}\Big(\frac{p(x)\int_{-\infty}^{\infty}\mu^2(y)dy}{n(p''(x))^2 K_2^2}\Big)^{-1/5} p(x)\int_{-\infty}^{\infty}\mu^2(y)dy+\\
&\quad \frac{1}{4}(p''(x))^2 K_2^2 \Big(\frac{p(x)\int_{-\infty}^{\infty}\mu^2(y)dy}{n(p''(x))^2 K_2^2}\Big)^{4/5} + \text{ infinitesimal of higher order}\\
&= n^{-4/5}\Big\{\big[(p''(x))^2 K_2^2\big]^{1/5} \big[p(x)\int_{-\infty}^{\infty}\mu^2(y)dy\big]^{4/5}+\\
&\quad \frac{1}{4}\big[(p''(x))^2 K_2^2\big]^{1/5} \big[p(x)\int_{-\infty}^{\infty}\mu^2(y)dy\big]^{4/5}\Big\}+ \text{ infinitesimal of higher order}\\
&= \frac{5}{4}\Big\{p^2(x)\Big(\int_{-\infty}^{\infty}\mu^2(y)dy\Big)^2 K_2 p''(x)\Big\}^{2/5} n^{-4/5} + o(n^{-4/5}). \qquad (5.70)
\end{aligned}
$$

From (5.70), we know that when sample size $n \to \infty$, then $MSE(\tilde{p}_n(x)) \to 0$ with rate $n^{-4/5}$, while $\Delta_n \to 0$ with rate $n^{-1/5}$. Naturally, Eq.(5.69) does not help us to choose Δ_n, because $p(x)$ is unknown.

5.7.4 Empirical distribution function, histogram and diffusion estimate

Order $x_1, x_2, \cdots, x_n$ from small to large, and we obtain $x_1^* \leq x_2^* \leq \cdots \leq x_n^*$. The empirical distribution function is then defined by

$$P_n(x) = \begin{cases} 0, & x < x_1^*, \\ \frac{k}{n}, & x_k^* \leq x < x_{k+1}^*, \\ 1, & x \geq x_n^*. \end{cases} \tag{5.71}$$

A Russian mathematician proved[40] that:

$$\lim_{n\to\infty} \sup_{-\infty<x<\infty} | P_n(x) - P(x) | = 0. \tag{5.72}$$

where $P(x)$ is cumulative distribution function. That is, when sample size tends to sufficient large, empirical distribution is progressing toward real probability distribution of the population.

Suppose x is a continuous random variable, and density function $p(x)$ is unknown. For any bounded interval $(a, b]$, with the following points to divide the interval into m subintervals, where $m < n$ and the lengths of subintervals may be differential.

$$a = u_0 < u_1 < \cdots < u_{m-1} < u_m = b.$$

Suppose there are ν_j sample points of X falling into interval $(u_j, u_{j+1}]$, then the frequency of event $u_j < x \leq u_{j+1}$ is ν_j/n. According to the strong law of large number, we obtain that:

$$\lim_{n\to\infty} \frac{\nu_j}{n} = \int_{u_j}^{u_{j+1}} p(x)dx. \tag{5.73}$$

Therefore, if $p(x)$ is continuous, and n is sufficient large, then:

$$\frac{\nu_j}{n} \approx \Delta_n p(u_j), (\Delta_n = u_{j+1} - u_j). \tag{5.74}$$

That is,

$$\frac{\nu_j}{n} \cdot \frac{1}{\Delta_n} \approx p(u_j).$$

So-called histogram in $(a, b]$ in fact is

$$\hat{p}_n(x) = \frac{\nu_j}{n\Delta_n}, \quad \text{for } u_j < v \leq u_{j+1}. \tag{5.75}$$

When n and m are sufficient large, the figure of the histogram approximate to the figure of $p(x)$ in $(a, b]$. According to (5.73), we have:

$$\lim_{n\to\infty} \frac{\nu_j}{n} = \int_{u_j}^{u_{j+1}} p(x)dx = p(u_{j+1}) - p(u_j). \tag{5.76}$$

Hence, When $\Delta_n \to 0 (n \to \infty)$, we obtain that:

$$\lim_{n\to\infty} \hat{p}_n(x) = \lim_{\Delta_n \to 0} \frac{p(u_{j+1}) - p(u_j)}{\Delta_n} = p(x). \tag{5.77}$$

Eq.(5.77) shows that histogram estimate converges with density function of population. But, Chen, Fang, Li and Tao (1989, see [6]) proved that $\hat{p}_n(x) \to p(x)$ with rate $n^{-2/3}$. In other words, the histogram estimate is slower than the diffusion estimate whose convergence rate is $n^{-4/5}$, and the diffusion method is better than the histogram approach. The basic difference between them reflects the fact that the diffusion estimate can utilize fuzzy information, which the histogram estimate does not. Naturally, the more information used, the more reliable the estimate.

In fact, the diffusion estimate in (5.39) is just a Parzen kernel estimate [28], but, there are many way to do diffusion estimate. On the other hand, the whole of the kernel theory focus on what properties the estimate possesses. Unfortunately, few people have discussed the reason why the kernel estimate is better in some cases. This may be difficult within the classical probability theory framework. Therefore, the principle of information diffusion provides us with a new approach to again know and deeply study kernel estimation theory from the fuzzy information point of view. In other words, the kernel estimation method used fuzzy information without being aware of it before fuzzy set theory appeared. Thus, in this book, the existence of the principle of information diffusion for incomplete-data set showed the reason — incompleteness of X — for the first time.

Table 5-2 shows the convergence rate of diffusion estimate (CSDE), $n^{-4/5}$, and the convergent rate of histogram estimate (CSHE), $n^{-2/3}$.

Table 5-2 Numerical Comparing CSDE with CSHE

Sample size n	SCDE $n^{-4/5}$	CSHE $n^{-2/3}$	Relative error $\frac{(n^{-2/3} - n^{-4/5})}{n^{-2/3}}$
1	1	1	0.0000000
4	$3.298770e^{-001}$	$3.968503e^{-001}$	0.1687621
27	$7.159933e^{-002}$	$1.111111e^{-001}$	0.3556060
256	$1.184154e^{-002}$	$2.480314e^{-002}$	0.5225792
3125	$1.600000e^{-003}$	$4.678427e^{-003}$	0.6580048
46656	$1.840238e^{-004}$	$7.716049e^{-004}$	0.7615052
823543	$1.851190e^{-005}$	$1.138175e^{-004}$	0.8373545
16777220	$1.660443e^{-006}$	$1.525879e^{-005}$	0.8911812
$3.874205e^{008}$	$1.347268e^{-007}$	$1.881678e^{-006}$	0.9284007
$1.000000e^{010}$	$9.999997e^{-009}$	$2.154435e^{-007}$	0.9535841
$2.853117e^{011}$	$6.850847e^{-010}$	$2.307384e^{-008}$	0.9703090
$8.916100e^{012}$	$4.363749e^{-011}$	$2.325680e^{-009}$	0.9812367
$3.028751e^{014}$	$2.600097e^{-012}$	$2.217300e^{-010}$	0.9882736
$1.111201e^{016}$	$1.456684e^{-013}$	$2.008189e^{-011}$	0.9927463
$1.048576e^{026}$	$1.525878e^{-021}$	$4.497113e^{-018}$	0.9996607
$6.156119e^{036}$	$3.702999e^{-030}$	$2.977109e^{-025}$	0.9999875

From the table we know that, $n^{-4/5}$ is considerably faster becoming zero than $n^{-2/3}$. Roughly speaking, for $n = 27$, the relative error=0.355606 means that, we can reduce the estimation error about 35.6% if we replace the histogram method by the diffusion method when estimating a probability distribution. This assumes we can find reasonable μ and Δ_n. It is not easy work. In the following two chapters we will discuss two special diffusion functions and give some suggestions for finding μ and Δ_n. In many cases, it is easier to choose μ and Δ_n for the diffusion estimate than choosing reasonable intervals for the histogram method.

5.8 Conclusion and Discussion

To use the method of information distribution for constructing a valid matrix, it is necessary to choose reasonable controlling points, $u_1, u_2, \cdots, u_m$, and distribution function, μ. This is hard work and requires some engineering experience. To overcome the difficulty, in this chapter, we generalized (expanded) μ to diffusion information over the universe.

The generalizing is based on the study investigating the fuzziness of an incomplete sample and some properties of random sets.

That the sample X is incomplete implies that the observations in X are the representatives of all possible observations. That is, there may be a group A_i behind each observation x_i. And, x_i cannot crisply indicates the boundary of A_i. The groups are fuzzy. Therefore, when we want to inverse every observation from X to estimate the population from which X is drawn, we have to regard X as fuzzy information. The fuzziness of an incomplete sample is due to that the groups empowering $x_1, x_2, \cdots, x_n$ are fuzzy.

On one side, the groups empower the observations. The group which plays the important role of agent in controlling the population would affect many more observations. For example, the observations most likely to appear in the range of higher density in terms of a probability distribution. On the other side, the observations diffuse their information through groups for estimating the population. For example, in the model of information distribution, in fact, the controlling points are selected to be agencies of the fuzzy intervals to receive the information from the observations. If the value of Q_j is large, the corresponding interval must belong to the higher density range of the population.

Mapping observations into groups, we obtain the random-set sample in the Euclidean space. According to a theorem of Wang (1985, see[39]), we infer that: for any fuzzy relationship R on $U \times V$, if the membership function $\mu_R(u, v)$ is Borel measurable, then there is a random set ξ such that μ_R is the falling shadow function of the random set.

Summing up the superiority of the diffusion estimates and analyzing the mechanism of the diffusion, we give the **principle of information diffusion, which asserts that, when we use an incomplete-data set to es-**

timate a relationship, there must exist reasonable diffusion means to change observations into fuzzy sets to partly fill the gap caused by incompleteness and improve non-diffusion estimate.

The principle holds, at least, to estimate a PDF, because the diffusion estimate is asymptotically unbiased, mean squared consistent , and convergence at a faster rate than the histogram estimate (when we know nothing about distribution shape and haven't other knowledge, the best non-diffusion estimate seems to be histogram estimate).

When we introduce fuzzy methods to study incomplete data problems concerning small samples, it is not to make a clever disguise for probability. We are interested in the incompleteness of a small sample given rather than whether or not an observation occurs.

Shortly after the concept of information diffusion has been started, there have been a number of claims that the information diffusion is nothing but the kernel estimate in disguise. Indeed, when the model of information diffusion is employed to estimate a probability density function with the operation as same as that in the kernel estimator, the diffusion functions are the kernel functions. And, needless to say, we scarcely employ the fuzzy terminology which is prevalent.

In fact, a kernel function and an information diffusion function are related but different concepts. The former is a 'bump' placed at the observation. The kernel estimate is the sum of 'bumps'. The latter is a 'membership function' which attains its supremum, at the observation. The information diffusion estimate is a falling shadow[39] of fuzzy random sets with 'membership functions'. The bumps with respect to a kernel estimate must satisfies some strict conditions. The membership functions is more flexible. In fact, the embryonic form of information diffusion function, linear information distribution, is without any relation to a kernel function.

From the point of view of diffusing, it is easy to understand the so-called monitoring space (see Definition 5.5). When the space is a part of the whole universe (e.g. taking a discrete universe to be the monitoring space from the whole universe), any diffusion function can be reformed to be conservative. Then, it becomes possible to propose the model of discrete regression (see Chapter 8).

It is important to note that neither the kernel model nor the information diffusion model govern the physical processes in Nature. They are introduced by humans to compensate for their own limitations.

References

1. Bezdek, J. C.(1994), The thirsty traveller visits gamont: a rejoinder to "comments on fuzzy set - what are they and why?" IEEE transactions on Fuzzy Systems, Vol.2, pp.43-45
2. Billingsley, P.(1979), Probability and Measure, Wiley, New York
3. Black, M. (1937), Vagueness: an exercise in logical analysis. Philosophy of Science, Vol.4, pp.427-455 (reprinted in Intern. J. of General Systems, vol.17, No.2-3, 1990 , pp.107-128).
4. Chaitin, G.J.(1982), Gödel's theorem and information.International Journal of Theoretical Physics, Vol.22, pp.941-954
5. Chaitin, G.J.(1990), Information, Randomness & Incompleteness. World Scientific, Singapor
6. Chen, X.R., Fang,Z.B, Li, G.Y. and Tao, B. (1989), Non-parametric Statistics, Shanghai Science and Technology Press, Shanghai, pp.284-292 (in Chinese)
7. Choquet, G.(1953-54), A theory of capacities. Ann. Inst. Fourier, Vol. 5, pp.131-295
8. Cover, T.M. and Hart, P.E.(1967), Nearest neighbor pattern classification. IEEE Trans. Information Theory, Vol.13, pp.21-27
9. Dubois, D. and Prade, H. (1988), Possibility Theory. Plenum Press, New York.
10. Fix, E. and Hodges, J.L.(1977), Discriminatory analysis, nonparametric discrimination: small sample performance. USAF School of Aviation Medicinen Report, 1951; also In Agrawala, A.K., Ed: Machine Recognition of patterns, IEEE Press, New York, pp.280-322
11. Goodman, I.R.(1987), Identification of fuzzy sets with random sets. Singh, M.G. (ed): Syastems and Control Encyclopedia, Pergamon Press, New York, pp.2293-2301
12. Goodman, I.R. and Nguyen, H.T.(1985), Uncertainty Models for Knowledge-based Systems, North-Holland, Amsterdam
13. Higashi, M. and Klir, G. J. (1982), On measures of fuzziness and fuzzy complements. Intern. J. of General Systems, Vol.8(3), pp.169-180
14. Hodges,J.L.and Fix, E. (1977), Discriminatory analysis, nonparametric discrimination: consistency properties. USAF School of Aviation Medicinen Report, 1951; also In Agrawala, A.K., Ed: Machine Recognition of patterns, IEEE Press, New York, pp.261-279
15. Huang, C.F.(1997), Principle of information diffusion. Fuzzy Sets and Systems, Vol.91, pp.69-90
16. Kandle, A.(1986), Fuzzy Mathematical Techniques with Applications. Addison-Wesley, Reading, Massachusetts
17. Kelly, J. L.(1955), General Topology, Van Nostrand, Princeton
18. Klir, G. J. and Folger, T. A. (1988), Fuzzy Sets, Uncertainty and Information. Prentice-Hall, Englewood Cliffs, NJ
19. Kong, Q. and Chen, G. (1995), On deductive databases with incomplete information. ACM Transactions on Information Systems, Vol.13, Iss.3, pp.354-369
20. Lin, C. T. and Lee, C. S. G.(1996), Neural Fuzzy Systems. Prentice Hall, Englewood Cliffs, New Jersey
21. Lu, Y.C, Shang H.J., Xu, X.M., and Chen, Q.(1999), Risk analysis and evaluation of some diseases (II), Proceedings of 18th International Conference of the North American Fuzzy Information Processing Society, New York, pp.308-312
22. Matheron, G.(1975), Random Sets and Integral Geometry, John Wiley & Sons, New York

23. Meier, W., Weber, R. and Zimmermann, H.-J. (1994), Fuzzy data analysis - methods and industrial applications. Fuzzy Sets and Systems, Vol.61(1), pp.19-28

24. Muriel,H., Nicotra, M.A. and Lambas, D.G.(1995), The luminosity function of elliptic galaxies. Astronomical Journal, Vol.110, Iss. 3, pp.1032-1038

25. Nguyen, H.T.(1984), On modeling of linguistic information using random sets. Information Sciences, Vol.35, pp.265-274

26. Nguyen, H.T.(1987), On entropy of random sets and possibility distribution. Bezdek, J. (ed): The Analysis of Fuzzy Information, Vol. I, CRC Press, Boca Raton, Florida

27. Otness, R.K. and Encysin, L.(1972), Digital Time series Analysis. John Wiley, New York

28. Parzen, E.(1962), On estimation of a probability density function and mode. Ann. Math. Statist., Vol.33, pp.1065-1076.

29. Peirce, C.S.(1931), Collected Papers of Charles Sanders Peirce. Hartshorne, C. And Weiss, P. (eds): Harvard University Press, Cambridge

30. Peng, X.T., Wang, P.Z., and Kandel, A.(1996), Knowledge acquisition by random sets. International Journal of Intelligent Systems, Vol.11, No.3, pp.113-147

31. Pratt,J.W.(1961), Length of confidence intervals. Journal of American Statistical Association, Vol. 56, pp.549-567

32. Robbins, H.E.(1944,1945), On the measure of a random set, I& II. Ann. Math. Statist., Vol.15, pp.70-74, Vol.16, pp.342-347

33. Russell,B.(1923), Vagueness. Australian J. of Psychology and Philosophy, Vol.1, No. 2, pp. 84-92

34. Sarkar, M. and Yegnanarayana, B. (1998), Rough-Fuzzy Membership Functions. Proceedings of FUZZ-IEEE'98, Anchorage, USA, pp.796-801

35. Shang, H.J., Lu, Y.C, Chen, Q., and Xu, X.M.(1999), Risk analysis and evaluation of some diseases (I), Proceedings of 18th International Conference of the North American Fuzzy Information Processing Society, New York, pp.304-307

36. Silverman, B.W.(1986), Density Estimation for Statistics and Data Analysis. Chapman and Hall, London

37. Viertl, R.(1995), Statistical Methods for Non-precise Data. CRC Press, Boca Raton, Florida

38. Votaw, D.F.(1946), The probability distribution of the measure of a random linear set. Ann. Math. Statist., Vol.17, pp.240-244

39. Wang, P.Z.(1985), Fuzzy Sets and Falling Shadows of Random Sets. Beijing Normal University Press, Beijing (in Chinese)

40. Wang, Z. K.(1976), Probability Theory Basis and Applications, Science Press, Beijing, pp.214-218 (in Chinese)

41. Yager, R. R. (1979), On the measure of fuzziness and negation. Part I: Membership in the unit interval. Intern. J. of General Systems, Vol.5(4), pp.189-200

42. Zadeh, L. A. (1978), Fuzzy sets as a basis for a theory of possibility. Fuzzy Sets and Systems, Vol.1, No.1, pp.3-28.

43. Zadeh, L. A. (1987), Fuzzy Sets as Applications: Selected Papers, Yager, R.R., Ovchinnikov, S., Ton, R.M. and Nguyen, H.T., Eds., John Wiley & Sons, New York

6. Quadratic Diffusion

This chapter proves that the Epanechnikov function (a quadratic function) is just the optimal diffusion function in theory. Some models from the kernel theory are introduced to choose the diffusion coefficients. The golden section method is employed to search for a diffusion coefficient for estimating unimodal distributions. We use computer simulation to estimate a lognormal distribution and compare quadratic diffusion with some other estimates.

6.1 Optimal Diffusion Function

If we use formula (5.39) to estimate a density function $p(x)$, in theory, the diffusion function $\mu(y)$ must satisfy:
(1) μ is non-negative (from Definition 5.5), i.e.,

$$\mu(y) \geq 0, \quad -\infty < y < \infty,$$

(2) μ is conservative (from Theorem 5.5), i.e.,

$$\int_{-\infty}^{\infty} \mu(y)dy = 1,$$

(3) The first order moment of μ is equal to zero (from asymptotically property of mean square error), i.e.,

$$\int_{-\infty}^{\infty} y\mu(y)dy = 0,$$

(4) The second order moment of μ is bounded (from asymptotically property of mean square error), i.e.,

$$\int_{-\infty}^{\infty} y^2\mu(y)dy < \infty.$$

According to Eq.(5.70), the optimal diffusion function $\mu(y)$ must have J_μ as the following taking minimum.

$$J_\mu = \left(\int_{-\infty}^{\infty} \mu^2(y)dy\right)^2 \left(\int_{-\infty}^{\infty} y^2\mu(y)dy\right). \tag{6.1}$$

The J_μ is called the *error coefficient* of μ with respect to the diffusion estimate.

Let diffusion function μ be satisfying the conditions (1)-(4). Taking notice of

$$\int_{-\infty}^{\infty} y^2 \mu(y) dy = 0 \Longleftrightarrow \mu(y) \equiv 0,$$

from condition (2), we know that it is impossible. Therefore, we can suppose:

$$\int_{-\infty}^{\infty} y^2 \mu(y) dy = \frac{1}{a^2}, \quad a > 0. \tag{6.2}$$

One easily verifies function

$$f(y) = \frac{1}{a}\mu\left(\frac{y}{a}\right), \tag{6.3}$$

satisfies conditions (1)-(3), and

$$\int_{-\infty}^{\infty} y^2 f(y) dy = a^2 \int_{-\infty}^{\infty} \left(\frac{y}{a}\right)^2 \mu\left(\frac{y}{a}\right) d\left(\frac{y}{a}\right) = 1. \tag{6.4}$$

On the other hand, we have

$$J_f = \left(\int_{-\infty}^{\infty} f^2(y) dy\right)^2 \left(\int_{-\infty}^{\infty} y^2 f(y) dy\right)$$

$$= \left(\int_{-\infty}^{\infty} \mu^2\left(\frac{y}{a}\right) d\left(\frac{y}{a}\right)\right)^2 \left(\int_{-\infty}^{\infty} \left(\frac{y}{a}\right)^2 \mu\left(\frac{y}{a}\right) d\left(\frac{y}{a}\right)\right)$$

$$= \left(\int_{-\infty}^{\infty} \mu^2(y) dy\right)^2 \left(\int_{-\infty}^{\infty} y^2 \mu(y) dy\right)$$

$$= J_\mu.$$

That is, what we have to do is looking for function $f(y)$ which satisfies the conditions (1)-(3) and

(4') $$\int_{-\infty}^{\infty} y^2 f(y) dy = 1,$$

and which has J_f taking minimum.

Because condition (4'), we have

$$J_f = \left(\int_{-\infty}^{\infty} f^2(y) dy\right)^2 \left(\int_{-\infty}^{\infty} y^2 f(y) dy\right) = \left(\int_{-\infty}^{\infty} f^2(y) dy\right)^2.$$

Hence, it is enough to have

$$C = \int_{-\infty}^{\infty} f^2(y) dy$$

taking minimum.

Summarizing:

$$\text{Min} \int_{-\infty}^{\infty} f^2(y)dy$$

s. t. $f(y) \geq 0$, $\int_{-\infty}^{\infty} f(y)dy = 1$, $\int_{-\infty}^{\infty} yf(y)dy = 0$, $\int_{-\infty}^{\infty} y^2 f(y)dy = 1$.   (6.5)

Theorem 6.1 The Epanechnikov function[2]

$$f_0(y) = \begin{cases} \frac{3}{20\sqrt{5}}(5 - y^2), & \text{for } |y| < \sqrt{5}, \\ 0, & \text{otherwise.} \end{cases}$$ (6.6)

is the optimal function.

Proof: $\forall f(y)$ satisfying constraint (6.5), we have

$$\int_{-\infty}^{\infty} f^2(y)dy$$

$$= \int_{-\infty}^{\infty} f_0^2(y)dy + \int_{-\infty}^{\infty} \left[f(y) - f_0(y)\right]^2 dy$$

$$+ 2\int_{-\infty}^{\infty} f_0(y)\left[f(y) - f_0(y)\right]dy.$$ (6.7)

Let

$$f_1(y) = \frac{3}{20\sqrt{5}}(5 - y^2), \quad (-\infty < y < \infty).$$

According to constraint (6.5) we obtain

$$\int_{-\infty}^{\infty} f_1(y)f(y)dy$$

$$= \frac{3}{20\sqrt{5}}\left(\int_{-\infty}^{\infty} 5f(y)dy - \int_{-\infty}^{\infty} y^2 f(y)dy\right)$$

$$= \frac{3}{20\sqrt{5}}(5 - 1)$$

$$= \frac{3}{5\sqrt{5}}.$$

On the other hand, we have

$$\int_{-\infty}^{\infty} f_0^2(y)dy$$

$$= \frac{9}{16 \times 5^3} \int_{-\sqrt{5}}^{\sqrt{5}} (25 - 10y^2 + y^4)dy$$

$$= \frac{9}{16 \times 5^3} \left[25 \times 2\sqrt{5} - \frac{10}{3} \times 2(\sqrt{5})^3 + \frac{1}{5} \times 2(\sqrt{5})^5\right]$$

$$= \frac{3}{5\sqrt{5}}.$$

That is,

$$\int_{-\infty}^{\infty} f_1(y)f(y)dy = \int_{-\infty}^{\infty} f_0^2(y)dy.$$

Because $f(y) \geq 0$, and when $|y| > \sqrt{5}$, $f_1(y) < 1$, hence we have

$$\int_{-\infty}^{\infty} f_1(y)f(y)dy \leq \int_{-\infty}^{\infty} f_0(y)f(y)dy.$$

That is,

$$\int_{-\infty}^{\infty} f_0^2(y)dy \leq \int_{-\infty}^{\infty} f_0(y)f(y)dy.$$

Therefore, for any f which satisfies constraint (6.5), we have

$$\int_{-\infty}^{\infty} f_0(y)\Big[f(y) - f_0(y)\Big]dy \geq 0.$$

So that, from Eq.(6.7), we know

$$\int_{-\infty}^{\infty} f^2(y) \geq \int_{-\infty}^{\infty} f_0^2(y).$$

□

From Eq. (6.3), when we take f_0 be f, the optimal diffusion function will be

$$\mu_0(y) = \begin{cases} \frac{3a}{20\sqrt{5}}(5 - a^2y^2), & \text{for } |y| < \frac{\sqrt{5}}{a}, \\ 0, & \text{otherwise.} \end{cases} \qquad a > 0 \qquad (6.8)$$

Then, J_μ takes minimum:

$$J_{\mu_0} = \left(\int_{-\infty}^{\infty} f_0^2(y)dy\right)^2 = \left(\frac{3}{5\sqrt{5}}\right)^2 = 0.072. \qquad (6.9)$$

Therefore, the diffusion based on Eq.(5.39) would take a function as Eq.(6.8) in theory. To be simple, meanwhile the accurate is same, naturally we let $a = 1$. So that, the final diffusion function is

$$\mu(y) = \begin{cases} \frac{3}{20\sqrt{5}}(5 - y^2), & \text{for } |y| < \sqrt{5}, \\ 0, & \text{otherwise.} \end{cases} \qquad a > 0 \qquad (6.10)$$

It is also denoted as

$$\mu(y) = \frac{3}{20\sqrt{5}}(5 - y^2)I_{|y|<\sqrt{5}}.$$

Because Eq.(6.10) is a quadratic function, the diffusion based on the function is called *quadratic diffusion*.

Some kernels as diffusion functions and their error coefficients are given in Table 6-1. It is quite remarkable that the error coefficients obtained are so close to the minimum, 0.072. Even the rectangular kernel used when constructing the naive estimate has an error coefficient of nearly 0.083.

Table 6-1 Some kernels as diffusion functions and their error coefficients

Kernel	Diffusion function $\mu(y)$		Error coefficient
Epanechnikov	$\frac{3}{4}(1 - \frac{1}{5}y^2)/\sqrt{5}$,	for $|y| < \sqrt{5}$,	0.072
	0,	otherwise	
Biweight	$\frac{15}{16}(1 - y^2)^2$,	for $|y| < 1$,	0.073
	0,	otherwise	
Triangular	$1 - |y|$,	for $|y| < 1$,	0.074
	0,	otherwise	
Gaussian	$\frac{1}{\sqrt{2\pi}}e^{-(1/2)y^2}$	$-\infty < y < \infty$	0.080
Rectangular	$\frac{1}{2}$,	for $|y| < 1$,	0.083
	0,	otherwise	

The message of Table 6-1 is really that there is very little to choose between the various kernels on the basis of integrated error.

6.2 Choosing Δ Based on Kernel Theory

The problem of choosing a diffusion coefficient is of crucial importance in diffusion estimate. Before discussing various methods in detail, it is worth pausing to make some remarks of a general nature. It should never be forgotten that the appropriate choice of a diffusion coefficient will always be influenced by the purpose for which the diffusion estimate is to be used. If the purpose is to illustrate the data in order to suggest possible models and hypotheses, then it will probably be quite sufficient, and indeed desirable, to choose the coefficient subjectively or experientially as we choose the step length between two controlling points in the model of information distribution. A natural method for subjectively choosing a diffusion coefficient is to plot out several curves and choose the estimate that is most in accordance with one's prior ideas about the relationship. For many applications this approach will be perfectly satisfactory. When using a corresponding diffusion estimate for presenting conclusions in terms of probability distributions, various methods studied in the kernel theory can be considered.

6.2.1 Mean integrated square error

The first (Rosenblatt, 1956, see[4]) and the most widely used way of placing a measure on the *global accuracy* of $\hat{p}(x)$ as an estimator of $p(x)$ is the *mean integrated square error* (abbreviated MISE) defined by

$$MISE(\hat{p}(x)) = E \int_{-\infty}^{\infty} [\hat{p}(x) - p(x)]^2 dx. \tag{6.11}$$

Though there are other global measures of discrepancy which may be more appropriate to one's intuitive ideas about what constitutes a globally good estimate, the MISE is by far the most tractable global measure. For diffusion estimate $\tilde{p}_n(x)$, we have

$$
\begin{aligned}
&MISE(\tilde{p}_n(x)) \\
&= \int_{-\infty}^{\infty} MSE(\tilde{p}_n(x))dx \\
&= \int_{-\infty}^{\infty} Var(\tilde{p}_n(x))dx + \int_{-\infty}^{\infty} [E\tilde{p}_n(x) - p(x)]^2 dx \\
&= (n\Delta_n)^{-1} \int_{-\infty}^{\infty} p(x)dx \int_{-\infty}^{\infty} \mu^2(y)dy + \int_{-\infty}^{\infty} \frac{1}{4}(p''(x))^2 K_2^2 \Delta_n^4 dx \\
&= (n\Delta_n)^{-1} \int_{-\infty}^{\infty} \mu^2(y)dy + \frac{1}{4} K_2^2 \Delta_n^4 \int_{-\infty}^{\infty} (p''(x))^2 dx \\
&\quad + \text{(infinitesimal of higher order)}
\end{aligned}
\tag{6.12}
$$

Denote the sum of them in $G(\Delta_n)$, then:

$$
G'(\Delta_n) = -n^{-1}\Delta_n^{-2} \int_{-\infty}^{\infty} \mu^2(y)dy + \int_{-\infty}^{\infty} (p''(x))^2 K_2^2 \Delta_n^3 dx. \tag{6.13}
$$

Let $G'(\Delta_n) = 0$, then, the ideal value of Δ_n, from the point of view of minimizing the approximate mean integrated square error, can be shown by simple calculus to be equal to $\Delta_{opt}(n)$, where

$$
\Delta_{opt}(n) = K_2^{-2/5} \left\{ \int_{-\infty}^{\infty} \mu^2(y)dy \right\}^{1/5} \left\{ \int_{-\infty}^{\infty} (p''(x))^2 dx \right\}^{-1/5} n^{-1/5}. \tag{6.14}
$$

6.2.2 Reference to a standard distribution

A very easy and natural approach is to use a standard family of distributions to assign a value to the term $\int_{-\infty}^{\infty} (p''(x))^2 dx$ in the expression (6.14) for the ideal diffusion coefficient. For example, when $p(x)$ is a normal density with variance σ^2, we have,

$$
\begin{aligned}
\int_{-\infty}^{\infty} (p''(x))^2 dx &= \sigma^{-5} \int_{-\infty}^{\infty} (\phi''(x))^2 dx \\
&= \frac{3}{8\sigma^5 \sqrt{\pi}} \\
&\approx 0.212\sigma^{-5}.
\end{aligned}
\tag{6.15}
$$

where $\phi(x)$ is the standard normal density.

If a Gaussian kernel is being used, then the diffusion coefficient obtained from (6.14) would be, substituting the value (6.15),

$$\Delta_{opt}(n) = (4\pi)^{-1/10}\frac{3}{8\sqrt{\pi}}\sigma n^{-1/5} = \left(\frac{4}{3}\right)^{1/5}\sigma n^{-1/5} = 1.06\sigma n^{-1/5}. \quad (6.16)$$

A quick way of choosing the diffusion coefficient, therefore, would be to estimate σ from the data and then to substitute the estimate into (6.16). Either the usual sample standard deviation or a more robust estimator of σ could be used.

While (6.16) will work well if the population really is normally distributed, it may oversmooth somewhat if the population is mutimodal, as a result of the value of $(\int_{-\infty}^{\infty}(p''(x))^2 dx)^{1/5}$ being larger relative to the standard deviation. One easily verifies this effect if the true $p(x)$ is an equal mixture of two unit normal distributions with means separated by varying amounts.

In the kernel theory, the lognormal and t family were studied[7] for investigating the sensitivity of the optimal smoothing parameter, to skewness and kurtosis in unimodal distributions. In these cases, for heavily skewed data, using (6.16) will again oversmooth, but the formula is remarkably insensitive to kurtosis within the t family of distributions.

6.2.3 Least-squares cross-validation

Another approach is to employ the so-called least-squares cross-validation, used in the kernel theory, to automatically choose the diffusion coefficient. The method was suggested by Rudemo (1982, see[5]) and Bowman (1984, see [1]). In this subsection, we introduce it to choose the diffusion coefficient.

Given any diffusion estimate $\tilde{p}(x)$ of a density function $p(x)$, the integrated square error can be written

$$\int_{-\infty}^{\infty}(\tilde{p}(x) - p(x))^2 dx = \int_{-\infty}^{\infty}\tilde{p}^2(x)dx - 2\int_{-\infty}^{\infty}\tilde{p}(x)p(x)dx + \int_{-\infty}^{\infty}p^2(x)dx.$$
$$(6.17)$$

Now the last term of (6.17) does not depend on $\tilde{p}(x)$, and so the ideal choice of a diffusion coefficient (in the sense of minimizing integrated square error) will correspond to the choice which minimizes the quantity C defined by

$$C(\tilde{p}(x)) = \int_{-\infty}^{\infty}\tilde{p}^2(x)dx - 2\int_{-\infty}^{\infty}\tilde{p}(x)p(x)dx. \quad (6.18)$$

The basic principle of least-squares cross-validation is to construct an estimate of $C(\tilde{p}(x))$ from the data themselves and then to minimize this estimate over Δ to give the choice of the coefficient. The term $\int_{-\infty}^{\infty}\tilde{p}^2(x)dx$ can be found from the estimate $\tilde{p}(x)$. Define $\tilde{p}_{-i}(x)$ to be the function estimate constructed from all the data point *except* x_i, that is to say,

$$\tilde{p}_{-i}(x) = \frac{1}{(n-1)\Delta}\sum_{j\neq i}\mu\left(\frac{x - x_j}{\Delta}\right). \quad (6.19)$$

Now define

$$M_0(\Delta) = \int_{-\infty}^{\infty} \tilde{p}^2(x)dx - \frac{2}{n}\sum_{i=1}^{n}\tilde{p}_{-i}(x_i). \tag{6.20}$$

The score M_0 depends only on the data (although it is not in a very suitable form for easy calculation). The idea of least-squares cross-validation is to minimize the score M_0 over Δ. We shall discuss below why this procedure can be expected to give good results and also obtain a computationally simpler approximation to M_0.

In order to understand why minimizing M_0 is a sensible way to proceed, we consider the expected value of $M_0(\Delta)$. The summation term in (6.19) has expectation

$$E\frac{1}{n}\sum_{i=1}^{n}\tilde{p}_{-i}(x_i) = E\tilde{p}_{-n}(x_n)$$

$$= E\int_{-\infty}^{\infty}\tilde{p}_{-n}(x)p(x)dx$$

$$= E\int_{-\infty}^{\infty}\tilde{p}(x)p(x)dx \tag{6.21}$$

since $E(\tilde{p}(x))$ depends only on the diffusion function and the coefficient, and not on the sample size. Substituting (6.21) back into the definition of $M_0(\Delta)$ shows that $EM_0(\Delta) = EC(\tilde{p}(x))$. It follows from (6.17) that $M_0(\Delta) - \int_{-\infty}^{\infty}p^2(x)dx$ is, for all Δ, an unbiased estimator of the mean integrated square error; since the term $\int_{-\infty}^{\infty}p^2(x)dx$ is the same for all Δ, minimizing $EM_0(\Delta)$ corresponds precisely to minimizing the mean integrated square error. Assuming that the minimizer of M_0 is close to the minimizer of $E(M_0)$ indicates why we might hope that minimizing M_0 gives a good choice of diffusion coefficient.

To express the score M_0 in a form which is more suitable for computation, first define $\mu^{(2)}(y)$ to be the convolution of the diffusion function with itself. If, for example, $\mu(y)$ is the standard Gaussian function, the $\mu^{(2)}(y)$ will be the Gaussian function with variance 2. Now, assuming $\mu(y)$ is symmetric, we have. substituting $u = \frac{x}{\Delta}$,

$$\int_{-\infty}^{\infty}\tilde{p}^2(x)dx = \int_{-\infty}^{\infty}\frac{1}{n\Delta}\sum_{i=1}^{n}\mu\left(\frac{x-x_i}{\Delta}\right) \times \frac{1}{n\Delta}\sum_{j=1}^{n}\mu\left(\frac{x-x_j}{\Delta}\right)dx$$

$$= \frac{1}{n^2\Delta}\sum_{i=1}^{n}\sum_{j=1}^{n}\int_{-\infty}^{\infty}\mu\left(\frac{x_i}{\Delta}-u\right)\mu\left(u-\frac{x_j}{\Delta}\right)du$$

$$= \frac{1}{n^2\Delta}\sum_{i=1}^{n}\sum_{j=1}^{n}\mu^{(2)}\left(\frac{x_i-x_j}{\Delta}\right). \tag{6.22}$$

Also

$$\frac{1}{n}\sum_{i=1}^{n}\check{p}_{-i}(x_i) = \frac{1}{n}\sum_{i=1}^{n}\frac{1}{n-1}\sum_{j\neq i}\frac{1}{\Delta}\mu\left(\frac{x_i-x_j}{\Delta}\right)$$

$$= \frac{1}{n(n-1)}\sum_{i=1}^{n}\sum_{j=1}^{n}\frac{1}{\Delta}\mu\left(\frac{x_i-x_j}{\Delta}\right) - \frac{1}{(n-1)\Delta}\mu(0). \qquad (6.23)$$

To find $M_0(\Delta)$, the expressions (6.22) and (6.23) can be substituted into the definition (6.20). A very closely related score function $M_1(\Delta)$, still easier to calculate, is given by changing the factors $\frac{1}{n-1}$ in (6.23) to the simpler $\frac{1}{n}$, and the substituting into (6.20) to give

$$M_1(\Delta) = \frac{1}{n^2\Delta}\sum_{i=1}^{n}\sum_{j=1}^{n}\mu^*\left(\frac{x_i-x_j}{\Delta}\right) + \frac{2}{n\Delta}\mu(0), \qquad (6.24)$$

where the function μ^* is defined by

$$\mu^*(y) = \mu^{(2)}(y) - 2\mu(y). \qquad (6.25)$$

We shall see below that the least-squares cross-validation score can be found remarkably quickly by Fourier transform methods. The algorithms developed in the kernel theory are an enhanced version of Silverman (1982, see [6]).

Given any function g, denote by $\check{g}$ its Fourier transform

$$\check{g}(s) = \frac{1}{\sqrt{2\pi}}\int e^{isy}g(y)dy.$$

Define $\psi(s)$ to be the Fourier transform of the data,

$$\psi(s) = \frac{1}{n\sqrt{2\pi}}\sum_{j=1}^{n}\exp(isx_j).$$

Let $\check{p}_n(s)$ be the Fourier transform of the diffusion estimate; taking Fourier transforms of the definition of the diffusion estimate yields

$$\check{p}_n(s) = \sqrt{2\pi}\check{\mu}(\Delta s)\psi(s) \qquad (6.26)$$

by the standard convolution formula for Fourier transforms. Here we have used the property that Fourier transform of the scaled diffusion $\frac{1}{\Delta}\mu(\frac{y}{\Delta})$ is $\check{\mu}(\Delta s)$. Formula (6.26) is particularly suitable for use when μ is the Gaussian kernel; in this case the Fourier transform of μ can be substituted explicitly to yield

$$\check{p}_n(s) = \exp(-\frac{1}{2}\Delta^2 s^2)\psi(s). \qquad (6.27)$$

The basic idea of the algorithm is to use the fast Fourier transform both to find the function ψ and also to invert $\check{p}_n$ to find the diffusion estimate $\tilde{p}$. Finding the least-squares cross-validation score $M_1(\Delta)$ from the Fourier transform is straightforward. Define

$$\phi(s) = \frac{1}{n^2\sqrt{2\pi}} \sum_{j=1}^{n}\sum_{k=1}^{n} \exp[is(x_j - x_k)] = \sqrt{2\pi}|\psi(s)|^2. \tag{6.28}$$

The Fourier transform of the function μ^* defined in (6.25) is

$$\begin{aligned}
\breve{\mu}^*(s) &= \breve{\mu}^{(2)}(s) - 2\breve{\mu}(s) \\
&= \sqrt{2\pi}\breve{\mu}^2(s) - 2\breve{\mu}(s) \\
&= \frac{1}{\sqrt{2\pi}}[\exp(-s^2) - 2\exp(-\tfrac{1}{2}s^2)]
\end{aligned} \tag{6.29}$$

in the special case of the Gaussian kernel. Define a function

$$\Phi(y) = \frac{1}{n^2}\sum_{i=1}^{n}\sum_{i=1}^{n}\frac{1}{\Delta}\mu^*\left(\frac{x_i - x_j}{\Delta} - y\right). \tag{6.30}$$

Then the least-squares cross-validation score is given by

$$M_1(\Delta) = \Phi(0) + \frac{2}{n\Delta}\mu(0). \tag{6.31}$$

Now

$$\breve{\Phi}(s) = \sqrt{2\pi}\breve{\mu}^*(\Delta s)\phi(s) = 2\pi\breve{\mu}^*(\Delta s)|\psi(s)|^2$$

and so

$$\begin{aligned}
\Phi(0) &= \frac{1}{\sqrt{2\pi}}\int \breve{\Phi}(s)ds \\
&= \sqrt{2\pi}\int \breve{\mu}^*(\Delta s)|\psi(s)|^2 ds \\
&= \int [\exp(-\Delta^2 s^2) - 2\exp(-\tfrac{1}{2}\Delta^2 s^2)]|\psi(s)|^2 ds
\end{aligned} \tag{6.32}$$

if the Gaussian kernel is used to be the diffusion function μ. Substituting (6.32) into (6.31) gives the least-squares cross-validation score; note that it is not necessary to invert the transform to find this score.

Since the fast Fourier transform gives the discrete Fourier transform of a sequence rather than the Fourier transform of a function, it is necessary to make some slight adjustments to the procedure. Consider an interval $[a, b]$ on which all the data points lie. The effect of the Fourier transform method we shall describe is to impose periodic boundary conditions, by identifying the end points a and b, and so the interval should be chosen large enough that this does not present any difficulties. Putting

$$a < \min_{1\le i\le n}\{x_i\} - 3\Delta \quad \text{and} \quad b > \max_{1\le i\le n}\{x_i\} + 3\Delta$$

is ample for this purpose if the Gaussian kernel is being used. Of course, if periodic boundary conditions on a given interval are actually required, for example in the case of directional data, then that interval should be used as the interval $[a, b]$.

Choose $M = 2^r$ for some integer r; the diffusion estimate will be found at M points in the interval $[a, b]$ and choosing $r = 7$ or 8 will give excellent results. Define

$$\delta = (b - a)/M$$
$$t_k = a + k\delta, \quad \text{for } k = 0, 1, \cdots, M - 1.$$

Discretize the data as follows. If a data point x falls in the interval $[t_k, t_{k+1}]$, it is split into a weight $n^{-1}\delta^{-2}(t_{k+1} - x)$ at t_k and a weight $n^{-1}\delta^{-2}(x - t_k)$ at t_{k+1}; these weights are accumulated over all the data points x_i to give a sequence (ξ_k) of weights summing to δ^{-1}. Now, for $-\frac{1}{2}M \leq l \leq \frac{1}{2}M$, define Y_l to be the discrete Fourier transform

$$Y_l = \frac{1}{M} \sum_{k=0}^{M-1} \xi_k \exp\left(\frac{i2\pi kl}{M}\right)$$

which may be found by fast Fourier transformation.

Define

$$s_l = \frac{2\pi l}{b - a}$$

and assume for the moment that $a = 0$. Then, using the definition of the weight ξ_k,

$$Y_l = \frac{1}{M} \sum_{k=0}^{M-1} \xi_k \exp(it_k s_l)$$

$$\approx \frac{1}{nM\delta} \sum_{j=1}^{n} \exp(is_l x_j) \tag{6.33}$$

$$= \frac{\sqrt{2\pi}}{b - a} \psi(s_l). \tag{6.34}$$

The approximation (6.34) will deteriorate as $|s_l|$ increases, but, since the next step in the algorithm multiplies all Y_l for larger $|l|$ by a very small factor, this will not matter in practice.

Define a sequence ζ_l^* by

$$\zeta_l^* = \exp\left(-\frac{1}{2}\Delta^2 s_l^2\right) Y_l \tag{6.35}$$

and let ζ_k be the inverse discrete Fourier transform of ζ_l^*. Then

$$\zeta_k = \sum_{l=-M/2}^{M/2} \exp\left(-\frac{2\pi ikl}{M}\right)\zeta_l^*$$

$$\approx \sum_{l=-M/2}^{M/2} \exp(-is_l t_k)\frac{\sqrt{2\pi}}{b-a}\exp\left(-\frac{1}{2}\Delta^2 s_l^2\right)\psi(s_l)$$

$$\approx \frac{1}{\sqrt{2\pi}}\int \exp(-ist_k)\exp\left(-\frac{1}{2}\Delta^2 s^2\right)\psi(s)ds \qquad (6.36)$$

$$= \tilde{p}(t_k)$$

since (6.36) is the inverse Fourier transform of $\tilde{p}$ as derived in (6.26). The case of general a yields algebra which is slightly more complicated, but the end result (6.35) is exactly the same. For almost all practical purposes the errors are negligible.

Thus the diffusion estimate can be found on the lattic t_k by the following algorithm:

Step 1 Discretize to find the weight sequence ξ_k.
Step 2 Fast Fourier transform to find the sequence Y_l.
Step 3 Use (6.35) to fine the sequence ζ_l^*.
Step 4 Inverse fast Fourier transform to find the sequence $\tilde{p}(t_k)$.
Step 5 If estimates with other coefficients are required for the same data, repeat step 3 and 4 only.

Turn now to the question of finding the least-squares cross-validation score. We have, approximating the integral (6.32) by a sum, and substituting (6.34),

$$\Phi(0) = (b-a)\sum_{l=-M/2}^{M/2}[\exp(-\Delta^2 s_l^2) - 2\exp(-\frac{1}{2}\Delta^2 s_l^2)]|Y_l|^2$$

$$= -1 + 2(b-a)\sum_{l=1}^{M/2}[\exp(-\Delta^2 s_l^2) - 2\exp(-\frac{1}{2}\Delta^2 s_l^2)]|Y_l|^2 \qquad (6.37)$$

since $Y_0 = M^{-1}\sum \xi_k = M^{-1}\delta^{-1} = (b-a)^{-1}$ and $|Y_l| = |Y_{-l}|$ for all l. Substituting (6.37) back into (6.31) gives

$$\frac{1}{2}[1 + M_1(\Delta)] = (b-a)\sum_{l=1}^{M/2}[\exp(-\Delta^2 s_l^2) - 2\exp(-\frac{1}{2}\Delta^2 s_l^2)]|Y_l|^2 + \frac{1}{n\Delta\sqrt{2\pi}}. \qquad (6.38)$$

This criterion is easily found for a range of values of Δ. For the values that will be of interest, the exponential terms will rapidly become negligible and so the sum actually calculated will have far fewer than $\frac{1}{2}M$ terms. Very little detailed work has been done on the form of the score function $M_1(\Delta)$ and so strong recommendations cannot be made about an appropriate minimization algorithm for $M_1(\Delta)$.

6.2.4 Discussion

So far in this chapter we have concentrated attention on the case of diffusion functions being just kernels that satisfy the condition

$$\int_{-\infty}^{\infty} \mu(y)dy = 1, \quad \int_{-\infty}^{\infty} y\mu(y)dy = 0, \quad \text{and} \quad \int_{-\infty}^{\infty} y^2\mu(y)dy = \frac{1}{a^2}, \quad a > 0,$$

usually being symmetric probability density function. In the case, the optimal diffusion function is the Epanechnikov kernel. However, with respect to choosing a diffusion coefficient, almost all results from the kernel theory serve for Gaussian kernel. The model of the least-squares cross-validation score is too complicated to be introduce for Epanechnikov kernel. In the next section we will give a new method to search for a diffusion coefficient for Epanechnikov kernel.

6.3 Searching for Δ by Golden Section Method

One easily verifies that, for any diffusion function $\mu(y)$, when its coefficient Δ is large, the corresponding diffusion estimate $\tilde{p}(x)$ from Eq.(5.39) must be oversmooth. On the other hand, for any probability distribution $p(x)$ from which the sample is drawn, when the coefficient Δ is much small, Eq.(5.39) gives the estimates a somewhat ragged character which is not only aesthetically undesirable, but, more seriously, could provide the untrained observer with a misleading impression.

With help from the computer, a very easy and natural approach is to search for a diffusion coefficient Δ such that the corresponding diffusion estimate satisfies some property of the density function, neither oversmooth nor ragged. In this section, we propose the so-called *golden section method*[3] to search for Δ in estimating unimodal distributions. The method is also called the *0.618 algorithm* employed to divide a line segment a into two parts for getting a point we want, where, the greater part, y, is the mean proportional to the whole segment a and the smaller part $a - y$, i.e.

$$a : y = y : (a - y) \tag{6.39}$$

To find y, one has to solve a quadratic equation,

$$y^2 + ay - a^2 = 0, \tag{6.40}$$

and the positive solution of which is

$$y = \frac{\sqrt{5} - 1}{2} \cdot a \approx 0.618a. \tag{6.41}$$

The golden section method for obtaining Δ runs as follows:

Step 1 Initialization.
   (a) Give a sufficiently small number $\epsilon > 0$.
   (b) Let $x0 = 0, x2 = b - a, x1 = x2/2.0, \Delta = x2, \Delta0 = x0$, where

$$b = \max_{1 \leq i \leq n} \{x_i\} \quad \text{and} \quad a = \min_{1 \leq i \leq n} \{x_i\}.$$

Step 2 Calculate $\tilde{p}(x)$ by

$$\tilde{p}(x) = \frac{1}{n\Delta} \sum_{i=1}^{n} \mu\left(\frac{x - x_i}{\Delta}\right).$$

Step 3 Judge $\tilde{p}(x)$ and adjust $x1$. If $\tilde{p}(x)$ has only one peak go to (a) else go to (b).
   (a) Cut down: $x2 = x1, x1 = x0 + (x1 - x0) \times 0.618$. Go to step 4.
   (b) Enlarge: $x0 = x1, x1 = x1 + (x2 - x1) \times 0.618$.
Step 4 Adjust Δ and judge if $|\Delta - \Delta| < \epsilon$
   (a) Let $\Delta0 = \Delta, \Delta = x1$.
   (b) If $|\Delta - \Delta0| < \epsilon$, stop, else do step 2 again.

We write Program 6-1 to choose Δ for quadratic diffusion.

Program 6-1   Choosing a diffusion coefficient of quadratic
diffusion by golden section method

```
      SUBROUTINE GOLDENSM(X,N,U,M,Q)
      INTEGER N,M,NP
      REAL X(100),U(100),EPSILON,Q(100),DELTA,DELTA0,X0,X1,X2
      AN=N
      ROOT5=SQRT(5.0)
      EPSILON=0.00001
      CALL MAXMIN(N,X,B,A)
      X0=0
      X2=B-A
      X1=X2/2.0
      DELTA=X2
      DELTA0=X0
10    CONTINUE
      C=1.0/(AN*DELTA)
      DO 30 J=1,M
      Q(J)=0
      DO 20 I=1,N
      A=ABS(U(J)-X(I))/DELTA
      IF(A.GT.ROOT5) GOTO 20
      Q(J)=Q(J)+C*3.0/(20.0*ROOT5)*(5.0-A**2)
20    CONTINUE
30    CONTINUE
```

```
        CALL PEAKS(M,Q,NP)
        IF(NP.GT.1) GOTO 40
        X2=X1
        X1=X0+(X1-X0)*0.618
        GOTO 50
40      X0=X1
        X1=X1+(X2-X1)*0.618
50      DELTA0=DELTA
        DELTA=X1
        A=ABS(DELTA0-DELTA)
        IF(A.LT.EPSILON) GOTO 60
        GOTO 10
60      CONTINUE
        WRITE(*,70)DELTA0
70      FORMAT(1X,'DELTA=',F12.5,' Diffusion estimation:')
        WRITE(*,80)(J,U(J),Q(J), J=1,M)
80      FORMAT(1X,I3,F12.8,F12.8)
        RETURN
        END
```

Here, X is a given sample whose size is N, and U is a given discrete universe with M points at which the diffusion estimate will be calculated. All of them are given by a main program. DELTA and Q(J) are employed to represent a diffusion coefficient, Δ, and the diffusion estimate, $\tilde{p}(x)$, respectively. NP is employed to show the number of peaks of $\tilde{p}(x)$. Program 6-1 is supported by other two subprograms, MAXMIN and PEAKS. The former can be obtained from Appendix 4.A, the latter is the following Program 6-2 by which we find how many peaks $\tilde{p}(x)$ has.

Program 6-2 Finding the number of peaks of a diffusion estimate

```
        SUBROUTINE PEAKS(M,Q,NP)
        INTEGER M,J1,J2,NP
        REAL Q(100)
        NP=0
        DO 10 J=1,M-2
        A=Q(J+1)-Q(J)
        J1=-1
        IF(A.GT.0) J1=1
        A=Q(J+2)-Q(J+1)
        J2=-1
        IF(A.GT.0) J2=1
        IF(J1.EQ.J2) GOTO 10
        NP=NP+1
10      CONTINUE
        RETURN
```

END

In this program we use the difference method to judge if the diffusion estimate $\tilde{p}(x)$ reaches a peak at point u_{j+1}. When two neighboring differences, $\tilde{p}(u_{j+1}) - \tilde{p}(u_j)$ and $\tilde{p}(u_{j+2}) - \tilde{p}(u_{j+1})$, take different sign (positive or negative), the relevant center point, u_{j+1}, is a peak point.

Example 6.1 Let X and U in subsection 4.4.4 (lognormal experiment) be a given sample (N=19) and a given discrete universe (M=11), respectively. That is, we are given

$$\begin{aligned} X = & \{x_1, x_2, \cdots, x_{19}\} \\ = & \{17.70, 13.93, 24.27, 14.46, 43.11, 26.35, 31.76, 17.48, 20.79, 16.38, \\ & 13.51, 31.32, 18.48, 9.32, 18.81, 21.85, 14.39, 22.84, 28.89\}. \end{aligned}$$

Now, we study choosing Δ for estimating the population of X by quadratic diffusion. The diffusion estimate will be calculated at

$$\begin{aligned} U = & \{u_1, u_2, \cdots, u_{11}\} \\ = & \{2.05, 6.15, 10.25, 14.35, 18.45, 22.55, 26.65, 30.75, 34.85, 38.95, 43.05\}. \end{aligned}$$

Step 1: Initialization.
  (a) Let $\epsilon = 0.00001$.
  (b) Because $b = 43.11$, $a = 9.32$, we obtain $x0 = 0, x2 = b - a = 33.79, x1 = x2/2.0 = 16.90, \Delta = 33.79, \Delta0 = 0$.
Step 2: Calculating $\tilde{p}(x)$, we obtain

$$\begin{aligned} & \{\tilde{p}(u_1), \tilde{p}(u_2), \cdots, \tilde{p}(u_{11})\} \\ = & \{0.00917, 0.00941, 0.00960, 0.00973, 0.00980, 0.00981, \\ & 0.00977, 0.00966, 0.00950, 0.00928, 0.00900\}. \end{aligned}$$

Step 3: $\tilde{p}(x)$ has just one peak, $\tilde{p}(u_6) = \tilde{p}(22.55) = 0.00981$. Hence, It needs to cut down Δ.
  (a) $x2 = x1 = 16.90, x1 = x0 + (x1 - x0) \times 0.618 = 16.90 \times 0.618 = 10.44$.
Step 4:
  (a) $\Delta0 = \Delta = 33.79, \Delta = x1 = 10.44$.
  (b) Because $|\Delta - \Delta0| = |10.44 - 33.79| = 23.35 > \epsilon = 0.00001$, we do step 2 again.

After 5 iterations (step 2 runs 5 times), the diffusion estimate $\tilde{p}(x)$ has two peaks, then $x0 = 0, x1 = 2.46440, x2 = 3.98771, \Delta = 2.46440$. In the case, It needs to enlarge Δ. In step 3 we go to (b), i.e., we let $x0 = x1 = 2.46440, x1 = x1 + (x2 - x1) \times 0.618 = 2.46440 + (3.98771 - 2.46440) \times 0.618 = 3.40581$. Now, $|\Delta - \Delta0| = |3.40581 - 2.46440| > \epsilon$, we have to do step 2 again.

After 22 iterations, we have $|\Delta - \Delta0| = 0.000009 < \epsilon$, then the last diffusion estimate is one we need, which is

$$\begin{aligned} & \{\tilde{p}(u_1), \tilde{p}(u_2), \cdots, \tilde{p}(u_{11})\} \\ = & \{0.00084, 0.00514, 0.02362, 0.04492, 0.05250, 0.04051, \\ & 0.02960, 0.01968, 0.01037, 0.00495, 0.00495\}. \end{aligned}$$

6.4 Comparison with Other Estimates

Based on a correct assumption for X, in Example 6.1, which is drawn from a lognormal population, we calculate the maximum likelihood estimate by the following formulae.

For lognormal assumption, we known (see subsection 3.3.4)

$$E(x) = \exp(\mu + \frac{\sigma^2}{2}), \quad \mathrm{Var}(x) = [\exp(\sigma^2) - 1]\exp(2\mu + \sigma^2).$$

Let $\bar{x}$ be sample mean and s be estimator of the standard deviation. we have

$$\begin{cases} \bar{x} = \exp(\hat{\mu} + \frac{\hat{\sigma}^2}{2}) \\ s^2 = [\exp(\hat{\sigma}^2) - 1]\exp(2\hat{\mu} + \hat{\sigma}^2) \end{cases}$$

$$\begin{cases} \log \bar{x} = \hat{\mu} + \frac{\hat{\sigma}^2}{2} \\ \log s^2 = 2\hat{\mu} + \hat{\sigma}^2 + \log[\exp(\hat{\sigma}^2) - 1] \end{cases}$$

$$\begin{cases} 2\log \bar{x} = 2\hat{\mu} + \hat{\sigma}^2 \\ \log s^2 = 2\hat{\mu} + \hat{\sigma}^2 + \log[\exp(\hat{\sigma}^2) - 1] \end{cases}$$

$$\log s^2 - 2\log \bar{x} = \log[\exp(\hat{\sigma}^2) - 1]$$

$$\log \frac{s^2}{\bar{x}^2} = \log[\exp(\hat{\sigma}^2) - 1]$$

$$\frac{s^2}{\bar{x}^2} = \exp(\hat{\sigma}^2) - 1$$

$$\exp(\hat{\sigma}^2) = \frac{s^2}{\bar{x}^2} + 1$$

$$\hat{\sigma}^2 = \log(\frac{s^2}{\bar{x}^2} + 1) \tag{6.42}$$

$$2\log \bar{x} = 2\hat{\mu} + \log(\frac{s^2}{\bar{x}^2} + 1)$$

$$\log \bar{x} = \hat{\mu} + \frac{1}{2}\log(\frac{s^2}{\bar{x}^2} + 1)$$

$$\hat{\mu} = \log \bar{x} - \frac{1}{2}\log(\frac{s^2}{\bar{x}^2} + 1) \tag{6.43}$$

Using X, we obtain

$$\bar{x} = \frac{1}{19}\sum_{i=1}^{19} x_i = 21.35, \quad s = \sqrt{\frac{1}{18}\sum_{i=1}^{19}(\bar{x} - x_i)^2} = 8.16.$$

Therefore,

$$\hat{\mu} = \log 21.35 - 0.5\log(\frac{8.16^2}{21.35^2} + 1) = 2.993$$

$$\hat{\sigma}^2 = \log(\frac{8.16^2}{21.35^2} + 1) = 0.1363$$

Finally, we obtain

$$\hat{p}_1(x) = \frac{1}{x\hat{\sigma}\sqrt{2\pi}} \exp[-\frac{(\log x - \hat{\mu})^2}{2\hat{\sigma}^2}] = \frac{1.08}{x} \exp[-\frac{(\log x - \hat{\mu})^2}{0.273}]$$

If we assume that X is drawn from a normal population or an exponential population, with the maximum likelihood method we obtain estimates as the following.

(1) For normal assumption, we have

$$\hat{p}_2(x) = \frac{1}{s\sqrt{2\pi}} \exp[-\frac{(\bar{x} - x_i)^2}{2s^2}] = 0.04888 \exp[-\frac{(\bar{x} - x_i)^2}{133.2526}].$$

(2) For exponential assumption, we have

$$\hat{\lambda} = \frac{1}{\bar{x}} = \frac{1}{21.35} = 0.0468,$$

we obtain another estimate,

$$\hat{p}_3(x) = \hat{\lambda}e^{-\hat{\lambda}x} = 0.0468e^{-0.0468x}, \quad x \geq 0,$$

Fig. 6.1 shows the four estimates. Comparing them, it seems diffusion estimate (diffusion coefficient Δ is calculated by the golden section method, see Section 6.3.) $\tilde{p}(x)$, is better.

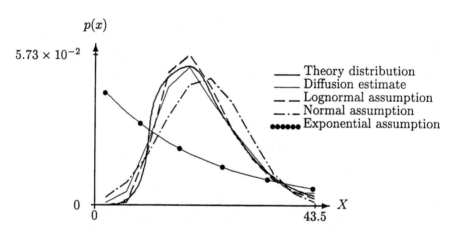

Fig. 6.1   Lognormal distribution estimation by quadratic diffusion, normal assumption, lognormal assumption and exponential assumption.

We let $n = 10, 12, \cdots, 30$, and run 90 simulation experiments with different seed numbers, respectively. The average non-log-divergences, $\tilde{\rho}$, $\hat{\rho}_1$, $\hat{\rho}_2$ and $\hat{\rho}_3$, in terms of the quadratic diffusion estimate $\tilde{p}(x)$, lognormal assumption estimate $\hat{p}_1(x)$, normal assumption estimate $\hat{p}_2(x)$ and exponential assumption estimate $\hat{p}_3(x)$, respectively, are shown in Fig. 6.2.

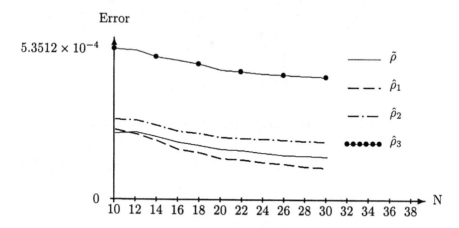

Fig. 6.2   Average non-log-divergence curves in terms of quadratic diffusion ($\tilde{\rho}$), lognormal assumption ($\hat{\rho}_1$), normal assumption ($\hat{\rho}_2$) and exponential assumption($\hat{\rho}_3$).

Obviously, from the point of view of statistics, if we can reach a correct assumption for X, the maximum likelihood estimate is the best. Otherwise, quadratic diffusion can be assessed as robust method for the population estimation, more reliable for the task at computer.

Employing the Gaussian kernel with $\Delta_{opt}(n) = 1.06\sigma n^{-1/5}$ (see Eq.(6.16)), we obtain a Gaussian diffusion estimate, denoted $\tilde{p}_G(x)$. For X in Example 6.1, the estimate is shown in Fig. 6.3(a). The corresponding average non-log-divergence curves, $\tilde{\rho}_G$, is given in Fig. 6.3(b).

Fig. 6.3 shows that the quadratic diffusion does not have the distinct advantage of Gaussian diffusion. Particularly, it is not easy to choose the diffusion coefficient for quadratic function if the population, from which a given sample is drawn, is not unimodal. In other words, although the quadratic diffusion is the optimal diffusion in theory, the Gaussian diffusion may be the more practical one for simplification in calculation. However, the kernel theory has provided no help in explaining the reason. In next chapter we will study it by the point of view of information diffusion.

density

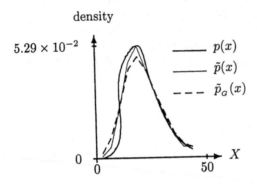

5.29 × 10⁻² → 5.29×10^{-2}

(a) Quadratic estimate $\tilde{p}(x)$ and Gaussian estimate $\tilde{p}_G(x)$

Error

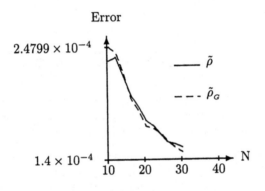

2.4799×10^{-4}

1.4×10^{-4}

(b) Average non-log-divergences

Fig. 6.3 Comparing of Quadratic diffusion and Gaussian diffusion

6.5 Conclusion

(1) A quadratic function, Epanechnikov function, is just the optimal diffusion function in theory.

(2) If a Gaussian kernel is being used, in theory, the optimal diffusion coefficient is $\Delta_{opt}(n) = 1.06\sigma n^{-1/5}$.

(3) The golden section method can be employed to search for a diffusion coefficient of quadratic diffusion to estimate unimodal distributions.

(4) If we cannot reach a correct assumption for X, quadratic diffusion can be assessed as robust method for the unimodal population estimation.

(5) The quadratic diffusion does not have the distinct advantage of Gaussian diffusion in practice. Thus Gaussian diffusion may be more practical for simplification in calculation.

References

1. Bowmanm A.W.(1984), An alternative method of cross-validation for the smoothing of density estimates. Biometrika, Vol.71, pp.353-360
2. Epanechnikov, V.A. (1969), Nonparametric estimation of a multidimensional probability density. Theor. Probab. Appl., Vol.14, pp.153-158
3. Govindarajulu, Z. (1981), The Sequential Statistical Analysis of Hypothesis Testing, Point and Interval Estimation, and Decision Theory, American Sciences Press, Columbus, Ohio
4. Rosenblatt, M.(1956), Remarks on some nonparametric estimates of a density function. Ann. Math. Statist., Vol.27, pp.832-837
5. Rudemo, M.(1982), Empirical choice of histograms and kernel density estimators. Scand. J. Statist., Vol.9, pp.65-78
6. Silverman, B.W.(1982), Kernel density estimation using the fast Fourier transform. Statistical Algorithm AS 176. Appl. Statist., Vol.31, pp.93-97
7. Silverman, B.W.(1986), Density Estimation for Statistics and Data Analysis. Chapman and Hall, London

7. Normal Diffusion

Many techniques model their manners on the human brain to process information. Fuzzy controllers use "if-then" rules. Neural networks learn. Genetic algorithms imitate biological genes. Along the novel line of thinking, in this chapter, we introduce the molecule diffusion theory to produce an information diffusion equation. Solving it, we obtain a simple and practical diffusion function, named normal diffusion. Some formulas are suggested to calculate the diffusion coefficient for the function, and some simulation experiments are done to show the quality of this type diffusion.

7.1 Introduction

Before introducing normal diffusion, it is worth pausing to make some remarks for the above chapters. It should never be forgotten that our overall goal is to find structure (information) about a given sample for function approximation. The information diffusion is a relatively new field of data analysis. In general, data analysis can be considered as a process in which starting from some given data sets information about the respective application is generated[6]. In this sense data analysis can be defined as search for structure in data [1].

When we study information matrixes for illustrating a given sample, a very interesting phenomenon occurs. A fuzzy-interval information matrix can collect all information from the given sample. Then, the information distribution is developed to change the crisp observations into fuzzy sets for finding a suitable structure. To avoid one of the "bottle neck" problems of the information distribution – choosing framework space $U = \{u_1, u_2, \cdots, u_m\}$, we study the fuzziness of a given sample and suggest the principle of information diffusion. The principle is of great interest when applied to probability estimation. In the case, the diffusion estimate is just the same as the kernel estimate if a kernel function is employed to be the diffusion function. A natural approach is to use the kernel theory to study information diffusion. Hence, we proved that the principle of information diffusion holds, at a minimum, in this case of estimating a probability density function. However, as with other statistical theories, the kernel theory is based on a sufficiently large sample. Therefore, the so-called optimal diffusion function, Epanechnikov function, does not have the distinct advantage of Gaussian diffusion. Nobody can explain the reason. Furthermore, in kernel theory, it is almost impossible to construct diffusion functions according to practical applications. It seems that the use of kernel theory cannot be completed for information diffusion, only for probability estimation.

Recent research in data analysis has produced a novel line of thinking highly different the classical applied mathematics. Many fuzzy techniques[5][7], neural networks[9][10] and genetic algorithms[3][8] are all from this novel line of thinking — an attempt to automate one of the capabilities of the human brain or a law of nature. Fuzzy techniques are closer to human thinking. Neural networks have so much learning power or non-linear modeling capabilities. Genetic algorithms imitate biological genes and are suitable for random optimization problems.

It seems that this line of thinking of imitation would be worthy of in-depth investigations of information diffusion. In this chapter, we analyze similarities of molecule diffusion and information diffusion and then produce the model of normal information diffusion.

7.2 Molecule Diffusion Theory

Molecule diffusion means the slow motion of molecules from one place to another. As solid as a pane of glass can seem, molecules of air can easily pass through the glass. This can readily be noticed on a cold winter night! The smell of cookies baking diffuses into the living room from the kitchen!

7.2.1 Diffusion

Diffusion is a very basic physicochemical process known as passive permeability and osmosis, and connects to the various concepts associated with mediated transport processes.

Diffusion can be qualitatively described as the net migration of molecules by means of their random motion from one region to another of lower concentration. Consider a container in which two solutions, one more dilute than the other, are initially separated along a hypothetical, sharply defined boundary. All the molecules are in random motion as a result of thermal energy. There is no preferred direction for the random movement of individual molecules, so molecules can move toward the region of low concentration or toward the region of high concentration. Consider now two equal volumes (A and B) on either side of the hypothetical boundary. Because of their random movement, during a unit of time a certain fraction of the molecules, say one-quarter, will move from side A to side B, and vice versa. If, for example, there are initially 1000 molecules on side A and 500 on side B, then after the specified time interval the ratio can be expected to have shifted from 1000:500 to 875:625.

The process can be repeated starting from the "new" ratio and, if repeated a number of times, it is evident that the ratio will approach 1:1; an equilibrium condition will eventually be reached.

The proportion of molecules which move across the boundary per unit time is a constant characteristic for the given molecules (e.g., fewer large molecules, which diffuse slowly, can be expected to cross the boundary, compared to small molecules). Thus, the number of molecules moving from a specified region across a unit area of the boundary per unit time is simply proportional to the concentration of the molecules in that region.

The rate of this movement is referred to as the flux of that compound and is designated by the letter "J." An important concept in cell physiology is that the net flux of molecules between two compartments is the difference between two oppositely oriented unidirectional fluxes. Recall the qualitative description of diffusion introduced above. The unidirectional fluxes of molecules from side A to side B (J_{AB}) and from side B to side A (J_{BA}) are proportional to the concentrations of the molecules on each side. The net flux from side A to B ($J_{AB_{net}}$) is the difference between the two unidirectional fluxes, i.e.,

$$J_{AB_{net}} = J_{AB} - J_{BA}. \tag{7.1}$$

In the simple system described, the net flux is zero when the concentration on side A is equal to the concentration on side B. This concept is important when one considers that the net rate of transport into and out of cells is the sum of all the transport processes occurring across cell membranes, including passive diffusion and carrier-mediated transport.

The difference in the unidirectional fluxes is proportional to the "driving force" acting on the molecule. For a nonelectrolyte the driving force is the chemical gradient, $\Delta C/\Delta x$, where ΔC is the difference in concentration between the regions of interest, and Δx is the distance separating these regions. The working relationship between net flux and the driving force for transport is:

$$J_{net} = D\frac{\Delta C}{\Delta x}. \tag{7.2}$$

where J_{net} has units of moles.time-1.area-1. The term "D" is the diffusion coefficient, a proportionality constant unique for each solute and set of experimental conditions (i.e., temperature, pressure, solvent, etc.); D has units of cm^2/sec and for small molecules is typically in the range of $1 \times 10^{-5} cm^2 sec^{-1}$. This relationship is commonly referred to as "Fick's law."

Two aspects of Fick's law should be stressed. First, simple diffusion, a process whereby a substance passes through a membrane without the aid of an intermediary, is directly proportional to the magnitude of the concentration gradient, $\Delta C/\Delta x$; if the gradient doubles, the net movement of molecules also doubles. This is an important diagnostic characteristic of passive diffusion (i.e., a major take-home message).

Second, the time required for a molecule to diffuse between two points is proportional to the square of the distance separating the two points. This is of extreme consequence in assessing the role of diffusion in the physiology of cells and organisms and warrants some discussion. Albert Einstein's examination of the theoretical aspects of diffusion resulted in development of what has become known as the "Einstein relationship", which relates the time of diffusion to the distance traveled (i.e., the "average molecular displacement").

The critical observation is that the time required for diffusion to occur is related to the square of the distance which the molecule must travel. Thus, though diffusion is comparatively fast over small, "cellular" distances (equivalent to 50 millisec when x is equivalent to 10 microns), it can represent a "limiting process" over large, "organismic" distances (e.g., > 8 minutes when x is > 1 millimeter). Cellular processes can typically rely on diffusion to provide substrates or remove wastes. In the brain, for example, few, if any, cells normally lie more than 100 microns from a capillary which serves as a source (and sink) for nutrients (and wastes). However, one notable example where diffusion would be limiting within a single cell is the case of nerve cells with long axonal or dendritic processes. Axons can be many cm long (up to a meter for motor axons running from the spine to the foot!). The synthetic machinery for making proteins, peptides and other cell components resides within the nerve cell body. Supplying the synaptic terminals at the end of axonal processes requires a form of convective (not diffusional) transport called "axonal transport." On a still larger scale, multicellular organisms must have non-diffusion dependent processes to move substrates over macroscopic distances. The requirement for an efficient circulatory system in large, multicellular organisms is a direct consequence of the relationship between diffusion, time, and distance.

The influence of distance on diffusion also has consequences in a number of pathological situations. For example, a consequence of stroke is a reduction or cessation of blood flow resulting in an ischemic condition in areas downstream of the obstruction. Cells in these areas can be several millimeters to several centimeters from a capillary source of nutrients and O_2. As noted above, diffusion over such distances can take from minutes to many hours, too slow to subserve metabolic demand.

7.2.2 Diffusion equation

For the case of diffusion in *one dimension* (the x-direction), Fig. 7.1 gives a diagram of molecule diffusion through a small unit.

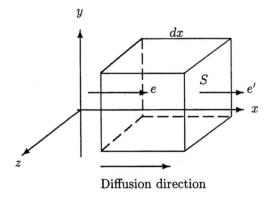

Fig. 7.1 Molecule diffusion through a small unit.

Let ψ be the molecules concentration function, e be the rate of molecule diffusion, and $\partial\psi/\partial x$ be the concentration gradient. In 1855, the physiologist Adolf Fick determined experimentally that e is directly proportional to the change in $\partial\psi/\partial x$:

$$e = -D\frac{\partial\psi}{\partial x} \tag{7.3}$$

D is a constant of proportionality called the diffusion constant, or diffusion coefficient, which is same as in Eq.(7.2). This relationship is called *Fick's first law*. Where the concentration is conventionally taken to be negative so that there is a positive flux of molecules along the x-axis.

The diffusion process must also satisfy the continuity equation describing the conservation of molecules, namely that any variation of the concentration with time must produce a concomitant spatially varying flux. For the unit in Fig. 7.1, length dx, section area S, its volume is

$$dV = \text{section area} \times \text{length} = Sdx.$$

At some time, the number of molecules of the type we study in the unit is $\psi dV = \psi Sdx$. The number of molecules which have come in left bottom is called *incidence flux*, being equal to eS. The number of molecules which have left right bottom is called *emission flux*, being equal to $e'S$. The conservation of molecules demands that the *cumulative speed* of the molecules in dV must be equal to the difference between the incidence flux and emission flux, i.e.,

$$\text{cumulative speed} = \text{incidence flux} - \text{emission flux}.$$

Hence, we obtain

$$\begin{aligned} \text{cumulative speed} \ &= eS - e's \\ &= -(e' - e)S \\ &= -(de)S \\ &= -\frac{\partial e}{\partial x}Sdx. \end{aligned} \tag{7.4}$$

But, the cumulative speed is equal to the number of added molecules per unit time per unit volume, $\partial\psi/\partial t$, by the volume, $S dx$, i.e.,

$$\text{cumulative speed} = \frac{\partial\psi}{\partial t} S dx. \qquad (7.5)$$

Combining Eq.(7.4) and (7.5), we obtain

$$-\frac{\partial e}{\partial x} S dx = \frac{\partial\psi}{\partial t} S dx,$$

therefore

$$\frac{\partial\psi}{\partial t} = -\frac{\partial e}{\partial x}. \qquad (7.6)$$

Substituting Eq.(7.3) into Eq. (7.6), we obtain

$$\frac{\partial\psi}{\partial t} = D\frac{\partial^2\psi}{\partial x^2} \qquad (7.7)$$

which is called the *diffusion equation*, or *Fick's second law*.

7.3 Information Diffusion Equation

7.3.1 Similarities of molecule diffusion and information diffusion

A molecule is an existence form of material, and our word is filled with molecules which are moving very slowly or quickly. Information can be regarded as an existence form of knowledge, which is useful for us and can be carried from here to there. Molecule and information both posses the property of diffusion.

According to Definition 5.5, we know that an information diffusion function $\mu(x, v)$ is decreasing: $\forall x \in X$, $\forall v', v'' \in V$, if $||v' - x|| \leq ||v'' - x||$ then $\mu(x, v') \geq \mu(x, v'')$. It implies that information must move toward the region of low concentration. Consider now two groups of people A and B on either side of the hypothetical boundary. Let us suppose that A is a team of sports reporters who has more information about a football match, and B is a group of the audience watching the match via TV. In this case, group A can be regarded as a region of high concentration and group B a region of low concentration. Certainly, the audiences' favor may affect the reporters' behavior. However, the audiences always receive net information from the reporters.

Although there is random movement when molecules diffuse, during a unit of time a certain fraction of the molecules will move from side A to side B if B is a region of low concentration, and vice versa, the last result is that molecules diffuse from one region to another of lower concentration.

In each case the diffusing substance moves from a region where its concentration is high to one where its concentration is low. Similarly, if there is a lack of information, and in some place there is information, information diffusion may be occurring. Both possess the property to fill in blanks. Molecules and information take diffusing direction to reduce concentration. It may be argued that, information diffusion obeys Fick's first law. Strictly speaking, the diffusion process we discuss is simple diffusion, whereby information passes through a medium without the aid of an intermediary.

According to Definition 5.7, $\mu(x, u)$ is conservative if and only if $\forall x \in X$, $\int_U \mu(x, u) du = 1$. It implies that some diffusion is conservative, i.e., without any birth or death phenomenon in the diffusion process. In other words, when we restrain ourselves to discuss information diffusion which is occurring in a seal system where the sum of information is set to 1, we say that information diffusion obeys the principle of information conserved.

From the principle of molecules conserved, we inferred that molecule diffusion obeys Fick's second law. Naturally, for information diffusion being conservative we suppose it obeys Fick's second law as we suppose atomic diffusion obeys Fick's second law[2] in the physics and chemistry of solids.

Therefore, it is possible to employ the diffusion equation (7.7) to study a special information diffusion which is a simple process without birth-death.

7.3.2 Partial differential equation of information diffusion

Let the information diffusion function be $\mu(y, t)$ where y is the distance between an observation x_i and a monitoring point u, t is time when diffusion just stops. One should note that this function differs from the one in Definition 5.5. Although both of them are information diffusion functions and use the Greek letter μ, their variables are different. In fact, $y = ||x_i - u||$, hence, $\mu(y, t)$ also can be written as $\mu(x_i, u, t)$. We use $\mu(y, t)$ to greatly simplify the following expressions. The diffusion function is similar to the molecules concentration function, i.e, the larger $\mu(y, t)$ is, the more information received by the monitoring point u.

Because the information diffusion we discuss obeys Fick's second law, we obtain

$$\frac{\partial \mu(y, t)}{\partial t} = D \frac{\partial^2 \mu(y, t)}{\partial y^2} \tag{7.8}$$

Obviously, $\mu(0, 0) = 1$, it is the boundary conditions of equation in (7.8). That is, before diffuse, information concentrates on the information injection point where observation x_i just falls. The condition also can be expressed as

$$\mu|_{t=0} = \mu(y, 0) = \delta(y),$$

where

$$\delta(x) = \begin{cases} 1, \text{for } y = 0, \\ 0, \text{otherwise.} \end{cases}$$

We use $\breve{\mu}(\lambda, t)$ to denote the function derived from $\mu(y, t)$ by Fourier transform, i.e., $\breve{\mu}(\lambda, t) = F[\mu(y, t)]$. According to Fourier transform theory, we know

$$F\left[\frac{\partial \mu(y,t)}{\partial t}\right] = \frac{d\breve{\mu}(\lambda,t)}{dt},$$

$$F\left[\frac{\partial^2 \mu(y,t)}{\partial y^2}\right] = \lambda^2 \breve{\mu}(\lambda,t),$$

$$F[\delta(y)] = 1.$$

Taking Fourier transform for two sides of Eq.(7.8), we obtain

$$\frac{d\breve{\mu}(\lambda,t)}{dt} + D\lambda^2 \breve{\mu}(\lambda,t) = 0 \tag{7.9a}$$

$$\breve{\mu}(0,t) = 1 \tag{7.9b}$$

Using method of separation of variables to solve it, we obtain that:

$$\breve{\mu}(y,t) = e^{-D\lambda^2 t}. \tag{7.10}$$

Therefore,

$$\begin{aligned}
\mu(y,t) &= F^{-1}[\breve{\mu}(\lambda,t)] \\
&= \frac{1}{2\pi} \int_{-\infty}^{\infty} \breve{\mu}(\lambda,t) e^{-i\lambda y} d\lambda \\
&= \frac{1}{2\pi} \int_{-\infty}^{\infty} e^{-D\lambda^2 t}(\cos \lambda y - i \sin \lambda y) d\lambda \qquad (7.11) \\
&= \frac{1}{2\pi} \int_{-\infty}^{\infty} e^{-D\lambda^2 t} \cos \lambda y d\lambda \\
&= \frac{1}{2\sqrt{D\pi t}} e^{-\frac{y^2}{4Dt}}
\end{aligned}$$

Denote $\sigma(t) = \sqrt{2Dt}$, then:

$$\mu(y,t) = \frac{1}{\sigma(t)\sqrt{2\pi}} \exp(-\frac{y^2}{2\sigma^2(t)}) \tag{7.12}$$

Because the process of information diffusion is an abstract process, we assume that diffusing will finish at t_0 which is not long. Denote $\sigma = \sigma(t_0)$, then at the time and the place, from where to the information injection point the length is y, we obtain diffused information:

$$\mu(y) = \frac{1}{\sigma\sqrt{2\pi}} \exp(-\frac{y^2}{2\sigma^2}). \tag{7.13}$$

The function is just the same as the normal probability distribution density function, hence we call it *normal diffusion function*. In short, $\mu(y)$ is also called *normal diffusion*.

If this normal diffusion is used to estimate a PDF, then the corresponding diffusion estimate obtained from (5.39) would be, regarding $\mu\left(\frac{x-x_i}{\Delta_n}\right)$ as $\mu(y)$,

$$\tilde{p}_n(x) = \frac{1}{n\Delta_n} \sum_{i=1}^{n} \mu\left(\frac{x - x_i}{\Delta_n}\right)$$

$$= \frac{1}{n\Delta_n} \sum_{i=1}^{n} \mu(y) \qquad (\text{let } y = \frac{x - x_i}{\Delta_n})$$

$$= \frac{1}{n\Delta_n} \sum_{i=1}^{n} \left(\frac{1}{\sigma\sqrt{2\pi}} \exp(-\frac{y^2}{2\sigma^2})\right)$$

$$= \frac{1}{n\Delta_n\sigma\sqrt{2\pi}} \sum_{i=1}^{n} \exp(-\frac{y^2}{2\sigma^2})$$

$$= \frac{1}{n\Delta_n\sigma\sqrt{2\pi}} \sum_{i=1}^{n} \exp\left(-\frac{(x - x_i)^2}{\Delta_n^2(2\sigma^2)}\right) \qquad (\text{substitute } \frac{x - x_i}{\Delta_n} \text{ back into } y)$$

$$= \frac{1}{n\sigma\Delta_n\sqrt{2\pi}} \sum_{i=1}^{n} \exp\left(-\frac{(x - x_i)^2}{2(\sigma\Delta_n)^2}\right)$$

Let $h = \sigma\Delta_n$, $x_i \in X$, $x \in \mathbb{R}$,

$$\tilde{p}(x) = \frac{1}{nh\sqrt{2\pi}} \sum_{i=1}^{n} \exp[-\frac{(x - x_i)^2}{2h^2}] \qquad (7.14)$$

is called *normal diffusion estimate* about $p(x)$ from which the given sample is drawn.

Obviously, the normal diffusion estimate is just the same as the Gaussian kernel estimate. It implies that Gaussian kernel connects to some simple diffusion without birth-death phenomenon. Furthermore we know that any one of diffusion functions cannot express all diffusion phenomena.

7.4 Nearby Criteria of Normal Diffusion

We know, $\forall x_i \in X$, if X is incomplete, then, x_i can be regarded as a representative of its "around." That is called nearby phenomenon.

For monitor u, we first introduce the concept of the nearest information injecting point from X.

Definition 7.1 For u, if $x' \in X$ and satisfies

$$|u - x'| = \min\{|u - x_i| \mid x_i \in X\}, \qquad (7.15)$$

x' is called the *first nearest information injecting point* of u from X.

For example, we suppose that John has only three friends: Dadid, William and Mary. John sees David every day, William every weekend, and Mary just at Christmas. In this case, David can be regarded as the first nearest information injecting point of John from his friends.

Definition 7.2 For u, let x' be its first nearest information injecting point from X. If $x'' \in X \setminus \{x'\}$ and satisfies

$$|u - x''| = \min\{|u - x_i| \mid x_i \in X \setminus \{x'\}\}, \qquad (7.16)$$

x'' is called the *second nearest information injecting point* of u from X.

Definition 7.3 For u, let x', x'' be its first and second nearest information injecting points, respectively. If $x''' \in X \setminus \{x', x''\}$ and satisfies

$$|u - x'''| = \min\{|u - x_i| \mid x_i \in X \setminus \{x', x''\}\}, \qquad (7.17)$$

x''' is called the *third nearest information injecting point* of u from X.

For monitor $x \in \mathbb{R}$, we let its first, second and third nearest information injecting points are x', x'', x''', respectively.

The h satisfying the *one-point criterion* is the one that makes the following inequality hold:

$$\exp[-\frac{(x - x')^2}{2h^2}] \geq \sum_{x_i \neq x'} \exp[-\frac{(x - x_i)^2}{2h^2}], x \in \mathbb{R} \qquad (7.18)$$

When h makes the following inequality hold:

$$\exp[-\frac{(x - x')^2}{2h^2}] + \exp[-\frac{(x - x'')^2}{2h^2}] \geq \sum_{x_i \neq x', x''} \exp[-\frac{(x - x_i)^2}{2h^2}], x \in \mathbb{R}$$
$$(7.19)$$

we say that the h satisfies the *two-points criterion*. Naturally, if

$$\sum_{x_k \in \{x', x'', x'''\}} \exp[-\frac{(x - x_k)^2}{2h^2}] \geq \sum_{x_i \neq x', x'', x'''} \exp[-\frac{(x - x_i)^2}{2h^2}], x \in \mathbb{R} \quad (7.20)$$

we can say that the h satisfies the *three-points criterion*. And so on.

One of the criteria may help us to choose h.

As it is well known from discussion in the last chapter the Gaussian kernel will work well if the population really is normal. Therefore, it is enough to study the standard normal density for assessing the performance of the criterion.

7.5 The 0.618 Algorithm for Getting h

Let $X = \{x_1, x_2, \cdots, x_n\}$ be a given sample. For monitor $x \in \mathbb{R}$, suppose its first, second and third nearest information injecting points are x', x'', x''', respectively. Note x', x'', $x''' \in X$. We let

$$X_1 = \{x'\}, \quad X_2 = \{x', x''\}, \quad X_3 = \{x', x'', x'''\}. \qquad (7.21)$$

For convenience sake, we write

$$L_s(h) = \sum_{x_k \in X_S} \exp[-\frac{(x - x_k)^2}{2h^2}], \tag{7.22}$$

and

$$R_s(h) = \sum_{x_i \in X \backslash X_S} \exp[-\frac{(x - x_i)^2}{2h^2}], \tag{7.23}$$

where $s \in \{1, 2, 3\}$. For example, when $s = 2$ we have

$$L_2(h) = \exp[-\frac{(x - x')^2}{2h^2}] + \exp[-\frac{(x - x'')^2}{2h^2}],$$

and

$$R_2(h) = \sum_{x_i \in X \backslash \{x', x''\}} \exp[-\frac{(x - x_i)^2}{2h^2}].$$

For any i, j, if $i \neq j$ implies $x_i \neq x_j$, we call X a set of *pairwise unequal elements*.

We restrict ourselves to the study of a set of pairwise unequal elements and suppose that $n > 2s + 1$ and $(\forall x \in X)(|x| < \infty)$. In this case, we know that:

$$0 \leq (x - x')^2 < \infty, \quad 0 \leq (x - x'')^2 < \infty, \quad 0 \leq (x - x''')^2 < \infty,$$

hence,

$$\lim_{h \to \infty} L_s(h) = s < s + 1 = 2s + 1 - s < \lim_{h \to \infty} R_s(h) = n - s. \tag{7.24}$$

Therefore, there exists a larger number $h^{(1)}$ satisfying

$$L_s(h^{(1)}) < R_s(h^{(1)}). \tag{7.25}$$

One can easily verify that any h' larger than $h^{(1)}$ also satisfies Eq.(7.25). In other words, a diffusion coefficient satisfying the s-points criterion must be smaller than $h^{(1)}$.

We can prove that there exists a real positive number $h^{(0)}$ satisfying:

$$L_s(h^{(0)}) > R_s(h^{(0)}). \tag{7.26}$$

Let $\psi_s(h) = L_s(h) - R_s(h)$. We know the interval $[h^{(0)}, h^{(1)}]$ containing the roots of the equation $\psi_s(h) = 0$. We can employ the 0.618-algorithm (see section 6.3) to obtain a positive root h of the equation.

The 0.618-algorithm for obtaining h runs as follows:

(1) Set $x0 = 0$, $x2 = 10^{10}$, $x1 = x2/2$;

(2) Set $h = x1$;

(3) If $L_s(h) > R_s(h)$, set $x0 = x1$, $x1 = x1 + 0.618(x2 - x1)$ and $x2 = x2$. If $L_s(h) < R_s(h)$, set $x0 = x0$, $x2 = x1$ and $x1 = x0 + 0.618(x1 - x0)$;

(4) Give a sufficiently small number $\delta > 0$. If $|h - x1| > \delta$, perform steps (2) and (3) again, else stop.

Since the diffusion model gives the discrete estimate rather than a function, it is necessary to make some slight adjustment to the procedure. Consider a discrete universe $U = \{u_1, u_2, \cdots, u_m\} \subset \mathbb{R}$ on which all the data points lie.

With the 0.618 algorithm, for every u_j, we can obtain a diffusion coefficient h_j. Obviously, their average value

$$\bar{h} = \frac{1}{n} \sum_{j=1}^{m} h_j \tag{7.27}$$

makes every h_j satisfy approximately the s-points criterion. $\bar{h}$ is called the *average diffusion coefficient* of X based on s-points criterion.

Example 7.1 Let

$$p(x) = \frac{1}{\sqrt{2\pi}} \exp(-\frac{x^2}{2}), -\infty < x < \infty,$$

x is a random variable obeying standard normal distribution $N(0, 1)$.

Running Program 3-3 in subsection 3.5.3 with MU=0, SIGMA=1, N=10, SEED=907690, we obtain 10 random numbers:

$$
\begin{aligned}
X = & \{x_1, x_2, \cdots, x_{10}\} \\
= & \{-0.02, 0.00, -0.93, 0.63, -0.70, 0.51, -0.81, 0.48, -0.05, -0.76\}.
\end{aligned}
$$

Given an origin $x_0 = -3$, step $\Delta = 0.3$, for positive integer $m = 21$, we obtain 21 monitoring points to construct a monitoring space U

$$U = \{u_1, u_2, \cdots, u_{21}\} = \{-3, -2.7, \cdots, 3\}.$$

We use above 0.618-algorithm to search for h satisfying the two-points criterion.

Obviously, for $u_1 = -3$, we have $x' = -0.93$ and $x'' = -0.81$. Doing steps (1) and (2), we obtain $h^{(1)} = 5 \times 10^9$. Then

$$L_2(h^{(1)}) = \exp\left[-\frac{(u_1 - x')^2}{2h^{(1)}h^{(1)}}\right] + \exp\left[-\frac{(u_1 - x'')^2}{2h^{(1)}h^{(1)}}\right] \approx 2$$

$$R_2(h^{(1)}) = \sum_{i=1}^{10} \exp\left[-\frac{(u_1 - x_i)^2}{2h^{(1)}h^{(1)}}\right] - L_2(h^{(1)}) \approx 8$$

That is, $L_2(h) < R_2(h)$. Set $x0 = x0 = 0$, $x2 = x1 = 5 \times 10^9$ and $x1 = x0 + 0.618(x1 - x0) = 0.618 \times 10^9 = 3.09 \times 10^9$.

We use $\delta = 0.00001$ to control the searching. Now, $|h^{(1)} - x1| > \epsilon$. We perform steps (2) and (3) again.

After 64 iterations, we have $|h^{(1)} - x1| < \delta$, then we obtain $h_1 = 1.042005$. For other u_j, by same algorithm, we can obtain corresponding h_j. Finally, we obtain $h = 0.711$. Using it to calculate normal diffusion estimate about $p(x)$, we obtain

$$\{\tilde{p}(u_1), \tilde{p}(u_2), \cdots, \tilde{p}(u_{21})\}$$
$$= \{0.002, 0.007, 0.019, 0.045, 0.092, 0.160, 0.242,$$
$$0.321, 0.382, 0.415, 0.414, 0.379, 0.315, 0.233,$$
$$0.152, 0.086, 0.042, 0.017, 0.006, 0.002, 0.000\}.$$

Because the true probability distribution of $N(0,1)$ on U is

$$\{p(u_1), p(u_2), \cdots, p(u_{21})\}$$
$$= \{0.004, 0.010, 0.022, 0.044, 0.079, 0.130, 0.194,$$
$$0.266, 0.333, 0.381, 0.399, 0.381, 0.333, 0.266,$$
$$0.194, 0.130, 0.079, 0.044, 0.022, 0.0100.004\}.$$

Therefore, we obtained the non-log-divergence $\tilde{\rho}_2 = 0.004677$. Plotting $p(u_j), j = 1, 2, \cdots, 21$ in a thick curve, and $\tilde{p}(u_j), j = 1, 2, \cdots, 21$ in a thin curve, we obtain Fig. 7.2 (b).

Using the one-point and three-points criteria to choose h for normal diffusion estimate about the same $p(x)$, we obtain $\tilde{\rho}_1 = 0.014337$, $\tilde{\rho}_3 = 0.008878$, respectively. Fig. 7.2 (a),(c) are estimates from the one-point and three-points criteria, respectively.

Employed result for the kernel theory,

$$h_{opt}(n) = 1.06\sigma n^{-1/5}.$$

For X in Example 7.1, $\sigma = 0.5973$, $n = 10$, hence $h_{opt}(n) = 0.3995$. Using it for kernel estimate, we have Fig. 7.2 (d) and the non-log-divergence $\rho_{opt} = 0.012823$.

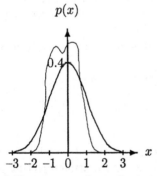

(a) From one-point criterion

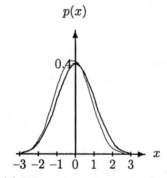

(a) From two-points criterion

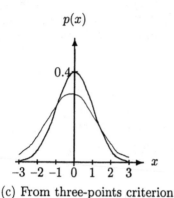

$p(x)$

0.4

-3 -2 -1 0 1 2 3 x

(c) From three-points criterion

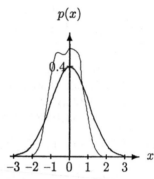

$p(x)$

0.4

-3 -2 -1 0 1 2 3 x

(d) From optimal Gaussian kernel

Fig. 7.2   Normal diffusion estimates (thin curves)based on the
s-points criterion and kernel theory.

It seems the two-points criterion is the best one for the normal diffusion.
We run 90 simulation experiments for every model with different seed
numbers, respectively. The average non-log-divergences are

$$\tilde{\rho}_1 = 0.01098, \tilde{\rho}_2 = 0.01020, \tilde{\rho}_3 = 0.01515, \rho_{opt} = 0.01177.$$

The result supports our inference, i.e., the two-points criterion would be
regarded as a reasonable criterion to control diffusion coefficient h.

7.6 Average Distance Model

The 0.618 algorithm requires us to calculate h for every monitor u. In general,
for more accurate estimation, the number of monitoring points would be more
large. A more simple formula calculating the diffusion coefficient would be
welcome by engineers. We use the two-points criterion and average distance
assumption to produce the formula.

Firstly, we arrange observations in X in ascending order, i.e. from small
to large, and still denote them by $x_1, x_2, \cdots, x_n$;

Let $d_i = x_{i+1} - x_i$ and define the average distance as:

$$d = \frac{1}{n-1} \sum_{i=1}^{n-1} d_i = \frac{x_n - x_1}{n-1}. \tag{7.28}$$

In general, the first monitor u_1 is just located at the place where x_1 gets
its value, i.e. $u_1 = x_1$. Obviously, its first nearest information injecting point
is just x_1. Since X has been ordered, then its second nearest information injecting point is just x_2. For this monitor, if h satisfies the two-points criterion,
we have:

$$\exp[-\frac{(u_1-x_1)^2}{2h^2}] + \exp[-\frac{(u_1-x_2)^2}{2h^2}] \geq \sum_{i=3}^{n} \exp[-\frac{(u_1-x_i)^2}{2h^2}]. \qquad (7.29)$$

Replacing the distances between observations in (7.29) by the average distance in (7.28), we obtain:

$$1 + \exp(-\frac{d^2}{2h^2}) \geq \sum_{j=1}^{n-2} \exp[-\frac{(j+1)^2 d^2}{2h^2}]. \qquad (7.30)$$

Let

$$z = \exp(-\frac{d^2}{2h^2}), \qquad (7.31)$$

then

$$1 + z \geq z^4 + z^9 + z^{16} + \cdots + z^{(n-2)^2}. \qquad (7.32)$$

To get h, we only need to solve the following equation:

$$z^4 + z^9 + z^{16} + \cdots + z^{(n-2)^2} - z - 1 = 0. \qquad (7.33)$$

For example, when $n = 5$, we have

$$z^4 - z^9 + z^{16} - z - 1 = 0.$$

Because $\exp(-\frac{d^2}{2h^2})$ must be a positive real number, it is enough to get efficient root of Eq.(7.33). Suppose its efficient root is z. From (7.31) we obtain

$$-\frac{d^2}{2h^2} = \log z,$$

$$h = \frac{1}{\sqrt{2(-\log z)}} d. \qquad (7.34)$$

We denote

$$c = \frac{1}{\sqrt{2(-\log z)}}, \qquad (7.35)$$

which is called *adjustment coefficient*.

Let

$$f(z) = z^4 + z^9 + z^{16} + \cdots + z^{(n-2)^2} - z - 1, \qquad (7.36)$$

we can use the Newton iterative method,

$$x_{k+1} = x_k - \frac{f(z_k)}{f'(z_k)}, \quad k = 0, 1, \cdots, \qquad (7.37)$$

to find the efficient root of Eq.(7.33). Table 7-1 gives the efficient roots z for $n = 5, 6, 7, 8, 9, 10$ and corresponding adjustment coefficients c. For any $n \geq 9$, efficient root is 0.933 and adjustment coefficient is 2.6851.

Table 7-1 The efficient root of $z^4 + z^9 + z^{16} + \cdots + z^{(m+1)^2} - 0.5 = 0$

Sample size n	5	6	7	8	9	10
Efficient root z	0.9354	0.9338	0.9332	0.9331	0.9330	0.9330
Adjustment coefficient c	2.7363	2.7019	2.6893	2.6872	2.6851	2.6851

Finally, we obtain

$$h = \begin{cases} 0.6841(b-a), & \text{for } n = 5; \\ 0.5404(b-a), & \text{for } n = 6; \\ 0.4482(b-a), & \text{for } n = 7; \\ 0.3839(b-a), & \text{for } n = 8; \\ 2.6851(b-a)/(n-1), & \text{for } n \geq 9. \end{cases} \qquad (7.38)$$

where

$$b = \max_{1 \leq i \leq n} \{x_i\}, \qquad a = \min_{1 \leq i \leq n} \{x_i\}.$$

For example, when $n = 10$, we have $h = 2.6851(b-a)/(10-1) = 0.2983(b-a)$. The h calculated by formula (7.38) is called *simple coefficient*.

The normal diffusion estimate with simple coefficient h is called the *simple normal diffusion estimate* (SNDE).

Now, with h from the formula (7.38), we study the diffusion estimate about $p(x)$ in Example 7.1. In the example, the given sample X has 10 observations, its maximum is 0.63, minimum -0.93. Hence, $h = 0.2983(0.63 + 0.93) = 0.4653$. Therefore, according to Eq.(7.14), the diffusion estimate is

$$\tilde{p}(x) = \frac{1}{0.4653 \times 10\sqrt{2\pi}} \sum_{i=1}^{10} \exp[-\frac{(x - x_i)^2}{2 \times 0.4653^2}]$$

$$= 0.0857 \sum_{i=1}^{10} \exp[-2.309(x - x_i)],$$

which is shown in Fig. 7.3 by the dotted curve. Its non-log-divergence is $\tilde{p} = 0.01066$. After 90 simulation experiments with different seed numbers, we obtain average non-log-divergences $\bar{p} = 0.01092$. The result shows that the simple coefficient is worth adopting.

Until now, all discussion, with respect to the normal diffusion, focus on that a give sample is drawn from the standard Gaussian distribution. In general, we know nothing about distribution shape, therefore, it is necessary to investigate the performance of the normal diffusion for other populations.

In Chapter 4 we have indicated that the function curves of probability distributions have performance between symmetry curve and monotone decreasing curve. Therefore, it is enough to study the two distributions, the normal distribution (Gaussinan distribution) and exponential distribution. As a transition curve, we also discuss the lognormal distribution in our simulation experiment.

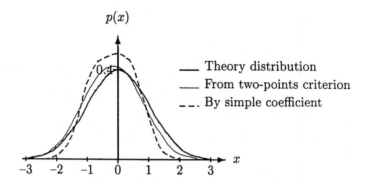

Fig. 7.3 Normal diffusion estimate about standard Gaussian distribution.

Firstly, we employ h calculated by the formula (7.38) and the normal estimate formula (7.14) to do the normal experiment described in subsection 4.4.2.

For X in (4.14), we have

$$n = 19, \quad b = \max_{1 \leq i \leq n} \{x_i\} = 7.5704, \quad a = \min_{1 \leq i \leq n} \{x_i\} = 6.1460,$$

hence,

$$h = 2.6851(b - a)/(n - 1) = 2.6851(7.5704 - 6.1460)/(19 - 1) = 0.2125.$$

Then, we obtain diffusion estimate

$$\tilde{p}(x) = \frac{1}{0.2125 \times 19\sqrt{2\pi}} \sum_{i=1}^{19} \exp[-\frac{(x - x_i)^2}{2 \times 0.2125^2}]$$

$$= 0.0988 \sum_{i=1}^{19} \exp[-11.0749(x - x_i)],$$

which is shown in Fig. 7.4 by a dotted curve. Let $n = 10, 12, \cdots, 30$ and respectively simulate 90 experiments with different seed numbers, we obtain Table 7-2 showing the average non-log-divergence $\tilde{\rho}_{SHE}$ of the soft histogram estimate, the average non-log-divergence $\tilde{\rho}_{SNDE}$ of the simple normal diffusion estimate, and the relative error of SHE and SNDE. The results show that SNDE is better then SHE. Roughly speaking, for a small sample, method of simple normal diffusion can improve a soft histogram estimator to reduce the error about 38%.

For X in (4.17) drawn from an exponential distribution, employing h calculated by the formula (7.38) and the normal estimate formula (7.14), we obtain SNDE shown in Fig. 7.5 by the dotted curve.

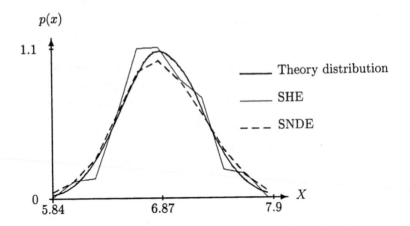

Fig. 7.4   Probability density estimators by SHE and SNDE
with respect to a normal distribution.

Table 7-2   Average non-log-divergences of SHE and SNDE,
and the relative error e for a normal distribution

n	10	12	14	16	18	20	22	24	26	28	30
$\tilde{\rho}_{SHE}$	0.157	0.148	0.140	0.128	0.122	0.114	0.109	0.103	0.097	0.091	0.087
$\tilde{\rho}_{SNDE}$	0.084	0.079	0.077	0.071	0.071	0.068	0.070	0.070	0.067	0.064	0.063
e	0.465	0.464	0.452	0.446	0.420	0.400	0.357	0.322	0.308	0.297	0.277

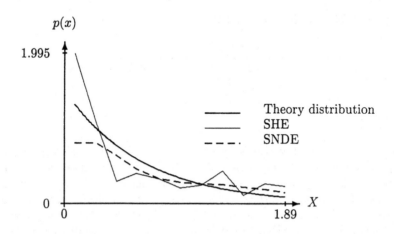

Fig. 7.5   Probability density estimators by SHE and SNDE
with respect to an exponential distribution.

Running 90 simulation experiments for $n = 10, 12, \cdots, 30$, we obtain Table 7-3 to show the average non-log-divergences of the soft histogram estimate and the simple normal diffusion estimate, respectively.

Table 7-3   Average non-log-divergences of SHE and SNDE, and the relative error e for an exponential distribution

n	10	12	14	16	18	20	22	24	26	28	30
$\tilde{\rho}_{SHE}$	0.272	0.213	0.174	0.138	0.127	0.117	0.106	0.098	0.084	0.079	0.072
$\tilde{\rho}_{SNDE}$	0.201	0.175	0.154	0.132	0.122	0.111	0.103	0.097	0.086	0.082	0.078
e	0.263	0.180	0.114	0.045	0.034	0.054	0.029	0.016	-0.025	-0.042	-0.081

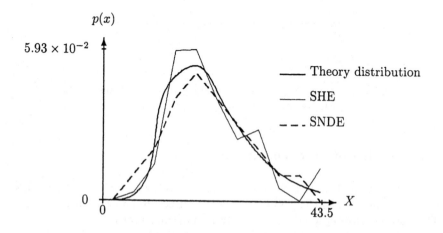

Fig. 7.6   Probability density estimators by SHE and SNDE with respect to a lognormal distribution.

For the exponential distribution we study, Table 7-3 shows that when n is less 24, SNDE is better then SHE, else do not use SNDE. The case reminds us that Gaussian assumption in the kernel theory faces serious risks.

For X in (4.19) drawn from a lognormal experiment, we obtain SNDE shown in Fig. 7.6 by the dotted curve.

Running 90 simulation experiments for $n = 10, 12, \cdots, 70$, respectively, we obtain Fig. 7.7 showing the average non-log-divergences of the soft histogram estimate and the simple normal diffusion estimate. Obviously, when n is small, SNDE is still better than SHE with respect to a lognormal distribution.

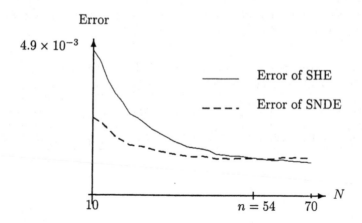

Fig. 7.7   Average non-log-divergence curves with respect to
a lognormal distribution.

7.7 Conclusion and Discussion

One application of fuzzy techniques, neural networks and genetic algorithms
is to automate one of the capabilities of the human brain or a law of nature.
The normal information diffusion imitates molecule diffusion.

In the case of information diffusion as a simple process (without the aid
of an intermediary) and without birth-death (in a seal system where the sum
of information is kept 1), information diffusion would obey the normal law,
i.e., *diffused information* from observation x to monitor u is

$$\mu(x, u) = \frac{1}{h\sqrt{2\pi}} \exp(-\frac{(u - x)^2}{2h^2}), \tag{7.39}$$

where h is called the diffusion coefficient of the normal diffusion.

Any observation x_i can be regarded as a representative of its "around."
This is called the nearby phenomenon. If h makes monitor u get the main
part of its sharing information from the nearest s-observations, we say that h
satisfies s-points criterion. Simulation shows that two-points criterion would
be regarded as a reasonable criterion to control the diffusion coefficient h.

Using an average distance to replace the distance between monitor and
observation, with the two-points criterion, we obtain a very simple formula
for calculating the diffusion coefficient of the normal diffusion. For $n \geq 9$, it
is

$$h = 2.6851(\max_{1 \leq i \leq n} \{x_i\} - \min_{1 \leq i \leq n} \{x_i\})(n - 1). \tag{7.40}$$

The normal diffusion estimate is just the same as the Gaussian kernel estimate. It implies that Gaussian kernel connects to some simple diffusion without birth-death phenomenon. However, the coefficient from the kernel theory is both non-explanatory and rough. When n is small, the method of normal diffusion is superior with respect to almost any distribution. Furthermore, for a given sample whose size is large and which is drawn from an exponential or lognormal distribution, the new method is not the best one. It means that any one of diffusion functions cannot express all diffusion phenomena.

Huang [4] suggested another formula

$$h = \begin{cases} 1.6987(b-a)/(n-1), & \text{for } 1 < n \le 5; \\ 1.4456(b-a)/(n-1), & \text{for } 6 \le n \le 7; \\ 1.4230(b-a)/(n-1), & \text{for } 8 \le n \le 9; \\ 1.4208(b-a)/(n-1), & \text{for } 10 \le n. \end{cases} \qquad (7.41)$$

to calculate h, resulted from the assumption $n = 2m + 3$ (m is a positive integer) and that h makes $u = x_{m+2}$ satisfy the two-points criterion. Then we obtain equation

$$z^4 + z^9 + z^{16} + \cdots + z^{(m+1)^2} - 0.5 = 0. \qquad (7.42)$$

For any m, by solving the equation, we can obtain formula (7.41). Further study shows that the assumption produces a larger h than we need.

References

1. Bezdek, J.C. and Pal, S.K. (Eds.)(1992), Fuzzy Models for Pattern Recognition, IEEE Press, New York
2. Elliott, S.R.(1998), The Physics and Chemistry of Solids, John Wiley & Sons, New York
3. Goldberg, D.E.(1989), Genetic Algorithms in Search, Optimization and Machine Learning, Addison-Wesley, New York
4. Huang, C.F.(1997), Principle of information diffusion. Fuzzy Sets and Systems, Vol.91, pp.69-90
5. Kandel, A. (1982), A Fuzzy Techniques in Pattern Recognition, John Wiley, New York
6. Meier, W., Weber, R. and Zimmermann,H.-J.(1994), Fuzzy data analysis - methods and industrial applications. Fuzzy Sets and Systems, Vol.61, pp.19-28
7. Pienkowski, A. E. K. (1989), Artificial Colour Perception Using Fuzzy Techniques in Digital Image Processing, Verlag TÜV Rheinland, Köln
8. Sanchez, E., Shibata, T. and Zadeh, L.A. (eds.)(1997), Genetic Algorithms and Fuzzy Logic Systems: Soft Computing Perspectives, World Scientific, Singapore
9. Simon Haykin (1994), Neural Networks: A Comprehensive Foundation. Prentice-Hall, Inc., Englewood Cliffs, New Jersey
10. Towell, G. G. and Shavlik, J. W.,(1994), Knowledge-based artificial neural networks. Artificial Intelligence, Vol.70, pp.119-16

Part II
Applications

8. Estimation of Epicentral Intensity

In this chapter, based on the principle of information diffusion, we suggest a new method, called self-study discrete regression, to construct a statistic relationship from a given sample. To understand this, a detail discussion develops around estimation of epicentral intensity. This chapter is organized as follows: in section 8.1 we introduce some basic concepts in seismology and earthquake engineering for studying estimation of epicentral intensity. In section 8.2, we review the linear regression and a fuzzy method for the estimation. Section 8.3 describes the method of self-study discrete regression. Section 8.4 and 8.5, respectively, give linear distribution self-study (LDSS) and normal diffusion self-study (NDSS) models to estimate epicentral intensity by magnitude. Then we conclude the chapter in section 8.6.

8.1 Introduction

Study on estimation of epicentral intensity is quite attractive in seismology and earthquake engineering. To help the readers who do not know seismology and earthquake engineering, we firstly introduce some basic concepts in the fields.

An earthquake is the vibration of the earth produced by the rapid release of energy. This energy radiates in all directions from its source, the *focus*, in the form of waves. Just as the impact of the stone sets water waves in motion, an earthquake generates seismic waves that radiate throughout the earth. Even though the energy dissipates rapidly with increasing distance from the focus, instruments located throughout the world can record the event.

Seismology is the science of earthquakes and related phenomena. Earthquake engineering is concerned with the design and construction of all kinds of civil and building engineering systems to withstand earthquake shaking. Earthquake engineers, in the course of their work, are faced with many uncertainties and must use sound engineering judgement to develop safe solutions to challenging problems. Both of them use two basic concepts, magnitude and intensity.

Early attempts to establish the size or strength of earthquakes relied heavily on subjective description. There was an obvious problem with this method because people's accounts are various. It is difficult to make an accurate classification of the quake's intensity. Then in 1902 a fairly reliable intensity scale based on the amount of damage caused to various types of structures was developed by Giuseppe Mercalli[7]. A modified form of this tool is presently used in China (see Appendix 4.B).

By definition, the earthquake *intensity* is a measure of the effects of a quake at a particular location. It is important to note that earthquake intensity depends not only on the strength of the earthquake, but also on other factors as well. These include the distance from the *epicenter* (the vertical projection of the focus on the surface), the nature on the surface materials, and building design. The intensity at the epicenter is called *epicentral intensity*.

In 1935, Charles Richter of the California Institute of Technology introduced the concept earthquake *magnitude* when attempting to rank earthquakes of southern California[1]. The earthquake magnitude is basically a relative scale[5]. It defines a standard size of earthquake and rates the others in a relative manner by their maximum amplitude under identical observational conditions. This is evident from Richter's definition:

$$M = \log[A(\Delta)/A_0(\Delta)] = \log A(\Delta) - \log A_0(\Delta), \qquad (8.1)$$

where Δ is an epicentral distance, and A_0 and A denote the maximum trace amplitudes, on a specified seismograph, of the standard event and the one to be measured, respectively. The standard earthquake, i.e. $M = 0$ (=log1) in Richter's formula, is such as to give the maximum trace amplitude of 0.001mm on a Wood-Anderson type seismograph at $\Delta = 100$ km.

The intensity scale differs from the Richter Magnitude Scale in that the effects of any one earthquake vary greatly from place to place, so there may be many intensity values measured from one earthquake. Each earthquake, on the other hand, should have just one magnitude, although the several methods of estimating it will yield slightly different values.

Study on the relationship between epicentral intensity, I_0, and magnitude, M, is very important because that

(1) A lot of historical earthquakes are recorded in the intensities. Using the relationship we can estimate their magnitudes;

(2) The earthquake prediction usually gives the magnitude of the future earthquake. To provide the design parameters with respect to withstand earthquake shaking, we must estimate intensity or acceleration from the given magnitude.

In general, we use a given sample to construct a statistic relationship between magnitude and intensity for the estimation. The relationship, in fact, is a regression result. The classical statistic tools are usually used to produce a continuous regression function. Employing the principle of information diffusion, in this chapter we suggest a new method, called *self-study discrete regression*, to estimate epicentral intensity. It has advantage over classical methods due to that we can avoid any assumption of the given sample and its accuracy may be higher. Particularly, the new method may have great advantage over classical statistical methods to model nonlinear functions.

The reason why we select, in this chapter, estimation of epicentral intensity as follows: The relationship of between epicentral intensity and magnitude is almost linear. Hence, if the estimation from the new method is better than one from the linear regression, we know the new method is effective. Strictly speaking, the universe of intensity consists of fuzzy concepts. However, in earthquake engineering, researchers use to map earthquake intensities into the real line $\mathbb{R}$. Therefore, it becomes easy to discuss variety of diffusion models.

8.2 Classical Methods

The linear regression is the simplest classical method to estimate epicentral intensity by magnitude. The method of fuzzy inference, based on the assumption that the earthquake intensities are normal fuzzy sets on magnitude, can be considered as one of classical methods.

8.2.1 Linear regression

For estimation of epicentral intensity, we can give a concrete model for the *linear regression* (LR) method as

$$I_0 = a + bM, \tag{8.2}$$

where a and b are regional constants, calculated by the method of least squares from a given sample.

Now let us study the given sample from Appendix 2.A, where we have 134 observations with magnitudes in the range 4.25-8.5 and epicentral intensity in the range VI-XII degrees.

Firstly, we employ f defined by (2.19) and (2.20) to map standard and nonstandard intensity into digit. Then, we have numerical sample shown in Table 8-1.

The linear regression based on the method of least squares can be obtained by solving simultaneous linear equations. Let

$$y = a + bx \tag{8.3}$$

be the function we need. For observations (x_i, y_i), $i = 1, 2, \cdots, n$, we calculate the square sum

$$F(a, b) = \sum_{i=1}^{n} (a + bx_i - y_i)^2. \tag{8.4}$$

To minimize the sum, a and b must satisfy the following equations

$$\frac{\partial F(a,b)}{\partial a} = 2\sum_{i=1}^{n}(a + bx_i - y_i) = 0,$$

$$\frac{\partial F(a,b)}{\partial b} = 2\sum_{i=1}^{n}(a + bx_i - y_i)x_i = 0.$$

That is,

$$an + b\sum_{i=1}^{n} x_i = \sum_{i=1}^{n} y_i, \tag{8.5a}$$

$$a\sum_{i=1}^{n} x_i + b\sum_{i=1}^{n} x_i^2 = \sum_{i=1}^{n} x_i y_i. \tag{8.5b}$$

Solving a, b, and substituting them into Eq.(8.3), we can obtain the linear repression result.

Table 8-1 Observations with magnitudes M and epicentral intensity I_0

No.	M	I_0	No.	M	I_0	No.	M	I_0	No.	M	I_0
1	5.75	7	35	5.2	6.2	69	6.2	7	103	5.5	7
2	5.75	7	36	5.4	7.2	70	6.8	9	104	5.8	7
3	6.4	8	37	5.2	7	71	5.5	7	105	6	8
4	6.4	8	38	5	6.2	72	5.5	7	106	5.8	7
5	6.5	9	39	6.4	9	73	6.8	9	107	5.6	7
6	6.5	9	40	5.2	6	74	6.3	9	108	4.8	6
7	6	8	41	7.80	10	75	5.8	7.2	109	6.2	7
8	6	8	42	5.3	6	76	5.8	7	110	5.6	7
9	6	7	43	5.5	6.2	77	7.90	10	111	6	7
10	7	8	44	5.5	7.2	78	8.50	12	112	6.75	9
11	7	9	45	5.5	7.2	79	8	11	113	5.5	8
12	6.25	8	46	6.5	8	80	6	8	114	4.75	7
13	5.75	8	47	5.5	7	81	7.30	10	115	5	7
14	5.8	8	48	5.5	7	82	5.4	6.5	116	5	6
15	6.3	9	49	5	6	83	5.3	6	117	5	6
16	6.5	8	50	5	6	84	5.7	7	118	5	6
17	5.8	7.2	51	5	6	85	5.4	7	119	5.75	7
18	5	7.2	52	4.75	5.8	86	5	6.2	120	5.5	7.2
19	5	6	53	5.4	6	87	4.8	7	121	4.25	6
20	5	6.2	54	6.3	8	88	7	8.2	122	5.2	7
21	6	6.8	55	7.50	10	89	5	6	123	6.8	9.2
22	4.8	6	56	6	8	90	4.8	6	124	5.6	6
23	5.5	6	57	6.8	9	91	5.5	7	125	7.20	10
24	5.8	8	58	6	8	92	6.8	8	126	6.2	7.2
25	6	8	59	5.5	7	93	7.5	9	127	6	8
26	5.5	7	60	7.25	10	94	8	11	128	5.2	7
27	6.2	7.8	61	5.75	7	95	7.25	9	129	6.3	7
28	6	6.8	62	5.5	6.5	96	7.25	9	130	5.5	6
29	5.4	7	63	6.8	9	97	7	9	131	5.7	7
30	5.2	6.2	64	5.3	7	98	6.8	9	132	5.4	6.2
31	6.1	7.2	65	7.5	9	99	6.4	8	133	4.6	6
32	5.1	7.2	66	6.8	9	100	6.5	8	134	7.3	9
33	6.5	8.8	67	5.5	7	101	6.6	8			
34	6.2	7	68	5.8	7	102	6.5	8			

In Table 8-1, M and I_0 can be regarded as x and y, respectively. For example, $x_1 = M_1 = 5.75$ and $y_1 = I_{01} = 7$. Hence, we have

$$\sum_{i=1}^{134} x_i = 797.65, \quad \sum_{i=1}^{134} y_i = 1010.8,$$

$$\sum_{i=1}^{134} x_i^2 = 4838.95, \quad \sum_{i=1}^{134} x_i y_i = 6142.16.$$

Employing

$$b = (n \sum_{i=1}^{n} x_i y_i - \sum_{i=1}^{n} x_i \sum_{i=1}^{n} y_i)/(n \sum_{i=1}^{n} x_i^2 - \sum_{i=1}^{n} x_i \sum_{i=1}^{n} x_i), \quad (8.6a)$$

and

$$a = \frac{1}{n}(\sum_{i=1}^{n} y_i - b \sum_{i=1}^{n} x_i), \quad (8.6b)$$

we have the linear regression result

$$I_0 = -0.66 + 1.38M. \quad (8.7)$$

By the formula, we calculate the epicentral intensities from the magnitudes in Table 8-1, which are shown in the column LR of Appendix 8.A. The column LR is divided tree sub-columns. The left sub-column shows the digital estimation calculated by formula (8.7) directly, the middle sub-column shows symbol estimation denoted by Roman numerals, and right sub-column shows the comment (C) with true – T and false – F. We take the nearest Roman numeral to be the symbol estimation. For example, for Gansu earthquake (No.79) occurred at May 23,1927, its magnitude is $M = 8.0$. The digital estimation of epicentral intensity we obtained by formula (8.7) is $\hat{y} = 10.38$. According to the mathematical mapping defied by (2.19) and (2.20), we know that the nearest Roman numeral is X, i.e., the symbolism estimation would be $\hat{I_0}$ =X. However, the real record is I_0=XI, i.e., the estimation is false. Hence, the comment is F.

In the column LR of Appendix 8.A there are 86 estimates being true, i.e., the correct rate of the linear regression is 64% (because 86/134=0.64179≈ 64%). The square error $\varepsilon_{LR} = 0.267$.

8.2.2 Fuzzy inference based on normal assumption

The earthquake intensity is a typical fuzzy scale based mainly on the macroscopic standards. Several methods of fuzzy mathematics for quantitative evaluation of earthquake intensity had been studied by the Chinese researchers. One of them was based on the concept of the degree of approaching for normal fuzzy sets and by means of fuzzy multifactorial evaluation[3]. In the method, the magnitude, M, is considered as the universe of the earthquake intensity, I, whose membership function is

$$\mu_I = \exp[-(\frac{m - a_I}{b_I})^2], \quad I = VI, VII, \cdots, XII, \quad (8.8)$$

where m is magnitude. Suppose there is n_I earthquake events for epicentral intensity $I_0 = I$, then

$$a_I = \frac{1}{n_I} \sum_{i=1}^{n_I} m_i, \quad b_I^2 = \frac{1}{n_I} \sum_{i=1}^{n_I} (m_i - a_I)^2, \quad (8.9)$$

where m_i is magnitude value of i-th observation. For using records of Appendix 2.A as more as possible, we regard I^- and I^+ as I. For the records involving two intensities such as 82-th record with VI—VII, we don't use them. Then, from Appendix 2.A we obtain a_I, b_I^2 shown in Table 8-2. Because the samples of XI and XII is too small, we take $b_{XII}^2 = b_{XI}^2 = b_X^2$.

Table 8-2 Parameters of normal fuzzy sets

Epicentral Intensity	n_I	a_I	b_I^2
VI	28	5.08	0.0885
VII	48	5.61	0.1314
VIII	26	6.24	0.1379
IX	21	6.84	0.1244
X	6	7.49	0.0737
XI	2	8.0	0.0737
XII	1	8.5	0.0737
Total	132		

Substituting these a_I, b_I^2 into Eq.(8.8), we obtain a fuzzy relation shown in Eq.(8.10).

$$R = \begin{array}{c} \\ m_1 \ (=4.0) \\ m_2 \ (=4.5) \\ m_3 \ (=5.0) \\ m_4 \ (=5.5) \\ m_5 \ (=6.0) \\ m_6 \ (=6.5) \\ m_7 \ (=7.0) \\ m_8 \ (=7.5) \\ m_9 \ (=8.0) \\ m_{10} \ (=8.5) \end{array} \begin{pmatrix} VI & VII & VIII & IX & X & XI & XII \\ 0 & 0 & 0 & 0 & 0 & 0 & 0 \\ 0.02 & 0 & 0 & 0 & 0 & 0 & 0 \\ 0.93 & 0.06 & 0 & 0 & 0 & 0 & 0 \\ 0.14 & 0.91 & 0.02 & 0 & 0 & 0 & 0 \\ 0 & 0.31 & 0.66 & 0 & 0 & 0 & 0 \\ 0 & 0 & 0.61 & 0.39 & 0 & 0 & 0 \\ 0 & 0 & 0.02 & 0.81 & 0.04 & 0 & 0 \\ 0 & 0 & 0 & 0.03 & 1 & 0.03 & 0 \\ 0 & 0 & 0 & 0 & 0.03 & 1 & 0.03 \\ 0 & 0 & 0 & 0 & 0 & 0.03 & 1 \end{pmatrix}$$

$$(8.10)$$

Then using fuzzy inference formula,

$$\tilde{I} = \tilde{M} \circ R, \quad (8.11)$$

we can calculate $\widetilde{I}$ from predicted magnitude $\widetilde{M}$, where operator "o" is the well known max-min fuzzy composition rule shown in Eq. (4.63), i.e.,

$$\mu_I(k) = \max_{1 \leq i \leq 10}\{\min\{\mu_M(m_i), \mu_R(m_i, k)\}\}, \quad k = VI, VII, \cdots, XII. \quad (8.12)$$

Due to that it is very difficult to accurately predict magnitude, Wang (1983, see [8]) suggested 4 types of predicted magnitudes for $\widetilde{M}$ shown in Table 8-3.

Table 8-3 Language magnitude prediction

Magnitude prediction	Mathematical expression ($i = 1, 2, \cdots, 10$)
will occur in m_i	$\widetilde{M} = \frac{1}{m_i}$
will occur around m_i	$\widetilde{M} = \frac{0.2}{m_{i-3}} + \frac{0.4}{m_{i-2}} + \frac{0.6}{m_{i-1}} + \frac{1}{m_i} + \frac{0.6}{m_{i+1}} + \frac{0.4}{m_{i+2}} + \frac{0.2}{m_{i+3}}$
possibly occur around m_i	$\widetilde{M} = \frac{0.04}{m_{i-3}} + \frac{0.16}{m_{i-2}} + \frac{0.36}{m_{i-1}} + \frac{1}{m_i} + \frac{0.36}{m_{i+1}} + \frac{0.16}{m_{i+2}} + \frac{0.04}{m_{i+3}}$
will occur from m_i to m_{i+s}	$\widetilde{M} = \frac{0.4}{m_{i-1}} + \frac{1}{m_i} + \frac{1}{m_{i+1}} + \cdots + \frac{1}{m_{i+s-1}} + \frac{1}{m_{i+s}} + \frac{0.4}{m_{i+s+1}}$

To calculate the epicentral intensities from the magnitudes in Table 8-1 (where a lot of magnitude records are not equal to one of m_i, $i = 1, 2, \cdots, 10$), firstly we change M into fuzzy set $\widetilde{M}$ with the method of information distribution as the following steps.

(1) Distributing. Let

$$\mu_A(m_i) = \begin{cases} 1 - \frac{|M - m_i|}{\Delta}, & \text{if } |M - m_i| \leq \Delta, \\ 0, & \text{otherwise.} \end{cases} \quad (8.13)$$

where $\Delta = m_2 - m_1 = 4.5 - 4.0 = 0.5$.

(2) Normalizing. Let

$$s = \max_{1 \leq i \leq 10}\{\mu_A(m_i)\},$$

then

$$\mu_M(m_i) = \frac{\mu_A(m_i)}{s}, \quad i = 1, 2, \cdots, 10.$$

For example, for $M = 5.75$ (No.1) which is between $m_4(= 5.5)$ and $m_5(= 6.0)$, we have

$$\mu_A(m_4) = 1 - \frac{|5.75 - 5.5|}{0.5} = 0.5, \quad \mu_A(m_6) = 1 - \frac{|5.75 - 6.0|}{0.5} = 0.5.$$

Hence $s = 0.5$, $\mu_A(m_4)/s = \mu_A(m_4)/s = 1$. Therefore

$$\widetilde{M} = \frac{0}{4.0} + \frac{0}{4.5} + \frac{0}{5.0} + \frac{1}{5.5} + \frac{1}{6.0} + \frac{0}{6.5} + \cdots + \frac{0}{8.5}.$$

Secondly, we use formula (8.12) and R to calculate $\widetilde{I}$. For the above $\widetilde{M}$, we have

$$\widetilde{I}_0 = \frac{0.14}{VI} + \frac{0.91}{VII} + \frac{0.66}{VIII} + \frac{0}{IX} + \frac{0}{X} + \frac{0}{XI} + \frac{0}{XII}.$$

Digitizing the intensities with the mapping in (2.19) and (2.20), $\widetilde{I}$ can also be represented as

$$\widetilde{I} = \frac{0.14}{6} + \frac{0.91}{7} + \frac{0.66}{8} + \frac{0}{9} + \frac{0}{10} + \frac{0}{11} + \frac{0}{12}.$$

Then, we calculate the gravity center of the fuzzy set, which is

$$\widetilde{y} = \left(\sum_{k=6}^{12} \mu_I(k)k \right) \Big/ \left(\sum_{k=6}^{12} \mu_I(k) \right).$$

For this $\widetilde{I}$, we have

$$\widetilde{y} = \frac{0.14 \times 6 + 0.91 \times 7 + 0.66 \times 8}{0.14 + 0.91 + 0.66} = \frac{12.49}{1.71} = 7.30.$$

Finally, taking the nearest Roman numeral to be the symbol estimation, we obtain the estimated intensity we need. From $\widetilde{y} = 7.30$, we obtain $\widetilde{I}_0 = VII$, which is just as same as the real record, so the comment is T.

The model is called *fuzzy inference based on normal assumption*(FINA). The FINA column of Appendix 8.A gives the estimated intensities by the model.

There are 87 estimates being true, i.e., the correct rate of the linear regression is also 65%. The square error $\varepsilon_{FINA} = 0.280$.

Both models of LR and FINA have almost same correct rate although some estimates are not same. However, the model of FINA shows that some estimated fuzzy sets have more than one supporting points, it just reflects separateness of the given sample. In some sense, FINA is better than LR.

8.3 Self-Study Discrete Regression

LR is useful only if we know that the relationship between input and output is an approximate linear function. FINA needs more experience from engineering practice. In this section we suggest a new method supported by the principle of information diffusion to construct a statistic relationship from a given sample.

8.3.1 Discrete regression

To resolve practical problems, we often want to know statistic relationships among factors by learning from the given samples. Least square procedure as a criterion is a method used by many people. In the first step for using it, the researcher is required to look for a reasonable analytical model. For example, the linear model, $y = a + bx$, is usually used. The second step is to estimate the parameters of the model by the help of least square fit. For linear model, the work is to estimate a and b by the method of least squares from a given sample. This approach has been so successful that it is now viewed as a standard tool for constructing statistic relationships.

However, in many cases, particularly for a new problem or a multivariate nonlinear system, it is very difficult or impossible to count upon a reasonable analytical model coming. In general, the model is unknown and the main work of analysis is to discover it. If the model is supposed firstly, the relationship we want to know must be limited by researcher's experience and knowledge.

As we have discussed in Chapter 2, during the last two decades, artificial neural network (ANN) has received extensive attentions. ANN is well known as a powerful tool to solve many practical problems including learning relationship from the given sample without any assumption of the model explaining where the sample is drawn from. The most popular neural networks for constructing statistic relationships are the multi-layer *back-propagation* (BP) networks whose training algorithm is the well-known gradient descent method. The algorithm is a supervised learning paradigm where a squared error criterion function[2] as a "teacher" provides the system with a detailed parallelism scheme for the update of the synaptic weights based on knowledge of the goal to be achieved. It has been shown that a BP network with one hidden layer can approximate any L^2 function f: $[0,1]^n \to R^m$ in any square error ε, a multiple layer network can approximate any function[4].

However, a neural network is a black box and we might meet the problem of getting trapped in local minima. Particularly, neural information processing models generally assume that the patterns used for training a neural network are compatible. In the real world, sometimes, the observations are strongly scattered and contradictory patterns do occur, so that, a neural network does not converge because the adjustments of weights and thresholds do not know where to turn due to the ambiguity brought forth by the contradictory patterns. In other words, in many cases, it is impossible to construct statistic relationships by neural networks directly.

Employing the principle of information diffusion, we can construct information matrixes for constructing statistic relationships, called regression relationships.

In the terminology of this book, any relationship learned from a given sample is called a *regression*. Here, word "regression" means repeating of the past procedure. The classical statistical regression usually gives a continuous function to represent the relationship between variables. A trained neural network is performing as a mapping from input space to output space. If the mapping is continuous, the trained neural network can be regarded as a continuous regression result. Both the classical statistical regression and the neural network regression use supervised learning such as least square methods to determine the parameters of the model or the output weights[6].

The information matrixes result from unsupervised learning, where we have no information available about the desired output of the matrixes. The overall goal of any regression is to find structure (information) about the given samples. One of the most common structures is the probability distribution of the population from which the sample was drawn. A common unsupervised learning algorithm, which is employed to searching for the structure, is the histogram. When we construct a histogram, there is no "teacher" who provides a criterion for the update of the histogram. Obviously, a histogram is a regression result. Particularly, it is impossible for a histogram to give a continuous function for estimation of a probability distribution. A histogram estimate of $p(x)$ calculated by formula (3.25) is a discrete function. According to the relation between the information matrixes and the histograms, we know that a statistic relationships which result from information matrixes must be discrete. That a histogram is a discrete regression means that an information matrix must be a discrete regression, too. The only difference between the histogram and the information matrix is that it is easier to soften an information matrix for obtaining a better fuzzy relationship among factors by learning from a given sample. To form the procedure, we give the following definitions.

Definition 8.1 Let R^r be the r-dimension Euclidean space. A set S consisting of more one points of R^r, $s_1, s_2, \cdots, s_n, \cdots$, is called a *discrete subset* of R^r.

For example, $S_1 = \{2, 6, 3\}$ is a discrete subset of R. $S_2 = \{(2, 1), (5, 6), (6, 4)\}$ is a discrete subset of R^2. But, $S_3 = [1, 0] \times [1, 10]$ is not a discrete set.

Definition 8.2 Let U be a set. If there is a discrete subset S of R^r such that we can construct an one-one mapping between U and S, we say that U is a *discrete set*.

For example, to make a fuzzy controller for an intelligent parking procedure, we use θ to represent the robot angle, $-90 \leq \theta \leq 270$. Let $U = \{$Right Bellow, Right Upper, Right Vertical, Vertical, Left Vertical, Left Upper, Left Below$\}$. It can be simplified to $U = \{$RB,RU,RV,V,LV,LU,LB$\}$. The elements of U are triangular fuzzy sets. Suppose their center-points are $\theta = -40, 25, 75, 90, 105, 155, 220$. Obviously, U is a discrete set because we can construct an one-one mapping from these fuzzy sets to their center-points that form a discrete subset $S = \{-40, 25, 75, 90, 105, 155, 220\}$ of $\mathbb{R}$.

Definition 8.3 Let U be a discrete set and V be a general set. The mapping $f : U \to V$ is called a *discrete mapping* of the set U into the set V. f is also called a *discrete function* from U to set V.

For example, let U consist of the intervals $I_1, I_2, \cdots, I_m$ which are employed to construct a relative frequency histogram, and $P = [0, 1]$, then the relative frequency histogram is a discrete function from U to range P of probability values.

Regression is a return to a former or earlier state. Hence, we have the following definition.

Definition 8.4 Let both U and V be discrete sets. If we can obtain a mapping f of the set U into the set V by learning from a given sample, we called f a *mapping discrete regression*.

For example, it is a generally accepted conclusion that domain experts can express if-then rules using their own fuzzy terms. We suppose that, integrating the knowledge of experts who know how to control an air condition, we obtain a rule-base:
IF temperature IS cold THEN force IS high;
IF temperature IS cool THEN force IS medium;
IF temperature IS warm THEN force IS low;
IF temperature IS hot THEN force IS zero.
Let

$$U = \{cold, cool, warm, hot\}, \quad V = \{high, medium, low, zero\}.$$

Then, above rule-base can be defined by the following mapping

$$
\begin{array}{rcl}
f: \quad U & \to & V \\
cold & \mapsto & high \\
cool & \mapsto & medium \\
warm & \mapsto & low \\
hot & \mapsto & zero
\end{array}
$$

Obviously, both U and V are discrete sets. Mapping f is obtained by learning from a sample given by domain experts. It is a regression result. According to Definition 8.4, mapping f is a mapping discrete regression.

Definition 8.5 Let both U and V be discrete sets. Any fuzzy relationship R on $U \times V$ by learning from the given sample X is called a *fuzzy-mapping discrete regression* of X on $U \times V$. In short, R is called a *discrete regression* of X.

When we obtain a discrete regression of X, using a relational approximate inference formula given in section 4.8, we can calculate output from a given input as same as that we calculate y_0 from x_0 by the linear regression $y = a + bx$.

A high quality discrete regression is the goal we look for. According to the discussion in part I of this book, one easily comes to the conclusions:

(1) The framework of information matrix is a reasonable approach to construct a discrete regression;
(2) The information matrixes constructed by the method of information distribution is better than ones by the classical histogram;
(3) Let the dimension of a given sample X be more than one. When X is incomplete, there must exist a reasonable diffusion function changing observation x_i into fuzzy set $\mu(x_i, u)$ for constructing an information matrix of quality.

In next subsection we suggest the so-called r-dimension diffusion for constructing information matrixes we need.

8.3.2 r-dimension diffusion

Let $X = \{x_1, x_2, \cdots, x_n\}$ be a r-dimension random sample. Then $\forall x_i \in X$, we know that it is a r-dimension vector, i.e.,

$$x_i = (x_{1i}, x_{2i}, \cdots, x_{ri}). \tag{8.14}$$

Denote $K = \{1, 2, \cdots, r\}$ and let

$$X_k = \{x_{ki} | i = 1, 2, \cdots, n\}, \quad k \in K \tag{8.15}$$

For example, $X = \{(5.75, 7), (6.4, 8), (6.5, 9)\}$ is a 2-dimension random sample. For it, we have $X_1 = \{5.75, 6.4, 9\}$ and $X_2 = \{7, 8, 9\}$.

$\forall k \in K$, we suppose that the chosen monitoring space of X_k is

$$U_k = \{u_{kj} | j = 1, 2, \cdots, m_k\}, \quad k \in K. \tag{8.16}$$

Then,

$$U = U_1 \times U_2 \times \cdots \times U_r = \prod_{k=1}^{r} U_k$$

is the chosen monitoring space of X. There are $m = m_1 m_2 \cdots m_r$ elements in U.

Let $\mu_{(k)}$ be a diffusion function of X_k on U_k, it can be written as

$$\mu_{(k)}(x_i, u_j) = \mu_{(k)}(x_{ki}, u_{kj}).$$

Definition 8.6 $\forall x_i \in X$, $\forall u_j \in U$,

$$\mu(x_i, u_j) = \prod_{k=1}^{r} \mu_{(k)}(x_{ki}, u_{kj}) \tag{8.17}$$

is called a r-dimension diffusion of X on U.

For example, the r-dimension linear distribution given in Eq.(4.27) is a r-dimension diffusion.

Definition 8.7 Let $X = \{x_1, x_2, \cdots, x_n\}$ be a r-dimension random sample, and $U = \{u_1, u_2, \cdots, u_m\}$ be the chosen monitoring space. For any $x_i \in X$, and $u_j \in U$, the following formula is called r-*dimension normal diffusion*:

$$\mu(x_i, u_j) = \prod_{k=1}^{r} \frac{1}{h_k \sqrt{2\pi}} \exp[-\frac{(u_{kj} - x_{ki})^2}{2h_k^2}]. \tag{8.18}$$

Where, h_k is called k-th diffusion coefficient.

8.3.3 Self-study discrete regression

Let $X = \{(x_1, y_1), (x_2, y_2), \cdots, (x_n, y_n)\}$ be a r-dimension random sample, the input and output universes of X are U and V, respectively.

Firstly, according to the accuracy we need, we choose two discrete universes from U and V, respectively, to be the monitoring spaces serving for diffusing. We still denote them by U and V. We denote the spaces as

$$U = \{u_1, u_2, \cdots, u_m\},$$

$$V = \{v_1, v_2, \cdots, v_t\}.$$

Secondly, for (x_i, y_i) and (u_j, v_k), we calculate $\mu((x_i, y_i), (u_j, v_k))$, which is called *diffused information* on (u_j, v_k) from (x_i, y_i), denoted as q_{ijk}.

Case 1: When $V \subset \mathbb{R}$, we use a r-dimension diffusion to calculate q_{ijk} as

$$q_{ijk} = \left[\prod_{l=1}^{r-1} \mu_{(l)}(x_{li}, u_{lj})\right]\mu(y_i, v_k), \tag{8.19}$$

where $\mu(y_i, v_k)$ is a 1-dimension diffusion of y_i on V.

Obviously, Eq.(8.19) is just the same as Eq.(8.17) if we have not separated the universe of X into input and output universes. In other words, we can employ a general r-dimension diffusion to calculate q if $V \subset \mathbb{R}$.

Case 2: When the elements of V are fuzzy concepts, we use the so-called pseudo-r-dimension diffusion to calculate q_{ijk}. That is

$$q_{ijk} = \begin{cases} \prod_{l=1}^{r-1} \mu_{(l)}(x_{li}, u_{lj}), & \text{if } y_i = v_k; \\ 0, & \text{otherwise.} \end{cases} \tag{8.20}$$

Thirdly, we combine q to produce an information matrix. Let

$$Q_{jk} = \sum_{i=1}^{n} q_{ijk}, \tag{8.21}$$

then,

$$Q = \begin{array}{c} \\ u_1 \\ u_2 \\ \cdots \\ u_m \end{array} \begin{array}{cccc} v_1 & v_2 & \cdots & v_t \\ \left(\begin{array}{cccc} Q_{11} & Q_{12} & \cdots & Q_{1t} \\ Q_{21} & Q_{22} & \cdots & Q_{2t} \\ \cdots & \cdots & \cdots & \cdots \\ Q_{m1} & Q_{m2} & \cdots & Q_{mt} \end{array} \right) \end{array} \tag{8.22}$$

is the information matrix we need. Q is called a *primary information matrix* of X on $U \times V$.

Obviously, a primary information distribution of X on U, defined in section 4.2, is a primary information matrix. Because information matrixes on discrete universes of discourse, crisp intervals and fuzzy intervals discussed in Chapter 2 are three special primary information distribution, we know that they are also primary information matrixes.

Then, from Q, employing or extending one of the models (R_f, R_m, R_c) given in section 4.7, we can obtain a fuzzy relation matrix R on $U \times V$.

Finally, with the approximate inference models given in section 4.8, we can estimate output y by input x.

All work we did depend on X, without any assumption. The R, acting as a function, is produced by learning from the given sample X, without any engineering experience. Obviously, R is a statistic relationship. Therefore, R is called a result of *self-study discrete regression*.

The estimation of epicentral intensity by magnitude based on a given sample is a problem of 2-dimension self-study discrete regression. In section 8.4 and 8.5 we will give two models to employ R_f for resolving the problem.

8.4 Linear Distribution Self-Study

The earthquake intensities, VI,VII,$\cdots$, XII, are typical fuzzy concepts. The simplest self-study model is to use the pseudo-2-dimension diffusion to calculate q_{ijk} (see Eq.(8.20)), then use

$$\begin{cases} R_f = \{r_{jk}\}_{m \times t}, \\ r_{jk} = Q_{jk}/s_k, \\ s_k = \max_{1 \le j \le m} Q_{jk}, \end{cases} \tag{8.23}$$

we can obtain a fuzzy relation matrix R_f.

For the given sample in Appendix 2.A, firstly we choose

$$U = \{u_1, u_2, \cdots, u_{10}\} = \{4.0, 4.5, \cdots, 8.5\},$$

$$V = \{v_1, v_2, \cdots, v_7\} = \{VI, VII, \cdots, XII\}$$

to be framework spaces.

Then, for (x_i, y_i), using the linear distribution

$$q_{ijk} = \begin{cases} 1 - \frac{|x_i - u_j|}{\Delta}, & \text{for } |x_i - u_j| \le \Delta \text{ and } \|y_i - v_k\| < 0.5, \\ 0, & \text{otherwise,} \end{cases} \tag{8.24}$$

where $\Delta = u_2 - u_1 = 0.5$, we can distribute (x_i, y_i) over $U \times V$.

For example, for No.1 record, $(x_1, y_1) = (5.75, VII)$, $u_4 < x_1 < u_5$, $\|y_1 - v_2\| = \|7 - VII\| = |7 - 7| = 0$, then we have

$$q_{142} = 1 - \frac{|5.75 - 5.5|}{0.5} = 0.5, \quad q_{152} = 1 - \frac{|5.75 - 6.0|}{0.5} = 0.5.$$

Distributing all records and using Eq.(8.21) we obtain a primary information matrix

$$Q = \begin{array}{c} \\ u_1 \ (=4.0) \\ u_2 \ (=4.5) \\ u_3 \ (=5.0) \\ u_4 \ (=5.5) \\ u_5 \ (=6.0) \\ u_6 \ (=6.5) \\ u_7 \ (=7.0) \\ u_8 \ (=7.5) \\ u_9 \ (=8.0) \\ u_{10} \ (=8.5) \end{array} \begin{array}{ccccccc} VI & VII & VIII & IX & X & XI & XII \\ 0.5 & 0.0 & 0.0 & 0.0 & 0.0 & 0.0 & 0.0 \\ 2.5 & 0.9 & 0.0 & 0.0 & 0.0 & 0.0 & 0.0 \\ 11.6 & 4.7 & 0.0 & 0.0 & 0.0 & 0.0 & 0.0 \\ 5.2 & 18.8 & 2.3 & 0.0 & 0.0 & 0.0 & 0.0 \\ 0.2 & 9.8 & 11.2 & 1.0 & 0.0 & 0.0 & 0.0 \\ 0.0 & 1.8 & 8.7 & 6.9 & 0.0 & 0.0 & 0.0 \\ 0.0 & 0.0 & 1.8 & 7.5 & 1.5 & 0.0 & 0.0 \\ 0.0 & 0.0 & 0.0 & 3.6 & 3.1 & 0.0 & 0.0 \\ 0.0 & 0.0 & 0.0 & 0.0 & 1.4 & 2.0 & 0.0 \\ 0.0 & 0.0 & 0.0 & 0.0 & 0.0 & 0.0 & 1.0 \end{array} \qquad (8.25)$$

Using Eq.(8.23) we obtain a fuzzy relation matrix

$$R_f = \begin{array}{c} \\ u_1 \ (=4.0) \\ u_2 \ (=4.5) \\ u_3 \ (=5.0) \\ u_4 \ (=5.5) \\ u_5 \ (=6.0) \\ u_6 \ (=6.5) \\ u_7 \ (=7.0) \\ u_8 \ (=7.5) \\ u_9 \ (=8.0) \\ u_{10} \ (=8.5) \end{array} \begin{array}{ccccccc} VI & VII & VIII & IX & X & XI & XII \\ 0.04 & 0 & 0 & 0 & 0 & 0 & 0 \\ 0.22 & 0.05 & 0 & 0 & 0 & 0 & 0 \\ 1.00 & 0.25 & 0 & 0 & 0 & 0 & 0 \\ 0.45 & 1.00 & 0.21 & 0 & 0 & 0 & 0 \\ 0.02 & 0.52 & 1.00 & 0.13 & 0 & 0 & 0 \\ 0 & 0.10 & 0.78 & 0.92 & 0 & 0 & 0 \\ 0 & 0 & 0.16 & 1.00 & 0.48 & 0 & 0 \\ 0 & 0 & 0 & 0.48 & 1.00 & 0 & 0 \\ 0 & 0 & 0 & 0 & 0.45 & 1.00 & 0 \\ 0 & 0 & 0 & 0 & 0 & 0 & 1.00 \end{array}$$

$$(8.26)$$

We use the max-min fuzzy composition rule shown in Eq. (8.12) to estimate epicentral intensity by magnitude.

For $M = 5.75$ (No.1), distributing it over U and normalizing the derived fuzzy set, we obtain

$$\widetilde{M} = \frac{0}{u_1} + \frac{0}{u_2} + \frac{0}{u_3} + \frac{1}{u_4} + \frac{1}{u_5} + \frac{0}{u_6} + \cdots + \frac{0}{u_{10}}.$$

Secondly, we use formula (8.12) and R_f to calculate $\widetilde{I}$. Then, the fuzzy estimation is

$$\widetilde{I}_0 = \frac{0.45}{VI} + \frac{1}{VII} + \frac{1}{VIII} + \frac{0.13}{IX} + \frac{0}{X} + \frac{0}{XI} + \frac{0}{XII}.$$

Digitizing the intensities with the mapping in (2.19) and (2.20), $\tilde{I}$ can also be represented as

$$\tilde{I} = \frac{0.45}{6} + \frac{1}{7} + \frac{1}{8} + \frac{0.13}{9} + \frac{0}{10} + \frac{0}{11} + \frac{0}{12}.$$

Then, we calculate the gravity center of the fuzzy set, which is

$$\tilde{y} = \left(\sum_{k=6}^{12} \mu_I(k)k \right) \bigg/ \left(\sum_{k=6}^{12} \mu_I(k) \right).$$

For this $\tilde{I}$, we have

$$\tilde{y} = \frac{0.45 \times 6 + 1 \times 7 + 1 \times 8 + 0.13 \times 9}{0.45 + 1 + 1 + 0.13} = \frac{18.87}{2.58} = 7.31.$$

Finally, taking the nearest Roman numeral to be the symbol estimation, we obtain the estimated intensity we need. From $\tilde{y} = 7.31$, we obtain $\tilde{I}_0 = VII$, which is just as same as the real record, so the comment is T.

The model is called *linear distribution self-study* (LDSS). The LDSS column of Appendix 8.A gives the estimated intensities by the model of self-study discrete regression.

There are 88 estimates being true, i.e., the correct rate of the linear regression is 66%. The square error $\varepsilon_{LDSS} = 0.265$.

$\varepsilon_{LDSS} < \varepsilon_{FINA}, \varepsilon_{LR}$, and we have successfully avoided any assumption. It may be argued that, the LDSS model is better than the classical models for estimation of epicentral intensity.

However, in LDSS, we need some engineering experience for choosing framework space U. The adjustment work must be done repeatedly. The number of T and the square error ε can be used to supervise learning. The result from LR shows that it is reasonable to map earthquake intensities into $\mathbb{R}$. Therefore we can use 2-dimension normal diffusion to construct Q as a complete self-study model.

8.5 Normal Diffusion Self-Study

The monitoring points for normal diffusion can be chosen as many as possible. We can do it based on the accuracy of the observations. For the data in Table 8-1, regarding M as x and I_0 as y, we have

$$\min_{x_i \neq x_j} \{|x_i - x_j| \mid i, j = 1, 2, \cdots 134\} = 0.05,$$

$$\min_{y_i \neq y_j} \{|y_i - y_j| \mid i, j = 1, 2, \cdots 134\} = 0.2.$$

Hence, it is enough to let

$$U = \{u_1, u_2, \cdots, u_{86}\} = \{4.25, 4.30, \cdots, 8.5\},$$

$$V = \{v_1, v_2, \cdots, v_{32}\} = \{5.8, 6, \cdots, 12\}$$

be the monitoring spaces, with steps $\Delta_u = 0.05$ and $\Delta_v = 0.2$.

Because

$$b_x = \max_{1 \leq i \leq 134} \{x_i\} = 8.5, \ a_x = \min_{1 \leq i \leq 134} \{x_i\} = 4.25,$$

$$b_y = \max_{1 \leq i \leq 134} \{y_i\} = 12, \ a_y = \min_{1 \leq i \leq 134} \{y_i\} = 5.8,$$

using formula (7.38) we obtain simple coefficients

$$h_x = 2.6851(b_x - a_x)/(n - 1) = 2.6851(8.5 - 4.25)/133 = 0.0858,$$

$$h_y = 2.6851(b_y - a_y)/(n - 1) = 2.6851(12 - 5.8)/133 = 0.1251.$$

We use 2-dimension normal diffusion

$$\mu((x_i, y_i), (u_j, v_k)) = \frac{1}{h_x \sqrt{2\pi}} \exp[-\frac{(u_j - x_i)^2}{2h_x^2}] \cdot \frac{1}{h_y \sqrt{2\pi}} \exp[-\frac{(v_j - y_i)^2}{2h_y^2}]$$

to calculate the diffused information on (u_j, v_k) from observation (x_i, y_i), denoted as q_{ijk}. Let

$$Q^{(i)} = \{q_{ijk}\}_{86 \times 32}.$$

It is an information matrix of single element sample $X^{(1)} = \{(x_i, y_i)\}$ on $U \times V$.

For example, for No.1 record in Table 8-1, $(x_1, y_1) = (5.75, 7)$, the corresponding matrix is given by (8.27), denoted as $Q^{(1)}$.

	v_1 (5.8)	$\cdots$	v_5 (6.6)	v_6 (6.8)	v_7 (7)	v_8 (7.2)	v_9 (7.4)	$\cdots$	v_{32} (12)
u_1 (4.25)	0.00	$\cdots$	0.00	0.00	0.00	0.00	0.00	$\cdots$	0.00
$\cdots$	$\cdots$	$\cdots$	$\cdots$	$\cdots$	$\cdots$	$\cdots$	$\cdots$	$\cdots$	$\cdots$
u_{24} (5.40)	0.00	$\cdots$	0.00	0.00	0.00	0.00	0.00	$\cdots$	0.00
u_{25} (5.45)	0.00	$\cdots$	0.00	0.01	0.03	0.01	0.00	$\cdots$	0.00
u_{26} (5.50)	0.00	$\cdots$	0.00	0.06	0.21	0.06	0.00	$\cdots$	0.00
u_{27} (5.55)	0.00	$\cdots$	0.01	0.27	0.98	0.27	0.01	$\cdots$	0.00
u_{28} (5.60)	0.00	$\cdots$	0.02	0.90	3.22	0.90	0.02	$\cdots$	0.00
u_{29} (5.65)	0.00	$\cdots$	0.05	2.10	7.51	2.10	0.05	$\cdots$	0.00
u_{30} (5.70)	0.00	$\cdots$	0.08	3.49	12.51	3.49	0.08	$\cdots$	0.00
u_{31} (5.75)	0.00	$\cdots$	0.09	4.13	14.82	4.13	0.09	$\cdots$	0.00
u_{32} (5.80)	0.00	$\cdots$	0.08	3.49	12.50	3.49	0.08	$\cdots$	0.00
u_{33} (5.85)	0.00	$\cdots$	0.05	2.10	7.51	2.10	0.05	$\cdots$	0.00
u_{34} (5.90)	0.00	$\cdots$	0.02	0.90	3.21	0.90	0.02	$\cdots$	0.00
u_{35} (6.95)	0.00	$\cdots$	0.01	0.27	0.98	0.27	0.01	$\cdots$	0.00
u_{36} (6.00)	0.00	$\cdots$	0.00	0.06	0.21	0.06	0.00	$\cdots$	0.00
u_{37} (6.05)	0.00	$\cdots$	0.00	0.01	0.03	0.01	0.00	$\cdots$	0.00
u_{38} (6.10)	0.00	$\cdots$	0.00	0.00	0.00	0.00	0.00	$\cdots$	0.00
$\cdots$	$\cdots$	$\cdots$	$\cdots$	$\cdots$	$\cdots$	$\cdots$	$\cdots$	$\cdots$	$\cdots$
u_{86} (8.50)	0.00	$\cdots$	0.00	0.00	0.00	0.00	0.00	$\cdots$	0.00

$Q^{(1)} =$ (the above matrix)

$$(8.27)$$

Obviously,

$$Q = \sum_{i=1}^{134} Q^{(i)}$$

is a primary information matrix of the given sample on $U \times V$. Using Eq.(8.23) we obtain a fuzzy relation matrix, denoted as R_f that is shown in (8.28).

	v_1 (5.8)	v_2 (6)	v_3 (6.2)	v_4 (6.4)	v_5 (6.6)	v_6 (6.8)	$\cdots$	v_{31} (11.8)	v_{32} (12)
u_1 (4.25)	0.12	0.11	0.05	0.00	0.00	0.00	$\cdots$	0.00	0.00
u_2 (4.30)	0.10	0.09	0.04	0.00	0.00	0.00	$\cdots$	0.00	0.00
u_3 (4.35)	0.06	0.06	0.03	0.00	0.00	0.00	$\cdots$	0.00	0.00
u_4 (4.40)	0.03	0.03	0.01	0.00	0.00	0.00	$\cdots$	0.00	0.00
u_5 (4.45)	0.03	0.03	0.01	0.00	0.00	0.00	$\cdots$	0.00	0.00
u_6 (4.50)	0.07	0.06	0.03	0.00	0.00	0.00	$\cdots$	0.00	0.00
u_7 (4.55)	0.13	0.10	0.05	0.00	0.00	0.01	$\cdots$	0.00	0.00
u_8 (4.60)	0.24	0.14	0.06	0.00	0.00	0.02	$\cdots$	0.00	0.00
u_9 (4.65)	0.40	0.18	0.08	0.01	0.00	0.06	$\cdots$	0.00	0.00
u_{10} (4.70)	0.60	0.25	0.11	0.01	0.01	0.12	$\cdots$	0.00	0.00
u_{11} (4.75)	0.77	0.34	0.16	0.02	0.01	0.16	$\cdots$	0.00	0.00
u_{12} (4.80)	0.79	0.42	0.22	0.04	0.01	0.16	$\cdots$	0.00	0.00
u_{13} (4.85)	0.73	0.50	0.34	0.12	0.01	0.13	$\cdots$	0.00	0.00
u_{14} (4.90)	0.76	0.66	0.57	0.27	0.01	0.11	$\cdots$	0.00	0.00
u_{15} (4.95)	0.92	0.89	0.85	0.44	0.02	0.10	$\cdots$	0.00	0.00
u_{16} (5.00)	1.00	1.00	1.00	0.54	0.02	0.11	$\cdots$	0.00	0.00
u_{17} (5.05)	0.85	0.86	0.91	0.50	0.02	0.13	$\cdots$	0.00	0.00
u_{18} (5.10)	0.57	0.59	0.71	0.43	0.02	0.18	$\cdots$	0.00	0.00
$\cdots$	$\cdots$	$\cdots$	$\cdots$	$\cdots$	$\cdots$	$\cdots$		$\cdots$	$\cdots$
u_{85} (8.45)	0.00	0.00	0.00	0.00	0.00	0.00	$\cdots$	0.84	0.84
u_{86} (8.50)	0.00	0.00	0.00	0.00	0.00	0.00	$\cdots$	1.00	1.00

$R_f = $ (the matrix above)

$$(8.28)$$

Any magnitude record in Table 8-1 must be equal to one of the monitoring points. It is unnecessary to change M into fuzzy set $\widetilde{M}$ with the method of information distribution for calculating the epicentral intensity. In other words, let $X_1 = \{M_1, M_2, \cdots, M_{134}\}$ from Table 8-1, $\forall M_i \in X_1$, $\exists u_j \in U$ such that $u_j = M_i$, hence $\widetilde{M} = \frac{1}{u_j}$. Using formula (8.12) we can calculate $\widetilde{I}$ directly. For example, $M = 4.8$ (No.22), it can be represented as $\widetilde{M} = \frac{1}{u_{12}} = \frac{1}{4.80}$. Using formula (8.12) we obtain

$$\widetilde{I}_0 = \frac{0.79}{5.8} + \frac{0.42}{6.0} + \frac{0.22}{6.2} + \frac{0.04}{6.4} + \frac{0.01}{6.6} + \frac{0.16}{6.8} + \frac{0.16}{7.0} + \frac{0.09}{7.2} + \frac{0.03}{7.4} + \frac{0.00}{7.6} + \cdots + \frac{0}{12}.$$

The gravity center of the fuzzy set is

$$\widetilde{y} = \left(\sum_{k=1}^{32} \mu_I(k) v_k \right) \Big/ \left(\sum_{k=1}^{32} \mu_I(k) \right) = 6.18.$$

Thence, we obtain $\tilde{I}_0 = VI$, which is just as same as the real record, so the comment is T.

Above model is called *normal diffusion self-study* (NDSS). The NDSS column of Appendix 8.A gives the estimated intensities by the model. Appendix 8.B gives a Fortran program to do the work.

There are 90 estimates being true, i.e., the correct rate of the linear regression is 67%. The square error $\varepsilon_{NDSS} = 0.226$.

$\varepsilon_{NDSS} < \varepsilon_{LDSS}, \varepsilon_{FINA}, \varepsilon_{LR}$, and we have successfully avoided to use any engineering experience . Therefore, the NDSS model is better than LDSS, FINA and LR for estimation of epicentral intensity.

Fig. 8.1 shows the estimations from the four models. The small circles are observations.

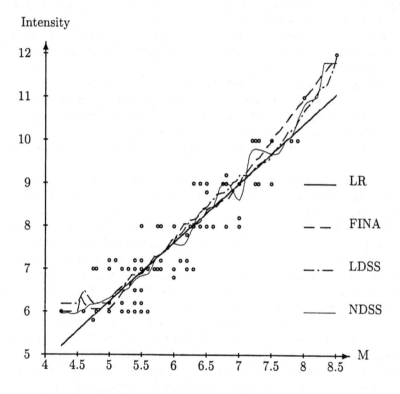

Fig. 8.1   Estimation by linear regression and fuzzy inference models.
            LR: linear regression;
            FINA: fuzzy inference based on normal assumption;
            LDSS: linear distribution self-study;
            NDSS: normal diffusion self-study.

Although NDSS gives the least square error, its curve is waving around M=6.5 to 7.5. An approach to smooth it is a hybrid fuzzy-neural-network, which will be discussed in next chapter.

8.6 Conclusion and Discussion

Many attempts have been made to realize input-output relationships by learning from known data. The traditional statistic regression is now viewed as a standard tool because of its simplicity, and the artificial neural networks are most popular because of their capability of nonlinear function approximation.

However, a conventional statistic relationship is limited by researcher's experience and knowledge, and the neural networks are not able to converge when the patterns used for training are incompatible.

With the self-study discrete regression based on r-dimension diffusion, it becomes easy to construct a statistic relationship from a given sample without any assumption.

The procedure includes the following 5 steps:

(1) Constructing a monitoring space for illustrating the given sample;

(2) Employing a r-dimension diffusion formula to produce a primary information matrix Q of the sample on the space;

(3) Changing Q into a fuzzy relation matrix R according to the property of Q;

(4) Using the fuzzy inference formula to calculate output by input;

(5) Analyzing the correct rate and the square error to judge if the regression is acceptable.

The results of the estimation of epicentral intensity show that both linear distribution self-study (LDSS) and normal diffusion self-study (NDSS) models are better than the traditional linear regression (LR) model. Because the monitoring points for NDSS can be chosen as many as possible and its square error is the least, we say that the NDSS model is the best model of self-study discrete regression we have found.

Strictly speaking, the relationship between epicentral intensity and magnitude is a nonlinear function. From Fig. 8.1 we know that the low part of the function ($M \in [4.5, 6.5]$) differs from the high part ($M > 6.5$). However, to minimize the square sum of LR, the linear function must be more compatible with the low part because the observations whose magnitude is not greater than 6.5 hold 79% of the whole sample. In other words, LR can not show the relationship with respect to the high part. All fuzzy models in this chapter can do that.

It is early to say that self-study discrete regression based on r-dimension diffusion is the best approach to construct a statistic relationship from a given sample. The reason is that, until now we have not found the best diffusion function and formula to calculate diffusion coefficient. Although NDSS gives the least square error, its curve is not smooth, and we need some tool to resolve the problem. In next chapter we suggest a hybrid fuzzy-neural-network to relax the problem.

References

1. Berlin, G.L.(1980), Earthquakes and the Urban Environment, Volume I, CRC Press, Boca Raton, Florida
2. Dracopoulos, D.C.(1997), Evolutionary Learning Algorithms for Neural Adaptive Control, Springer-Verlag, Heidelberg
3. Feng, D.Y., Lin, M.Z., Wu, G.Y. and Jiang, C.(1985), A study on fuzzy evaluation of earthquake intensity. Fen Deyi and Liu Xihui (eds): Fuzzy Mathematics in Earthquake Researches. Seismological Press, Beijing, pp.149-161
4. Irie, B. and Miyake, S.(1988), Capabilities of three-layered perceptrons, Proc. of the International Conference on Neural Networks, pp. 641-648
5. Kasahara, K.(1981), Earthquake Mechanics, Cambridge University Press, Cambridge, UK
6. Moody, J. and Darken, C. J. (1989), Fast learning in networks of locally tuned processing units. Neural Computation, Vol.1, pp.281-194
7. Tarbuck, E.J. and Lutgens, F.K.(1991), Earth Science (Sixth Edition), Macmillan Publishing Company, New York
8. Wang, F.(1983), Fuzzy recognition of relations between epicentral intensity and magnitude, Earthquake Engineering and Engineering Vibration, Vol.3, No.3, pp.84-96

Appendix 8.A: Real and Estimated Epicentral Intensities

Using Four Models to Estimate Epicentral Intensities

No.	M	Real intensity		LR			FINA			LDSS			NDSS		
		I_0	Digit	$\hat{y}$	$\hat{I_0}$	C	$\tilde{y}$	$\tilde{I_0}$	C	$\tilde{y}$	$\tilde{I_0}$	C	$\tilde{y}$	$\tilde{I_0}$	C
1	5.75	VII	7	7.28	VII	T	7.30	VII	T	7.31	VII	T	7.29	VII	T
2	5.75	VII	7	7.28	VII	T	7.30	VII	T	7.31	VII	T	7.29	VII	T
3	6.4	VIII	8	8.17	VIII	T	8.11	VIII	T	8.32	VIII	T	8.33	VIII	T
4	6.4	VIII	8	8.17	VIII	T	8.11	VIII	T	8.32	VIII	T	8.33	VIII	T
5	6.5	IX	9	8.31	VIII	F	8.39	VIII	F	8.46	VIII	F	8.41	VIII	F
6	6.5	IX	9	8.31	VIII	F	8.39	VIII	F	8.46	VIII	F	8.41	VIII	F
7	6	VIII	8	7.62	VIII	T	7.68	VIII	T	7.74	VIII	T	7.59	VIII	T
8	6	VIII	8	7.62	VIII	T	7.68	VIII	T	7.74	VIII	T	7.59	VIII	T
9	6	VII	7	7.62	VIII	F	7.68	VIII	F	7.74	VIII	F	7.59	VIII	F
10	7	VIII	8	9	IX	F	9.02	IX	F	9.2	IX	F	8.61	IX	F
11	7	IX	9	9	IX	T	9.02	IX	T	9.2	IX	T	8.61	IX	T
12	6.25	VIII	8	7.97	VIII	T	8.06	VIII	T	8.15	VIII	T	7.84	VIII	T
13	5.75	VIII	8	7.28	VII	F	7.30	VII	F	7.31	VII	F	7.29	VII	F
14	5.8	VIII	8	7.34	VII	F	7.35	VII	F	7.36	VII	F	7.36	VII	F
15	6.3	IX	9	8.03	VIII	F	8.06	VIII	F	8.16	VIII	F	8.04	VIII	F
16	6.5	VIII	8	8.31	VIII	T	8.39	VIII	T	8.46	VIII	T	8.41	VIII	T
17	5.8	VII$^+$	7.2	7.34	VII	T	7.35	VII	T	7.36	VII	T	7.36	VII	T
18	5	VII$^+$	7.2	6.24	VI	F	6.06	VI	F	6.2	VI	F	6.31	VI	F
19	5	VI	6	6.24	VI	T	6.06	VI	T	6.2	VI	T	6.31	VI	T
20	5	VI$^+$	6.2	6.24	VI	T	6.06	VI	T	6.2	VI	T	6.31	VI	T
21	6	VII$^-$	6.8	7.62	VIII	F	7.68	VIII	F	7.74	VIII	F	7.59	VIII	F
22	4.8	VI	6	5.96	VI	T	6.06	VI	T	6.2	VI	T	6.18	VI	T
23	5.5	VI	6	6.93	VII	F	6.89	VII	F	6.86	VII	F	6.83	VII	F
24	5.8	VIII	8	7.34	VII	F	7.35	VII	F	7.36	VII	F	7.36	VII	F
25	6	VIII	8	7.62	VIII	T	7.68	VIII	T	7.74	VIII	T	7.59	VIII	T
26	5.5	VII	7	6.93	VII	T	6.89	VII	T	6.86	VII	T	6.83	VII	T
27	6.2	VIII$^-$	7.8	7.9	VIII	T	8.06	VIII	T	8.05	VIII	T	7.67	VIII	T
28	6	VII$^-$	6.8	7.62	VIII	F	7.68	VIII	F	7.74	VIII	F	7.59	VIII	F
29	5.4	VII	7	6.79	VII	T	6.81	VII	T	6.86	VII	T	6.73	VII	T
30	5.2	VI$^+$	6.2	6.52	VII	F	6.44	VI	T	6.58	VII	F	6.51	VII	F
31	6.1	VII$^+$	7.2	7.76	VIII	F	7.95	VIII	F	7.83	VIII	F	7.55	VIII	F
32	5.1	VII$^+$	7.2	6.38	VI	F	6.24	VI	F	6.46	VI	F	6.45	VI	F
33	6.5	IX	8.8	8.31	VIII	F	8.39	VIII	F	8.46	VIII	F	8.41	VIII	F
34	6.2	VII	7	7.90	VIII	F	8.06	VIII	F	8.05	VIII	F	7.67	VIII	F
35	5.2	VI$^+$	6.2	6.52	VII	F	6.44	VI	T	6.58	VII	F	6.51	VII	F

36	5.4	VII$^+$	7.2	6.79	VII	T	6.81	VII	T	6.86	VII	T	6.73	VII	T
37	5.2	VII	7	6.52	VII	T	6.44	VI	F	6.58	VII	T	6.51	VII	T
38	5	VI$^+$	6.2	6.24	VI	T	6.06	VI	T	6.2	VI	T	6.31	VI	T
39	6.4	IX	9	8.17	VIII	F	8.11	VIII	F	8.32	VIII	F	8.33	VIII	F
40	5.2	VI	6	6.52	VII	F	6.44	VI	T	6.58	VII	F	6.51	VII	F
41	7.8	X	10	10.10	X	T	10.60	XI	F	10.24	X	T	10.11	X	T
42	5.3	VI	6	6.65	VII	F	6.59	VII	F	6.76	VII	F	6.54	VII	F
43	5.5	VI$^+$	6.2	6.93	VII	F	6.89	VII	F	6.86	VII	F	6.83	VII	F
44	5.5	VII$^+$	7.2	6.93	VII	T	6.89	VII	T	6.86	VII	T	6.83	VII	T
45	5.5	VII$^+$	7.2	6.93	VII	T	6.89	VII	T	6.86	VII	T	6.83	VII	T
46	6.5	VIII	8	8.31	VIII	T	8.39	VIII	T	8.46	VIII	T	8.41	VIII	T
47	5.5	VII	7	6.93	VII	T	6.89	VII	T	6.86	VII	T	6.83	VII	T
48	5.5	VII	7	6.93	VII	T	6.89	VII	T	6.86	VII	T	6.83	VII	T
49	5	VI	6	6.24	VI	T	6.06	VI	T	6.2	VI	T	6.31	VI	T
50	5	VI	6	6.24	VI	T	6.06	VI	T	6.2	VI	T	6.31	VI	T
51	5	VI	6	6.24	VI	T	6.06	VI	T	6.2	VI	T	6.31	VI	T
52	4.75	VI$^-$	5.8	5.89	VI	T	6.06	VI	T	6.2	VI	T	6.16	VI	T
53	5.4	VI	6	6.79	VII	F	6.81	VII	F	6.86	VII	F	6.73	VII	F
54	6.3	VIII	8	8.03	VIII	T	8.06	VIII	T	8.16	VIII	T	8.04	VIII	T
55	7.5	X	10	9.69	X	T	10	X	T	9.68	X	T	9.69	X	T
56	6	VIII	8	7.62	VIII	T	7.68	VIII	T	7.74	VIII	T	7.59	VIII	T
57	6.8	IX	9	8.72	IX	T	8.61	IX	T	8.83	IX	T	9.01	IX	T
58	6	VIII	8	7.62	VIII	T	7.68	VIII	T	7.74	VIII	T	7.59	VIII	T
59	5.5	VII	7	6.93	VII	T	6.89	VII	T	6.86	VII	T	6.83	VII	T
60	7.25	X	10	9.35	IX	F	9.56	X	T	9.39	IX	F	9.8	X	T
61	5.75	VII	7	7.28	VII	T	7.30	VII	T	7.31	VII	T	7.29	VII	T
62	5.5	VI–VII	6.5	6.93	VII	T	6.89	VII	T	6.86	VII	T	6.83	VII	T
63	6.8	IX	9	8.72	IX	T	8.61	IX	T	8.83	IX	T	9.01	IX	T
64	5.3	VII	7	6.65	VII	T	6.59	VII	T	6.76	VII	T	6.54	VII	T
65	7.5	IX	9	9.69	X	F	10	X	F	9.68	X	F	9.69	X	F
66	6.8	IX	9	8.72	IX	T	8.61	IX	T	8.83	IX	T	9.01	IX	T
67	5.5	VII	7	6.93	VII	T	6.89	VII	T	6.86	VII	T	6.83	VII	T
68	5.8	VII	7	7.34	VII	T	7.35	VII	T	7.36	VII	T	7.36	VII	T
69	6.2	VII	7	7.90	VIII	F	8.06	VIII	F	8.05	VIII	F	7.67	VIII	F
70	6.8	IX	9	8.72	IX	T	8.61	IX	T	8.83	IX	T	9.01	IX	T
71	5.5	VII	7	6.93	VII	T	6.89	VII	T	6.86	VII	T	6.83	VII	T
72	5.5	VII	7	6.93	VII	T	6.89	VII	T	6.86	VII	T	6.83	VII	T
73	6.8	IX	9	8.72	IX	T	8.61	IX	T	8.83	IX	T	9.01	IX	T
74	6.3	IX	9	8.03	VIII	F	8.06	VIII	F	8.16	VIII	F	8.04	VIII	F
75	5.8	VII$^+$	7.2	7.34	VII	T	7.35	VII	T	7.36	VII	T	7.36	VII	T
76	5.8	VII	7	7.34	VII	T	7.35	VII	T	7.36	VII	T	7.36	VII	T
77	7.9	X	10	10.24	X	T	10.79	XI	F	10.44	X	T	10.48	X	T
78	8.5	XII	12	11.07	XI	F	11.97	XII	T	12	XII	T	11.8	XII	T

79	8	XI	11	10.38	X	F	11	XI	T	10.69	XI	T	10.83	XI	T
80	6	VIII	8	7.62	VIII	T	7.68	VIII	T	7.74	VIII	T	7.59	VIII	T
81	7.3	X	10	9.41	IX	F	9.61	X	T	9.46	IX	F	9.79	X	T
82	5.4	VI–VII	6.5	6.79	VII	T	6.81	VII	T	6.86	VII	T	6.73	VII	T
83	5.3	VI	6	6.65	VII	F	6.59	VII	F	6.76	VII	F	6.54	VII	F
84	5.7	VII	7	7.21	VII	T	7.30	VII	T	7.21	VII	T	7.18	VII	T
85	5.4	VII	7	6.79	VII	T	6.81	VII	T	6.86	VII	T	6.73	VII	T
86	5	VI$^+$	6.2	6.24	VI	T	6.06	VI	T	6.2	VI	T	6.31	VI	T
87	4.8	VII	7	5.96	VI	F	6.06	VI	F	6.2	VI	F	6.18	VI	F
88	7	VIII$^+$	8.2	9	IX	F	9.02	IX	F	9.2	IX	F	8.61	IX	F
89	5	VI	6	6.24	VI	T	6.06	VI	T	6.2	VI	T	6.31	VI	T
90	4.8	VI	6	5.96	VI	T	6.06	VI	T	6.2	VI	T	6.18	VI	T
91	5.5	VII	7	6.93	VII	T	6.89	VII	T	6.86	VII	T	6.83	VII	T
92	6.8	VIII	8	8.72	IX	F	8.61	IX	F	8.83	IX	F	9.01	IX	F
93	7.5	IX	9	9.69	X	F	10	X	F	9.68	X	F	9.69	X	F
94	8	XI	11	10.38	X	F	11	XI	T	10.69	XI	T	10.83	XI	T
95	7.25	IX	9	9.35	IX	T	9.56	X	F	9.39	IX	T	9.8	X	F
96	7.25	IX	9	9.35	IX	T	9.56	X	F	9.39	IX	T	9.8	X	F
97	7	IX	9	9	IX	T	9.02	IX	T	9.2	IX	T	8.61	IX	T
98	6.8	IX	9	8.72	IX	T	8.61	IX	T	8.83	IX	T	9.01	IX	T
99	6.4	VIII	8	8.17	VIII	T	8.11	VIII	T	8.32	VIII	T	8.33	VIII	T
100	6.5	VIII	8	8.31	VIII	T	8.39	VIII	T	8.46	VIII	T	8.41	VIII	T
101	6.6	VIII	8	8.45	VIII	T	8.45	VIII	T	8.64	IX	F	8.47	VIII	T
102	6.5	VIII	8	8.31	VIII	T	8.39	VIII	T	8.46	VIII	T	8.41	VIII	T
103	5.5	VII	7	6.93	VII	T	6.89	VII	T	6.86	VII	T	6.83	VII	T
104	5.8	VII	7	7.34	VII	T	7.35	VII	T	7.36	VII	T	7.36	VII	T
105	6	VIII	8	7.62	VIII	T	7.68	VIII	T	7.74	VIII	T	7.59	VIII	T
106	5.8	VII	7	7.34	VII	T	7.35	VII	T	7.36	VII	T	7.36	VII	T
107	5.6	VII	7	7.07	VII	T	7.08	VII	T	7.03	VII	T	6.88	VII	T
108	4.8	VI	6	5.96	VI	T	6.06	VI	T	6.2	VI	T	6.18	VI	T
109	6.2	VII	7	7.90	VIII	F	8.06	VIII	F	8.05	VIII	F	7.67	VIII	F
110	5.6	VII	7	7.07	VII	T	7.08	VII	T	7.03	VII	T	6.88	VII	T
111	6	VII	7	7.62	VIII	F	7.68	VIII	F	7.74	VIII	F	7.59	VIII	F
112	6.75	IX	9	8.65	IX	T	8.61	IX	T	8.79	IX	T	9	IX	T
113	5.5	VIII	8	6.93	VII	F	6.89	VII	F	6.86	VII	F	6.83	VII	F
114	4.75	VII	7	5.89	VI	F	6.06	VI	F	6.2	VI	F	6.16	VI	F
115	5	VII	7	6.24	VI	F	6.06	VI	F	6.2	VI	F	6.31	VI	F
116	5	VI	6	6.24	VI	T	6.06	VI	T	6.2	VI	T	6.31	VI	T
117	5	VI	6	6.24	VI	T	6.06	VI	T	6.2	VI	T	6.31	VI	T
118	5	VI	6	6.24	VI	T	6.06	VI	T	6.2	VI	T	6.31	VI	T
119	5.75	VII	7	7.28	VII	T	7.30	VII	T	7.31	VII	T	7.29	VII	T
120	5.5	VII$^+$	7.2	6.93	VII	T	6.89	VII	T	6.86	VII	T	6.83	VII	T

121	4.25	VI	6	5.2	V	F	6	VI	T	6.19	VI	T	5.95	VI	T
122	5.2	VII	7	6.52	VII	T	6.44	VI	F	6.58	VII	T	6.51	VII	T
123	6.8	IX$^+$	9.2	8.72	IX	T	8.61	IX	T	8.83	IX	T	9.01	IX	T
124	5.6	VI	6	7.07	VII	F	7.08	VII	F	7.03	VII	F	6.88	VII	F
125	7.2	X	10	9.28	IX	F	9.46	IX	F	9.28	IX	F	9.78	X	T
126	6.2	VII$^+$	7.2	7.90	VIII	F	8.06	VIII	F	8.05	VIII	F	7.67	VIII	F
127	6	VIII	8	7.62	VIII	T	7.68	VIII	T	7.74	VIII	T	7.59	VIII	T
128	5.2	VII	7	6.52	VII	T	6.44	VI	F	6.58	VII	T	6.51	VII	T
129	6.3	VII	7	8.03	VIII	F	8.06	VIII	F	8.16	VIII	F	8.04	VIII	F
130	5.5	VI	6	6.93	VII	F	6.89	VII	F	6.86	VII	F	6.83	VII	F
131	5.7	VII	7	7.21	VII	T	7.30	VII	T	7.21	VII	T	7.18	VII	T
132	5.4	VI$^+$	6.2	6.79	VII	F	6.81	VII	F	6.86	VII	F	6.73	VII	F
133	4.6	VI	6	5.69	VI	T	6.19	VI	T	6.50	VII	F	6.02	VI	T
134	7.3	IX	9	9.41	IX	T	9.61	X	F	9.46	IX	T	9.79	X	F
Number of T				86			87			88			90		
Square error				$\varepsilon_{LR} = 0.267$			$\varepsilon_{FINA} = 0.280$			$\varepsilon_{LDSS} = 0.265$			$\varepsilon_{NDSS} = 0.226$		

Appendix 8.B: Program of NDSS

Fortran program of normal diffusion self-study for estimation of epicentral intensity

```
C ********************************************************
C *    2-dimension normal diffusion self-study model        *
C ********************************************************
      PROGRAM MAIN
      INTEGER N,NR,MR
      REAL X(134),Y(134),U(86),V(32),R(86,32),DI(134)
      CHARACTER*26 ROM(12),S(134),T(134)
C N - Size of sample, NR - Rows, MR - Columns
C X,Y - Magnitude and intensity
C R - Fuzzy relation between magnitude and intensity
C DI - Digital estimations
C U,V - Discrete universes of magnitude and intensity
C ROM - Roman numerals, S - Symbolical estimations
C T - Comment of estimation with "T"(True) and "F"(Fault)
      DATA ROM/'I','II','III','IV','V','VI','VII','VIII',
     1 'IX','X','XI','XII'/
      DATA X/5.75,5.75,6.40,6.40,6.50,6.50,6.00,6.00,6.00,7.00,
     1 7.00,6.25,5.75,5.80,6.30,6.50,5.80,5.00,5.00,5.00,
     2 6.00,4.80,5.50,5.80,6.00,5.50,6.20,6.00,5.40,5.20,
     3 6.10,5.10,6.50,6.20,5.20,5.40,5.20,5.00,6.40,5.20,
     4 7.80,5.30,5.50,5.50,5.50,6.50,5.50,5.50,5.00,5.00,
     5 5.00,4.75,5.40,6.30,7.50,6.00,6.80,6.00,5.50,7.25,
     6 5.75,5.50,6.80,5.30,7.50,6.80,5.50,5.80,6.20,6.80,
     7 5.50,5.50,6.80,6.30,5.80,5.80,7.90,8.50,8.00,6.00,
     8 7.30,5.40,5.30,5.70,5.40,5.00,4.80,7.00,5.00,4.80,
     9 5.50,6.80,7.50,8.00,7.25,7.25,7.00,6.80,6.40,6.50,
     1 6.60,6.50,5.50,5.80,6.00,5.80,5.60,4.80,6.20,5.60,
     2 6.00,6.75,5.50,4.75,5.00,5.00,5.00,5.00,5.75,5.50,
     3 4.25,5.20,6.80,5.60,7.20,6.20,6.00,5.20,6.30,5.50,
     4 5.70,5.40,4.60,7.30/
      DATA Y/7.0,7.0,8.0,8.0,9.0,9.0,8.0,8.0,7.0,8.0,
     1 9.0,8.0,8.0,8.0,9.0,8.0,7.2,7.2,6.0,6.2,
     2 6.8,6.0,6.0,8.0,8.0,7.0,7.8,6.8,7.0,6.2,
     3 7.2,7.2,8.8,7.0,6.2,7.2,7.0,6.2,9.0,6.0,
     4 10.0,6.0,6.2,7.2,7.2,8.0,7.0,7.0,6.0,6.0,
     5 6.0,5.8,6.0,8.0,10.0,8.0,9.0,8.0,7.0,10.0,
     6 7.0,6.5,9.0,7.0,9.0,9.0,7.0,7.0,7.0,9.0,
     7 7.0,7.0,9.0,9.0,7.2,7.0,10.0,12.0,11.0,8.0,
     8 10.0,6.5,6.0,7.0,7.0,6.2,7.0,8.2,6.0,6.0,
```

```
      9  7.0,8.0,9.0,11.0,9.0,9.0,9.0,9.0,8.0,8.0,
      1  8.0,8.0,7.0,7.0,8.0,7.0,7.0,6.0,7.0,7.0,
      2  7.0,9.0,8.0,7.0,7.0,6.0,6.0,6.0,7.0,7.2,
      3  6.0,7.0,9.2,6.0,10.0,7.2,8.0,7.0,7.0,6.0,
      4  7.0,6.2,6.0,9.0/
         N=134
         CALL DIFFUSION(N,X,Y,NR,MR,U,V,R)
         CALL ESTIMATE(N,X,Y,NR,MR,U,V,R,DI,S,T,ROM)
         CALL RESULT(N,X,Y,DI,S,T,ROM)
         STOP
         END

         SUBROUTINE DIFFUSION(N,X,Y,NR,MR,U,V,R)
         INTEGER N,NR,MR
         REAL X(134),Y(134),U(86),V(32),R(86,32),HX,HY
C HX - Diffusion coefficient of X
C HY - Diffusion coefficient of Y
         CALL DISCRETE(N,X,Y,NR,MR,U,V)
         CALL SIMPLEH(N,X,HX)
         CALL SIMPLEH(N,Y,HY)
         CX=2.0*HX*HX
         CY=2.0*HY*HY
         DO 10 J=1,NR
         DO 10 K=1,MR
         R(J,K)=0
10       CONTINUE
         DO 50 I=1,N
         DO 20 J=1,NR
         DO 20 K=1,MR
         A=EXP(-(U(J)-X(I))*(U(J)-X(I))/CX)
         B=EXP(-(V(K)-Y(I))*(V(K)-Y(I))/CY)
         R(J,K)=R(J,K)+A*B
20       CONTINUE
50       CONTINUE      CALL COLUMN(NR,MR,R)
         WRITE(*,*)'Fuzzy Relation Matrix R:'
         WRITE(*,60)(V(K),K=1,MR)
60       FORMAT(7X,32F5.1)
         DO 90 J=1,NR
         WRITE(*,100)U(J),(R(J,K),K=1,MR)
90       CONTINUE 100    FORMAT(1X,F5.2,32(F5.2))
         RETURN
         END

         SUBROUTINE ESTIMATE(N,X,Y,NR,MR,U,V,R,DI,S,T,ROM)
```

```
      INTEGER N,NR,MR,RY,EY
      REAL X(134),Y(134),U(86),V(32),R(86,32),XX,YY,DI(134)
      CHARACTER*26 AT,AF,TF,S(134),T(134),ROM(12)
      AT='T'
      AF='F'
      DO 10 I=1,N
      XX=X(I)
      CALL FINFER(NR,MR,U,V,R,XX,YY)
      RY=INT(Y(I)+0.5)
      EY=INT(YY+0.5)
      TF=AF
      IF(RY.EQ.EY) TF=AT
      DI(I)=YY
      S(I)=ROM(EY)
      T(I)=TF
10    CONTINUE
      RETURN
      END

      SUBROUTINE FINFER(N,M,U,V,R,X,Y)
      INTEGER N,M
      REAL X,Y,U(86),V(32),R(86,32),FM(86),FI(32),DELTA
      DELTA=(U(2)-U(1))/100.0
      DO 10 I=1,N
      FM(I)=0
      A=ABS(X-U(I))
      IF(A.LT.DELTA) FM(I)=1
10    CONTINUE
20    FORMAT(1X,2(F5.2))
      DO 40 J=1,M
      B=0
      DO 30 I=1,N
      A=FM(I)
      IF(A.GT.R(I,J)) A=R(I,J)
      IF(B.LT.A) B=A
30    CONTINUE
      FI(J)=B
40    CONTINUE
      CALL CENTER(M,V,FI,Y)
      RETURN
      END

      SUBROUTINE CENTER(N,V,W,C)
      INTEGER N
```

```
      REAL W(32),V(32),C
      A=0
      B=0
      DO 20 I=1,N
      A=A+W(I)*V(I)
      B=B+W(I)
20    CONTINUE
      IF(B.EQ.0) B=1
      C=A/B
      RETURN
      END

      SUBROUTINE RESULT(N,X,Y,DI,S,T,ROM)
      INTEGER N,TT,PN
      REAL X(134),Y(134),DI(134)
      CHARACTER*26 S(134),T(134),AT,C1,C2,RI(134),ROM(12)
      C1='Number of T:'
      C2='Square error:'
      AT='T'
      DO 10 I=1,N
      PN=(Y(I)+0.5)    RI(I)=ROM(PN)
10    CONTINUE
      WRITE(*,*)' '
      WRITE(*,20)
20    FORMAT(1X,'No.',' Magnitude',' Real Intensity ',
     1 7X,'Normal diffusion estimation ',' Comment')
      WRITE(*,25)
25    FORMAT(15X,'Symbol',3X,'Digit ',11X,'Digit',4X,'Symbol ')
      WRITE(*,30)(I,X(I),RI(I),Y(I),DI(I),S(I),T(I),I=1,N)
30    FORMAT(1X,I3,4X,F5.2,6X,A5,F6.1,12X,F7.2,7X,A5,13X,A3)
      AN=N
      TT=0
      A=0
      DO 40 I=1,N
      IF(T(I).EQ.AT) TT=TT+1
      B=Y(I)-DI(I)
      A=A+B*B
40    CONTINUE
      A=A/AN
      WRITE(*,60)C1,TT,C2,A
50    CONTINUE
60    FORMAT(1X,A13,I2,',',5X,A15,F6.3)
      RETURN
      END
```

```
      SUBROUTINE SIMPLEH(N,X,H)
      INTEGER N,PN
      REAL X(100),H,C(4),C9
      DATA C/0.6841,0.5404,0.4482,0.3839/
      C9=2.6851
      CALL MAXMIN(N,X,B,A)
      IF(N.LT.9) GOTO 10
      AN=N
      H=C9*(B-A)/(AN-1.0)
      GOTO 20
10    CONTINUE
      PN=N-4
      H=C(PN)*(B-A)
20    CONTINUE
      RETURN
      END

      SUBROUTINE MAXMIN(N,X,A,B)
      INTEGER N
      REAL X(134)
      A=0
      B=1E+10
      DO 10 I=1,N
      IF(A.LT.X(I)) A=X(I)
      IF(B.GT.X(I)) B=X(I)
10    CONTINUE
      RETURN
      END

      SUBROUTINE COLUMN(N,M,R)
      INTEGER N,M
      REAL R(86,32)
      DO 40 J=1,M
      A=0
      DO 10 I=1,N
      IF(A.LT.R(I,J)) A=R(I,J)
10    CONTINUE
      IF(A.EQ.0) A=1
      DO 20 I=1,N
      R(I,J)=R(I,J)/A
20    CONTINUE
40    CONTINUE
      RETURN
```

```
        END

        SUBROUTINE DISCRETE(N,X,Y,NR,MR,U,V)
        INTEGER N,NR,MR
        REAL X(134),Y(134),U(86),V(32),A,B,C,D
        CALL MAXMIN(N,X,B,A)
        CALL MIN(N,X,C)
        D=(B-A)/C
        NR=INT(D+1.5)
        DO 10 I=1,NR
        U(I)=A+(I-1)*C
10      CONTINUE
        CALL MAXMIN(N,Y,B,A)
        CALL MIN(N,Y,C)
        D=(B-A)/C
        MR=INT(D+1.5)
        DO 20 I=1,MR
        V(I)=A+(I-1)*C
20      CONTINUE
        RETURN
        END

        SUBROUTINE MIN(N,X,C)
        INTEGER N
        REAL X(134),C
        C=1E+10
        DO 20 I=1,N-1
        DO 10 J=I+1,N
        A=ABS(X(I)-X(J))
        IF(A.EQ.0) GOTO 10
        IF(C.GT.A) C=A
10      CONTINUE
20      CONTINUE
        RETURN
        END
```

9. Estimation of Isoseismal Area

In this chapter[1], based on the normal diffusion and the feedforward neural network with backpropagation algorithm (BP), we suggest a hybrid fuzzy neural network to estimate isoseismal area by earthquake magnitude. In section 9.1 we give the outline of estimation of isoseismal area. In section 9.2, we give a brief review of current methods for the construction of fuzzy relationships. Section 9.3 suggests the information diffusion function to produce if-then rules from observations. In section 9.4, we propose a model for pattern smoothing to assist a BP neural network to acquire knowledge from the data. In section 9.5, we give the architecture of the hybrid model which consists of an information-diffusion approximate reasoning and a conventional BP neural network. In section 9.6, we use the model to estimate isoseismal area by earthquake magnitude. The chapter is then summarized with a conclusion in section 9.7.

9.1 Introduction

Earthquake is one of the natural disasters which can cause tremendous damage to lives and properties. To have a better understanding, various aspects of earthquake have been studied over the years. Accompanied by the intensity attenuation analysis, the relationship between isoseismal area and earthquake magnitude has been a major area of concern which requires thorough investigation[8].

Approximately two decades ago, engineers in earthquake engineering[6][9] were active in the search for the following (or similar) expression relating intensity, I, to magnitude, M, and hypocentral distance, R (in kilometers):

$$I = aM - b\log_{10} R + c \tag{9.1}$$

where a, b, and c are empirical constants. Since there is a 60% probability that an observed intensity is more than one degree greater or smaller than

---
[1] Prof. Yee Leung given some distribution for the work and this project is supported by the earmarked grant CUHK 8/93H of Hong Kong Research Grant Council and the grant of the Croucher Foundation.

its predicted value[20], a more appropriate expression[14] relating isoseismal area, S (in square kilometers), with intensity to magnitude was developed:

$$\log_{10} S(I) = a + bM \tag{9.2}$$

where a and b are empirical constants.

However, several studies[19] have demonstrated that the linear relationship does not fit the seismicity of any region. In the western United States of America, many destructive earthquakes are controlled by one huge fault, the San Andreas Fault, and the relationship between isoseismal area and earthquake magnitude is approximately linear. Nevertheless, in regions, as Yunnan Province, China, where earthquakes are not controlled by one single fault, the linear relationship generally fails.

In principle, if there were a lot of observations recording historical earthquakes, researchers could use powerful statistical tools such as regression analysis[1][2][5] to reveal the nonlinear relationship. Nevertheless, destructive earthquakes are infrequent events with very small probability of occurrence. Therefore, to satisfy the conditions of most statistical tools, the researchers, in general, must collect enough observations from a wide zone, even the whole country. However, samples collected from a wide zone will miss out the intricate effects of seismotectonic structures on earthquakes. The relationship so obtained is of little use. Obviously, we cannot employ the traditional statistical tools to estimate the relationship between isoseismal area and earthquake magnitude.

In general, seismotectonic structures are very complex and the relationship between isoseismal area and earthquake magnitude is strongly nonlinear. Since we do not know which nonlinear function best describe the relationship, it is profitable to employ neural networks to search for the mapping from the input, which is earthquake magnitude, to the output, which is isoseismal area by observations. A structural systems test[27] in earthquake engineering showed that neural networks can generate more reliable results than they could be obtained using any other traditional methods.

However, the tests in laboratory can usually be controlled, and the observations are smoother. In the real world, the observations are strongly scattered and contradictory patterns do occur. Since neural information processing models largely assume that the learning patterns for training a neural network are compatible, then a neural network does not converge because the adjustments of weights and thresholds do not know where to turn due to the ambiguity brought forth by the contradictory patterns. That is, when earthquake magnitudes are the same but isoseismal areas are very different, the directions of change in weights and thresholds are ambiguous if we use these observations as patterns to train a neural network. Hence, to estimate the relationship, we need to handle the fuzziness and granularity of the observations.

Fuzzy models have been formulated to study the uncertainty and ambiguity of isoseismal maps[25]. However, different researchers usually obtain different results based on the same observations because their analyses depend on personal engineering experience rather than the observations. It is thus necessary to introduce a new fuzzy model to process the observations before we use them in the neural networks. We can do it by the information diffusion.

Obviously, if we want to estimate the relationship between isoseismal area and earthquake magnitude using the data on past seismic events, we must simultaneously consider the following three problems: small-sample, scattering, and ambiguity of the isoseismal maps.

In this chapter, we propose a hybrid fuzzy neural network to deal with the problems.

9.2 Some Methods for Constructing Fuzzy Relationships

9.2.1 Fuzzy relation and fuzzy relationship

Unless stated otherwise, we assume that we are given a sample of n real-valued observations, x_i $(i = 1, 2, \cdots, n)$, which have two components, earthquake magnitude, m_i, and isoseismal area, S_i, whose underlying relationship is to be estimated. An observation is also called a pattern. And, a given sample is also represented as follows:

$$X = \{x_1, x_2, \cdots, x_n\} = \{(m_1, S_1), (m_2, S_2), \cdots, (m_n, S_n)\}.$$

To reduce scattering in the sample, we generally consider logarithmic isoseismal area, $g = \log_{10} S$, instead and the given sample can be rewritten as:

$$X = \{x_1, x_2, \cdots, x_n\} = \{(m_1, g_1), (m_2, g_2), \cdots, (m_n, g_n)\}. \tag{9.3}$$

Let U be the universe of "magnitude" and V be the universe of " logarithmic isoseismal area ". Obviously, $U, V \subset \mathbb{R}$.

As we said in Chapter 1, any fuzzy set on $U \times V$ can be called a fuzzy relation from U to V. It is a relationship if and only if it shows some cause-effect connection for two variables, and U, V are the domain and the range, respectively. Formally, we give the following definitions.

Definition 9.1 Let U be the domain of variable x, and V be the domain of variable y. Any mapping R from $U \times V$ to $[0, 1]$

$$\mu_R : \quad U \times V \to [0, 1]$$
$$(x, y) \mapsto \mu_R(x, y), \forall (x, y) \in U \times V$$

is called a *fuzzy relation* between x and y.

For example, let us consider an image showing a hill. The researcher may have informed that the vegetation type is lawn, and that the hill is of medium height. An example of a similarity table where the height of hills was defined, is shown in Table 9-1[2]. It can be observed that two hills with "high" heights have a similarity degree equal to 1, so they are completely similar. On the other hand, the similarity degree between a hill with "low" height and another with "very high" height is equal to 0, so they are not similar.

Table 9-1 Similarity table among hill height values

	Height(Hills)			
	very high	high	medium	low
very high	1	0.92	0.4	0
high	0.92	1	0.9	0.38
medium	0.38	0.9	1	0.7
low	0	0.38	0.7	1

Let $U = V = \{very\ high, high, medium, low\}$. We know that Table 9-1 defines a fuzzy relation μ_R of $U \times V$.

In fact, μ_R defines a fuzzy subset on the universe of discourse $U \times V$. In other words, a binary fuzzy relation is a fuzzy membership function in a 2-dimension Cartesian space.

Obviously, it is useful if an fuzzy relation can approximately reveal the information structure implied by the sample X. Particularly, when the structure can show some cause-effect connection, the relation is called a relationship, which embodies the family of connections among variables.

Definition 9.2 A fuzzy relation R is called a *fuzzy relationship* if and only if it shows some cause-effect connection for input and output variables.

For example, a weekend party will be held Saturday. 10-15 people will attend. Suppose one person would drink about two bottles of Coca-Cola. The domain of input variable is $U = \{10, 11, 12, 13, 14, 15\}$. Let $V = \{10, 15, 20, 25, 30, 35, 40\}$ be the domain of output variable. The cause-effect connection would be a fuzzy relationship shown as Table 9-2.

Table 9-2 Cause-effect connection between the numbers of people attending and bottles of Coca-Cola

people	Coca-Cola bottles						
	10	15	20	25	30	35	40
10	0.17	0.58	1	0.58	0.17	0	0
11	0	0.42	0.83	0.73	0.33	0	0
12	0	0.25	0.67	0.92	0.5	0.08	0
13	0	0.08	0.50	0.92	0.67	0.25	0
14	0	0	0.33	0.75	0.83	0.42	0
15	0	0	0.17	0.58	1	0.58	0.17

[2] From
http://www.computer.org/proceedings/meta/1999/papers/32/MVieira.html

There are two common models in standard fuzzy system theory that one can use them to construct fuzzy relationships. They are respectively the employment of fuzzy rules and fuzzy graphs.

Mamdani[21] and Togai[28] independently suggested the fuzzy-rule model as a multivalued logical-implication operator. Zadeh[29][31], on the other hand, suggested the use of a fuzzy graph to characterize a coarse approximation to a function or a relationship.

9.2.2 Multivalued logical-implication operator

In the Mamdani-Togai model, several fuzzy rules, $A_i \rightarrow B_i$ (If A_i then B_i), must be given. For convenience, let the universe of discourse of A_i be U, and B_i be V. Let $\mu_{A_i}(u)$ and $\mu_{B_i}(v)$ denote fuzzy subsets A_i and B_i respectively. Thus, the relationship R_i based on $A_i \rightarrow B_i$ is a fuzzy subset in the universe of discourse $U \times V$, and μ_{R_i} defined by

$$\mu_{R_i}(u,v) = \min\{\mu_{A_i}(u), \mu_{B_i}(v)\}, \quad u \in U, v \in V. \tag{9.4}$$

Each rule $A_i \rightarrow B_i$ is translated into a relationship R_i. The overall protocol is then a relationship R formed by applying the connective " or " to R_i's as follows:

$$R = R_1 \vee R_2 \vee \cdots \vee R_i \vee \cdots \vee R_n. \tag{9.5}$$

Now, if an approximate value A' (the antecedent) is known, then one needs a way to infer the consequent B' from A' and R; say $B' = A' \circ R$ where A' is to be composed with R. This has to have the effect of reducing the dimensionality of the support set of R to that of B'. In Mamdani-Togai model, the compositional rule of reasoning used to relate $\mu_{B'}$ to μ_R and $\mu_{A'}$ is

$$\mu_{B'}(v) = \max_u \min\{\mu_{A'}(u), \mu_R(u,v)\}. \tag{9.6}$$

In fact, the min operator is a symmetric truth operator where $\mathrm{truth}(u \rightarrow v) = \min(u,v)$. So, it does not properly generalize the classical implication $P \rightarrow Q$, which is false if and only if the antecedent P is true and the consequent Q is false, i.e. $t(P) = 1$ and $t(Q) = 0$. To define a "conditional-possibility" matrix pointwise with continuous implication values, Zadeh[30] chooses the Lukasiewicz implication operator where $\mathrm{truth}(u \rightarrow v) = \min(1, 1-u+v)$. Unfortunately the Lukasiewicz operator usually equals or approximates unity, for $\min(1, 1 - u + v) < 1$ iff $u > v$.

9.2.3 Fuzzy associative memories

The purpose of Zadeh's fuzzy graph model is to obtain an approximation of a function. The concept of a fuzzy graph, in fact, is a natural generalization of crisp graphs using fuzzy subsets, and in many cases the extension principle. For example, a fuzzy graph which describes the dependence of a variable v on a variable u may be expressed as a disjunction of Cartesian products of fuzzy subsets. In this model, a fuzzy graph may be interpreted as a collection of if-then rules. Conversely, a collection of fuzzy if-then rules may be represented as a fuzzy graph. Meanwhile, the membership function of a Cartesian product based on fuzzy if-then rule "If u is A then v is B" is defined by

$$\mu_{A \times B}(u, v) = \mu_A(u) \wedge \mu_B(v). \tag{9.7}$$

In other words, when we use fuzzy rules to produce a fuzzy graph, the result must be the same as from the Mamdani-Togai model because formula (9.7) actually is the same as Eq.(9.4) and disjunction of Cartesian products of fuzzy subsets can be represented by formula (9.5).

To obtain if-then rules from a sample X, a common way is to employ the adaptive binary input-output fuzzy associative memories (BIOFAM) clustering[15][16]. Actually, it is the use of a frequency histogram to weight fuzzy if-then rules.

Let $x_1, x_2, \cdots, x_n$ denote n observations which are input-output data pairs $(m_i, g_i) \in R^2$, also called quantization vectors in the input-output product space.

Suppose the r fuzzy subsets $A_1, A_2, \cdots, A_r$ quantize the input universe of discourse U, and the s fuzzy subsets $B_1, B_2, \cdots, B_s$ quantize the output universe of discourse V. When A_i and B_j are triangular fuzzy subsets, the covariable method[3][4] projects the ellipse E_{ij}. Let K_{ij} be the number of x_k's in the ellipse E_{ij}. Let

$$N = \sum_{i=1}^{r} \sum_{j=1}^{s} K_{ij}, \quad w_{ij} = \frac{K_{ij}}{N}, \tag{9.8}$$

where w_{ij} is the weight of the fuzzy if-then rule "If u is A_i, then v is B_j". $w_{ij} > 0$ means that the sample X does support the relationship based on the fuzzy Cartesian product $A_i \times B_j$. Each fuzzy rule defines a Cartesian product as shown in Fig. 9.1.

This kind of fuzzy function approximation is equivalent to unsupervised competitive learning. In principle, tightly clustered data give smaller ellipses or more precise rules, and sparse or noisy data give larger ellipses or less certain rules. Nevertheless, in practical cases, the rule given by this model might be unreliable when the data in the ellipse is not dense even though the weight is large. Only when the sample size n is large, the density of the data can be high enough to have a reliable approximation. Therefore, for incomplete and sparse data, we need a new model for constructing fuzzy relationships.

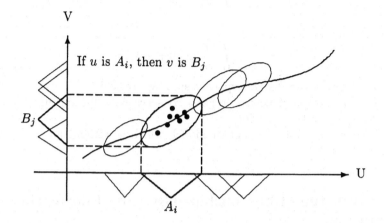

Fig. 9.1   The fuzzy rule " If u is A_i, then v is B_j" can be supported
by the ellipse which contains observations.

9.2.4 Self-study discrete regression

We have given the model in the last chapter. In fact, it is an improved
probability model. Let $Q = \{q_{ij}\}$ be a primary information matrix, and
$S = \sum_i \sum_j q_{ij}$. Dividing a primary information matrix by the S, we can
changed Q into a discrete probability estimation. In other word, if the sample
size n is large, the self-study discrete regression and the traditional regression
must tend to the same conclusion.

When n is small and we don't know the shape of the population of the
given sample, the model of the self-study discrete regression can give a better
estimation of output by input. The reason is that a reasonable information
diffusion mines the fuzzy information which hides in incomplete data. The
added information partly fills the gaps caused by incompleteness and improve
non-diffusion estimator.

It can be understand if someone claims that the self-study discrete regression is nothing else but the kernel estimator in disguise. However, as we have
stated in chapter 5, a kernel function and an information diffusion function
are related but different concepts. In fact, it is almost impossible to construct
any primary information matrix by a kernel function, because when U and V
are discrete, we know that, for any kernel function K, the following expression
is true.

$$\int_U \int_V K(x)dx < 1.$$

It means that K is not a kernel function on $U \times V$. In other words, there are
kernel functions on R^2, but not on a discrete space.

In the model of the self-study discrete regression, if $f(x_i, x)$ is employed
to diffuse x_i to $x \in U \times V$, we let

$$S = \int_U \int_V f(x_i, x)dx, \quad \text{and} \quad \mu(x_i, x) = \frac{f(x_i, x)}{S},$$

then the μ satisfies

$$\int_U \int_V \mu(x)dx = 1.$$

When the step of a discrete space is small we regard $S = 1$.

In other words, it is easy to get a conservative diffusion function for the self-study discrete regression.

Next section we give another model, which is without any relation to a kernel estimate.

9.3 Multitude Relationships Given by Information Diffusion

In the main existing models, fuzzy subsets used to construct fuzzy rules and fuzzy relationships are given by researchers. Different choices of fuzzy subsets lead to different results. In spite of that the fuzzy subsets can be adjusted according to the results, the approach in which first impressions are strongest does not fit incomplete data problems where there is insufficient information to support any first impression.

Usual methods are to look for evidence to support fuzzy relationships. The tracing curve of sample-data clusters is regarded as input-output fuzzy relationship function. Cluster analysis techniques can be developed to choose reasonable fuzzy subsets. When a sample is small or scattered, there does not exist any tracing curve of clusters. Clustering approach would be replaced by the information diffusion method[10][11][12] which helps us to change an observation into a fuzzy subset to fill the gap caused by incomplete data.

Let $\mu(m_i, u)$ be an information diffusion function of m_i on U. Normalizing it, denoted by $\mu_{m_i}(u)$, we can obtain a fuzzy classifying function indicating the degree of that u belongs to fuzzy set "around m_i". When the universe of discourse has been separated into many parts, information diffusion can be depicted by Fig. 9.2 where observation m_i is diffused to every point of U with different values. For continuous universe of discourse, information diffusion of m_i can be shown by a fuzzy membership function $\mu_{m_i}(u)$ as that depicted in Fig. 9.3.

Using normal diffusion functions

$$\mu(m_i, u) = \frac{1}{h_m\sqrt{2\pi}} \exp[-\frac{(u - m_i)^2}{2h_m^2}],$$

and

$$\mu(g_i, v) = \frac{1}{h_g\sqrt{2\pi}} \exp[-\frac{(v - g_i)^2}{2h_g^2}],$$

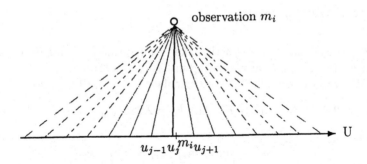

observation m_i

$u_{j-1}\, u_j^{m_i} u_{j+1}$

Fig. 9.2   Diffusion of a piece of information to all elements of a
universe discourse by a fuzzy set from an information
diffusion function

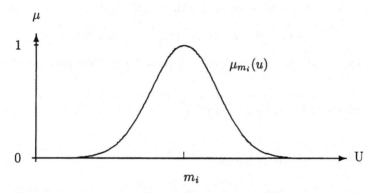

Fig. 9.3   Fuzzy set from the information diffusion function of
m_i on a continuous universe of discourse U.

and dividing them by

$$\frac{1}{h_m\sqrt{2\pi}}, \quad \text{and} \quad \frac{1}{h_g\sqrt{2\pi}},$$

respectively, to normalize, we can change anyone input-output observation
$(m_i, g_i) \in X$ in (9.3) into two fuzzy subsets

$$A_i = \int_U \mu_{m_i}(u)/u, \quad B_i = \int_V \mu_{g_i}(v)/v. \tag{9.9}$$

That is

$$A_i : \quad \mu_{m_i}(u) = \exp[-\frac{(u - m_i)^2}{2h_m^2}], \quad u \in U, \tag{9.10}$$

$$B_i: \quad \mu_{g_i}(v) = \exp[-\frac{(v-g_i)^2}{2h_g^2}], \quad v \in V. \tag{9.11}$$

Obviously, (m_i, g_i) means

$$A_i \rightarrow B_i, \tag{9.12}$$

translated as that "If an earthquake in magnitude around m_i occurs, then logarithmic isoseimal area relating intensity I is around g_i."

We can use formula (7.38) to calculate h for Eq.(9.10) and (9.11), i.e,

$$h_m = 2.6851(b_m - a_m)/(n-1), \quad h_g = 2.6851(b_g - a_g)/(n-1), \quad (n \geq 9)$$

where

$$b_m = \max_{1 \leq i \leq n} \{m_i\}, \ a_m = \min_{1 \leq i \leq n} \{m_i\},$$

$$b_g = \max_{1 \leq i \leq n} \{g_i\}, \ a_g = \min_{1 \leq i \leq n} \{g_i\}.$$

In order to preserve more information, we employ the correlation-product encoding[16] to produce fuzzy relationship based on $A_i \rightarrow B_i$, instead of the correlation-minimum encoding in the Mamdani-Togai model. Therefore,

$$\mu_{R_i}(u,v) = \mu_{A_i}(u)\mu_{B_i}(v); \quad u \in U, v \in V. \tag{9.13}$$

So, we can get n fuzzy relationships from n historical earthquake observations.

In the new model, the weights of all fuzzy if-then rules "If u is A_i, then v is B_i" are same. It can be regarded as 1.

9.4 Patterns Smoothening

Suppose we have n observations $(m_1, g_1), (m_2, g_2), \cdots, (m_n, g_n)$. Using information diffusion technique, we can get n fuzzy if-then rules $A_1 \rightarrow B_1, A_2 \rightarrow B_2, \cdots, A_n \rightarrow B_n$ as shown in Fig. 9.4.

Now if a crisp input value m_o (the antecedent) is known, then one needs a way, such as Eq.(9.6), to derive the consequent g_o from m_o and R_i.

In practical calculation, U is generally discrete so that m_o is not just equal to some value in U. We can employ the information distribution formula shown in Eq.(9.14) to get a fuzzy subset as an input:

$$\mu_{m_o}(u_j) = \begin{cases} 1 - |m_o - u_j|/\Delta, & \text{if } |m_o - u_j| \leq \Delta, \\ 0, & \text{if } |m_o - u_j| > \Delta; \end{cases} \tag{9.14}$$

where $\Delta = u_{j+1} - u_j$.

A fuzzy consequent $\underset{\sim}{g_{oi}}$ from $\underset{\sim}{m_o}$ and R_i is then obtained as

$$\mu_{g_{oi}}(v) = \sum_u \mu_{m_o}(u)\mu_{R_i}(u,v), \quad v \in V. \tag{9.15}$$

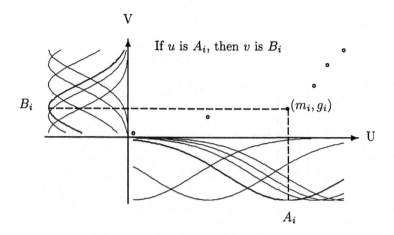

Fig. 9.4   Production of the fuzzy rule " If u is A_i, then v is B_i" through observation (m_i, g_i) by the information diffusion technique.

Using the so-called "max" principle, we defuzzify $\underset{\sim}{g_{oi}}$ into a crisp output value g_i while we obtain a weight value w_i. Suppose

$$\mu_{g_{oi}}(v') = \max_{v \in V}\{\mu_{g_{oi}}(v)\}, \tag{9.16}$$

hence the max-defuzzified value is $g_i = v'$ and the weight is $w_i = \mu_{g_{oi}}(v')$.

Then, to integrate all results coming from $R_1, R_2, , \cdots, R_n$, the relevant output value g_o becomes

$$g_o = \left(\sum_{i=1}^{n} w_i g_i\right) \Big/ \left(\sum_{i=1}^{n} w_i\right). \tag{9.17}$$

The procedure comprising (9.10)-(9.17) is called *information-diffusion approximate reasoning* (IDAR) whose system architecture is depicted in Fig. 9.5.

Any observation (m_i, g_i) of the sample can thus be changed into a new pattern $(m_i, \widetilde{g}_i)$ via information-diffusion approximate reasoning. In the study of a practical problem in section 9.6, it is apparent that the new patterns must be smoother.

9.5 Learning Relationships by BP Neural Networks

Using neural networks for automatic learning by examples has been a common approach employed in the construction of artificial systems. Backpropagation(BP) neural networks[23][26], a class of feedforward neural networks, are models commonly used for learning and reasoning.

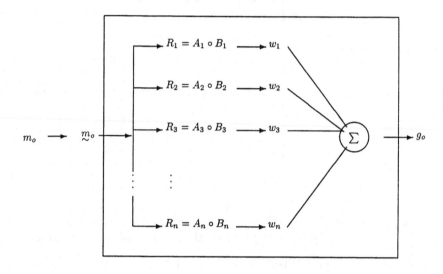

Fig. 9.5   System architecture of information-diffusion approximate
reasoning. Compositional rule "∘" is defined by formula (9.13).

However, neural information processing models generally assume that
the patterns used for training a neural network are compatible. In some
research[7][22], fuzzy neural networks are developed with stronger, compared
with conventional BP networks, nonlinear and imprecision mapping abilities.
They appear to have promising application prospects in nonlinear modeling,
fuzzy identification and self-organizing fuzzy control for complex systems.

There are two common architectures to process fuzzy information based
on the conventional BP neural network. One of them is to use fuzzy numbers
instead of real values as inputs and weights. This architecture[17][18] is called
fuzzy BP network and can be shown in Fig. 9.6.

Fuzzy BP networks perform nonlinear mapping between fuzzy input vectors and crisp outputs. This fuzzy type of BP networks, however, does not
have the ability of processing contradictory patterns. It usually encounters
the problem of convergence.

Another common fuzzy BP model is to generate fuzzy rules by the BP
technique. In this architecture [24], as depicted in Fig. 9.7, an input node represents a fuzzy antecedent A_i and an output node means a fuzzy consequent

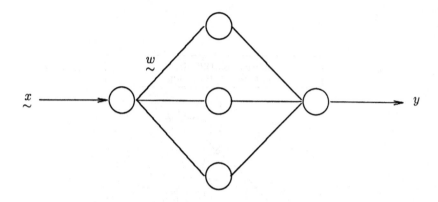

Fig. 9.6 BP network with fuzzy numbers.

B_j. Pattern (u, v) fires A_i and B_j in value $\mu_{A_i}(u)$ and $\mu_{B_j}(v)$. In other words, real values can be replaced by fuzzy memberships to train a conventional BP neural network. The fuzzy rules correspond to the final weights which are larger than zero.

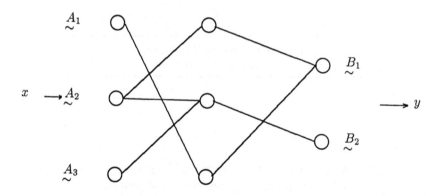

Fig. 9.7 Fuzzy-rule neural network model.

Though the generation of fuzzy rule from training data by neural networks is more automatic, the method does not ensure that we can always find the rules with a given pool of experts or with a fixed set of data. To effectively process incomplete and contradictory data to unravel relationships between inputs and outputs, we propose in here a *hybrid model* (HM) which integrates information-diffusion approximate reasoning and conventional BP neural network. The architecture of the hybrid model is depicted in Fig. 9.8.

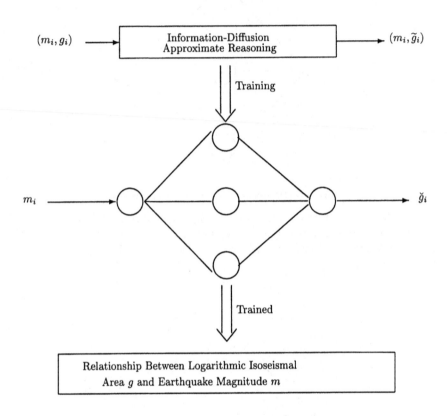

Fig. 9.8   Architecture of the hybrid model integrating information-diffusion approximate reasoning and conventional BP neural network.

   In the hybrid model, observations $(m_1, g_1), (m_2, g_2), \cdots, (m_n, g_n)$ are first changed, via the information-diffusion technique, into new patterns $(m_1, \tilde{g}_1)$, $(m_2, \tilde{g}_2), \cdots, (m_n, \tilde{g}_n)$. A conventional BP neural network is then employed to learn the relationship between isoseismal area, g, and earthquake magnitude, m. As discussed in section 9.4, the information-diffusion approximate reasoning serves as a pattern smoothing mechanism for the set of fuzzy if-then rules. Its results are then used for training the BP neural network.

9.6 Calculation

In Yunnan Province of China, there is a data set of strong earthquakes consisting of 25 records from 1913 to 1976 with magnitude, M, and isoseismal area, S, of intensity, $I \geq VII$ (Table 9-3).

Table 9-3 Magnitudes and isoseismal areas

Date	M	$S_{I \geq VII}$	Date	M	$S_{I \geq VII}$
1913.12.21	6.5	2,848	1965.7.3	6.1	733
1917.7.31	6.5	3,506	1966.1.31	5.1	19
1925.3.16	7	4,758	1966.2.5	6.5	1,703
1930.5.15	5.75	779	1966.9.19	5.4	261
1941.5.16	7	2,593	1966.9.28	6.4	404
1941.12.26	7	1,656	1970.1.5	7.7	8,176
1951.12.21	6.25	3,385	1970.2.7	5.5	100
1952.6.19	6.25	1,345	1971.4.28	6.7	212
1952.12.28	5.75	190	1973.3.22	5.5	18
1955.6.7	6	88	1973.8.6	6.8	200
1961.6.12	5.8	47	1974.5.11	7.1	837
1961.6.27	6	3,582	1976.2.16	5.7	99
1962.6.24	6.2	449			

M is the Richter magnitude scale,
S is square kilometers.

The given sample with magnitude m and logarithmic isoseismal area $g = \log_{10} S$ is

$$
\begin{aligned}
X &= \{x_1, x_2, \cdots, x_{25}\} \\
&= \{(m_1, g_1), (m_2, g_2), \cdots, (m_{25}, g_{25})\} \\
&= \{(6.5, 3.455), (6.5, 3.545), (7, 3.677), (5.75, 2.892), (7, 3.414), \\
&\quad (7, 3.219), (6.25, 3.530), (6.25, 3.129), (5.75, 2.279), (6, 1.944), \\
&\quad (5.8, 1.672), (6, 3.554), (6.2, 2.652), (6.1, 2.865), (5.1, 1.279), \\
&\quad (6.5, 3.231), (5.4, 2.417), (6.4, 2.606), (7.7, 3.913), (5.5, 2.000), \\
&\quad (6.7, 2.326), (5.5, 1.255), (6.8, 2.301), (7.1, 2.923), (5.7, 1.996)\}.
\end{aligned}
$$

(9.18)

Linear regression (LR) method has conventionally been used to estimate the relationship between g and m. Regressing g on m with data in the given sample, the regression line is obtained as:

$$g = -2.61 + 0.85m \tag{9.19}$$

whose linear relevance coefficient $r^2 = 0.503$ is relatively small.

To establish the relationship by the neural network approach, we constructed a conventional BP neural network with one node in the input layer, 15 nodes in the hidden layer, and one node in the output layer.

On the principle of the BP algorithm, we can, directly, use X in (9.18) to train a BP network. For that, we have to take

$$f(x) = \frac{\beta}{1 + \exp(-\gamma x)} \tag{9.20}$$

to be the activation function. It needs so much time to learn β and γ. To save time, we use the sigmoidal function

$$f(x) = \frac{1}{1 + \exp(-x)} \tag{9.21}$$

to be the activation function. Obviously, the output of the function is always in [0,1]. In this case, firstly we normalize the given sample X by

$$m' = \frac{m - a_m}{b_m - a_m}, \quad \text{and} \quad g' = \frac{g - a_g}{b_g - a_g}. \tag{9.22}$$

The normalized sample is

$$\begin{aligned}
X' &= \{x'_1, x'_2, \cdots, x'_{25}\} \\
&= \{(m'_1, g'_1), (m'_2, g'_2), \cdots, (m'_{25}, g'_{25})\} \\
&= \{(0.538, 0.8276), (0.538, 0.8616), (0.731, 0.9115), (0.250, 0.6158), \\
&\quad (0.731, 0.8123), (0.731, 0.7390), (0.442, 0.8559), (0.442, 0.7050), \\
&\quad (0.250, 0.3852), (0.346, 0.2594), (0.269, 0.1569), (0.346, 0.8651), \\
&\quad (0.423, 0.5257), (0.385, 0.6058), (0.000, 0.0088), (0.538, 0.7436), \\
&\quad (0.115, 0.4371), (0.500, 0.5085), (1.000, 1.0000), (0.154, 0.2803), \\
&\quad (0.615, 0.4031), (0.154, 0.0000), (0.654, 0.3935), (0.769, 0.6275), \\
&\quad (0.231, 0.2786)\}.
\end{aligned} \tag{9.23}$$

Setting the momentum rate $\eta = 0.9$ and learning rate $\alpha = 0.7$, we use X' to train the BP network. After 600,000 iterations, the normalized system error is 0.015594. For results from the trained network, we inverse them into the primary universe by

$$g = a_g + g'(b_g - a_g). \tag{9.24}$$

The results obtained by the regression and neural network methods are shown in Fig. 9.9. Apparently, the BP curve does not quite adequately capture the observed relationship. Neither does the regression method.

A careful examination of X shows that it is, in fact, relatively small and data contained are incomplete and sometimes contradictory. Therefore, any observation in X can be regarded as a piece of fuzzy information which represents partially the relationship between the two variables. Furthermore, an observation can produce a simple fuzzy relationship via the diffusion method. The integration of all fuzzy relationships will, in turn, produce a more appropriate description. For the analysis of data in Table 9-3, the procedure for our proposed approach is described as follows:

Firstly, let the discrete universe of discourse of earthquake magnitudes be

$$U = \{u_1, u_2, \cdots, u_{30}\} = \{5.010, 5.106, \cdots, 7.790\} \qquad (9.25)$$

where step length $\Delta_m = 0.096$, $u_1 = a_m - \Delta_m$ and $u_{30} = b_m + \Delta_m$.

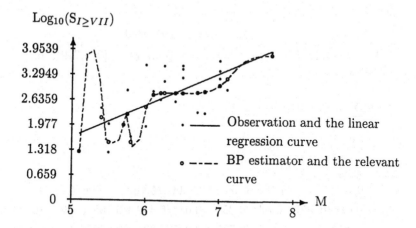

Fig. 9.9   Relationship between earthquake magnitude and logarithmic isoseismal area estimated by linear regression and BP network.

Let the discrete universe of discourse of logarithmic isoseismal area be

$$V = \{v_1, v_2, \cdots, v_{30}\} = \{1.164, 1.262, \cdots, 4.004\} \qquad (9.26)$$

where step length Δ_g is 0.098. $v_1 = a_g - \Delta_g$ and $v_{30} = b_g + \Delta_g$.

Then, for X in (9.18), using the formula in (7.38), we obtain the normal diffusion coefficient as:

$$h_m = 2.6851(b_m - a_m)/(n - 1) = 2.6851(7.7 - 5.1)/24 = 0.2909. \qquad (9.27)$$

and

$$h_g = 2.6851(b_g - a_g)/(n - 1) = 2.6851(3.913 - 1.255)/24 = 0.2974. \qquad (9.28)$$

Applying the normal information diffusion formulae (9.10) and (9.11), observation (m_i, g_i) can be changed into two fuzzy subsets as

$$\mu_{m_i}(u) = \exp[-\frac{(u - m_i)^2}{0.169}], \quad u \in U, \qquad (9.29)$$

and

$$\mu_{g_i}(v) = \exp[-\frac{(v - g_i)^2}{0.177}], \quad v \in V. \qquad (9.30)$$

Therefore, an observation $x_i = (m_i, g_i)$ can be transformed into a fuzzy relationship matrix:

$$R_i = \begin{array}{c} \\ u_1 \\ u_2 \\ \cdots \\ u_{30} \end{array} \begin{pmatrix} v_1 & v_2 & \cdots & v_{30} \\ r_{1,1} & r_{1,2} & \cdots & r_{1,30} \\ r_{2,1} & r_{2,2} & \cdots & r_{2,30} \\ \cdots & \cdots & \cdots & \cdots \\ r_{30,1} & r_{30,2} & \cdots & r_{30,30} \end{pmatrix}, \qquad (9.31)$$

where

$$r_{j,k} = \mu_{m_i}(u_j)\mu_{g_i}(v_k). \qquad (9.32)$$

For example, let $i = 6$, then $x_6 = (m_6, g_6) = (7, 3.219)$, and the fuzzy subsets are

$$A_6 = \sum_{j=1}^{30} \mu_{m_6}(u_j)/u_j$$

$$= \sum_{j=1}^{30} \exp[-\frac{(u-m_i)^2}{0.169}]/u_j$$

$= 0/5.010 + \cdots + 0.001/5.873 + 0.002/5.969 + 0.006/6.065 +$
$\quad 0.016/6.160 + 0.038/6.256 + 0.084/6.352 + 0.165/6.448 + 0.292/6.544 +$
$\quad 0.464/6.640 + 0.661/6.735 + 0.845/6.831 + 0.969/6.927 + 0.997/7.023 +$
$\quad 0.920/7.119 + 0.762/7.215 + 0.566/7.310 + 0.377/7.406 + 0.225/7.502 +$
$\quad 0.121/7.598 + 0.058/7.694 + 0.025/7.790, \qquad (9.33)$

$$B_6 = \sum_{j=1}^{30} \mu_{g_6}(v_j)/v_j$$

$$= \sum_{j=1}^{30} \exp[-\frac{(v-g_i)^2}{0.177}]/v_j$$

$= 0/1.164 + \cdots + 0.001/2.143 + 0.004/2.241 + 0.013/2.339 +$
$\quad 0.031/2.437 + 0.071/2.535 + 0.143/2.633 + 0.260/2.731 + 0.422/2.829 +$
$\quad 0.617/2.927 + 0.808/3.025 + 0.949/3.123 + 1.000/3.221 + 0.946/3.319 +$
$\quad 0.802/3.416 + 0.610/3.514 + 0.417/3.612 + 0.255/3.710 + 0.140/3.808 +$
$\quad 0.069/3.906 + 0.031/4.004. \qquad (9.34)$

Hence, using Eq.(9.32) we obtain a fuzzy relationship matrix:

$$R_6 = \begin{array}{c} \\ \\ u_1(5.010) \\ \cdots \\ u_{16}(6.448) \\ u_{17}(6.544) \\ u_{18}(6.640) \\ u_{19}(6.735) \\ u_{20}(6.831) \\ u_{21}(6.927) \\ u_{22}(7.023) \\ u_{23}(7.119) \\ u_{24}(7.215) \\ \cdots \\ u_{30}(7.790) \end{array} \begin{array}{ccccccccc} v_1 & \cdots & v_{20} & v_{21} & v_{22} & v_{23} & v_{24} & \cdots & v_{30} \\ 1.164 & \cdots & 3.025 & 3.123 & 3.221 & 3.319 & 3.416 & \cdots & 4.004 \\ \left(\begin{array}{ccccccccc} 0.000 & \cdots & 0.000 & 0.000 & 0.000 & 0.000 & 0.000 & \cdots & 0.000 \\ \cdots & \cdots & \cdots & \cdots & \cdots & \cdots & \cdots & \cdots & \cdots \\ 0.000 & \cdots & 0.133 & 0.157 & 0.165 & 0.156 & 0.132 & \cdots & 0.005 \\ 0.000 & \cdots & 0.236 & 0.277 & 0.292 & 0.276 & 0.234 & \cdots & 0.009 \\ 0.000 & \cdots & 0.375 & 0.440 & 0.464 & 0.439 & 0.372 & \cdots & 0.014 \\ 0.000 & \cdots & 0.534 & 0.627 & 0.661 & 0.625 & 0.530 & \cdots & 0.020 \\ 0.000 & \cdots & 0.683 & 0.802 & 0.845 & 0.799 & 0.678 & \cdots & 0.026 \\ 0.000 & \cdots & 0.783 & 0.919 & 0.969 & 0.916 & 0.777 & \cdots & 0.030 \\ 0.000 & \cdots & 0.805 & 0.946 & 0.997 & 0.943 & 0.800 & \cdots & 0.030 \\ 0.000 & \cdots & 0.743 & 0.873 & 0.920 & 0.870 & 0.738 & \cdots & 0.028 \\ 0.000 & \cdots & 0.615 & 0.723 & 0.762 & 0.720 & 0.611 & \cdots & 0.023 \\ \cdots & \cdots & \cdots & \cdots & \cdots & \cdots & \cdots & \cdots & \cdots \\ 0.015 & \cdots & 0.020 & 0.024 & 0.025 & 0.024 & 0.020 & \cdots & 0.001 \end{array}\right) \end{array} \quad (9.35)$$

When magnitude m_o is given, suppose $m_o = 6.5$, we can change it into a fuzzy subset on U through the information distribution formula in (9.14) as follows:

$$\underset{\sim}{m}_o = \sum_{j=1}^{30} \mu_{m_o}(u_j)/u_j$$
$$= 0/5.010 + \cdots + 0.457/6.448 + 0.543/6.544 + \cdots + 0/7.790. \quad (9.36)$$

Using sum-product composition shown in Eq.(9.15), based on R_6 we obtain a fuzzy Consequent

$$\underset{\sim}{g}_{o6} = \sum_{j=1}^{30} \mu_{g_{o6}}(v_j)/v_j$$
$$= 0/1.16 + \cdots + 0.144/2.927 + 0.189/3.025 + 0.222/3.123 + 0.234/3.221 +$$
$$0.221/3.319 + 0.188/3.416 + 0.143/3.514 + \cdots + 0/4.00 \quad (9.37)$$

Using the "max" principle, see Eq.(9.16), we obtain estimated value $g_6 = 3.221$ and the weight $w_6 = 0.234$. For $m_o = 6.5$, employing $R_1, R_2, \cdots, R_{25}$, we obtain a estimated vector, G, with weight vector, W, as

$$\begin{aligned} G &= (g_1, g_2, \cdots, g_{25}) \\ &= (3.416, 3.514, 3.710, 2.927, 3.416, 3.221, 3.514, 3.123, 2.241, \\ &\quad 1.947, 1.653, 3.514, 2.633, 2.829, 1.262, 3.221, 2.437, 2.633, \\ &\quad 3.906, 2.045, 2.339, 1.262, 2.339, 2.927, 1.947), \end{aligned} \quad (9.38)$$

$$\begin{aligned} W &= (w_1, w_2, \cdots, w_{25}) \\ &= (0.979, 0.981, 0.233, 0.039, 0.234, 0.234, 0.688, 0.688, 0.038, \\ &\quad 0.234, 0.059, 0.232, 0.587, 0.390, 0.000, 0.986, 0.001, 0.928, \\ &\quad 0.000, 0.003, 0.783, 0.003, 0.583, 0.124, 0.025). \end{aligned} \quad (9.39)$$

Then, using the integration formula in (9.17), the relevant output value is

$$g_o = \left(\sum_{i=1}^{25} w_i g_i\right) \Big/ \left(\sum_{i=1}^{25} w_i\right) = 27.233/9.053 = 3.008. \qquad (9.40)$$

It means that, in Yunnan Province of China, according to historical earthquake experience, if an earthquake of Richter magnitude scale M=6.5 occurs, the isoseismal area of intensity $I \geq VII$ caused by this earthquake is about $S = 10^{3.008} = 1,019$ square kilometers. Therefore, the building in this area must resist earthquakes of intensity $I \geq VII$.

Let $m_o = m_i \in X_m$ consisting of all inputs of X, we can obtain a new sample as follows:

$$
\begin{aligned}
\widetilde{X} &= \{\widetilde{x}_1, \widetilde{x}_2, \cdots, \widetilde{x}_{25}\} \\
&= \{(m_1, \widetilde{g}_1), (m_2, \widetilde{g}_2), \cdots, (m_{25}, \widetilde{g}_{25})\} \\
&= \{(6.5, 3.008), (6.5, 3.008), (7, 3.103), (5.75, 2.329), (7, 3.103), \\
&\quad (7, 3.103), (6.25, 2.906), (6.25, 2.906), (5.75, 2.329), (6, 2.643), \\
&\quad (5.8, 2.387), (6, 2.643), (6.2, 2.863), (6.1, 2.762), (5.1, 1.757), \\
&\quad (6.5, 3.008), (5.4, 2.004), (6.4, 2.984), (7.7, 3.742), (5.5, 2.085), \\
&\quad (6.7, 3.027), (5.5, 2.085), (6.8, 3.042), (7.1, 3.145), (5.7, 2.272)\}.
\end{aligned}
$$
$$(9.41)$$

It is used to train a conventional BP neural network with one unit in the input layer, one hidden layer with 15 units, and one unit in the output layer. With 0.9 as momentum rate and 0.7 learning rate, after only 34405 iterations, the normalized system error is 0.00001, then we obtain the weights and thresholds as Table 9-4 and Table 9-5, respectively.

Table 9-4 Weights of the trained BP network

Hidden nodes	Input node 1	Output node 1
1	-2.993257	-0.720176
2	-11.014281	-3.188434
3	-3.093669	2.189297
4	-0.882442	0.932863
5	12.790363	9.158985
6	-13.381745	-9.071100
7	-1.002514	1.058568
8	0.907408	0.386380
9	0.064322	-0.043502
10	-1.293816	1.388809
11	-0.469254	0.438369
12	-1.313313	1.218095
13	-1.121203	1.151400
14	0.911799	-0.498471
15	0.946231	-0.049769

Table 9-5 Thresholds of hidden layer and output layer

	Nodes	1	2	3	4	5
Hidden layer	Threshold	-2.285321	3.622231	0.446015	-1.343085	-12.601024
	Nodes	6	7	8	9	10
	Threshold	-0.523689	-1.129807	-3.957527	-2.070541	-0.557413
	Nodes	11	12	13	14	15
	Threshold	-1.792893	-0.816823	-0.916131	-2.659005	-3.500129
Output layer	Node	1				
	Threshold	-0.723701				

Using the hybrid model consisting of the information-diffusion approximate reasoning method and a BP neural network, we can get a better estimator depicted in Fig. 9.10.

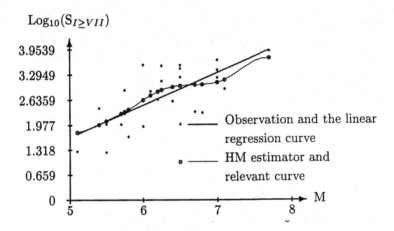

Fig. 9.10   Relationship between earthquake magnitude and logarithmic isoseismal area estimated by the hybrid model (HM) consisting of the IDAR method and a BP network.

For comparison, the average sums of squared errors ϵ of the two estimators: $\widehat{g}$, the linear-regression estimator (LR); and, $\breve{g}$, the hybrid-model (HM) are computed as follows:

$$\begin{cases} \epsilon_{LR} = \frac{1}{25}\sum_{i=1}^{25}(g_i - \widehat{g}_i)^2 = 0.2734 \\ \epsilon_{HM} = \frac{1}{25}\sum_{i=1}^{25}(g_i - \breve{g}_i)^2 = 0.2452 \end{cases} \tag{9.42}$$

The advantage of the hybrid-model is clearer identification of curve that nonlinear due to different part characteristics.

Obviously, the HM estimator is better than the linear-regression estimator, because the new estimator is more precise than old estimator. HM is also better than the conventional BP-neural-network, because the new one is more stable.

9.7 Conclusion and Discussion

Most systems in the real world are nonlinear. Neural networks provide a non-parametric approach for the nonlinear estimation of data, more specifically nonlinear, feedforward multilayer networks including the multilayer percep-tron and the radial basis function RBF networks. However, when the patterns used for training are incompatible, the traditional neural networks do not con-verge because the adjustments of weights and thresholds do not know where to turn.

Earthquake data with magnitudes and isoseismal areas are strongly in-compatible patterns. Two earthquakes with same magnitude never produce same isoseismal area, and usually are very different.

In this chapter we propose an approach to resolve the problem. It is a hybrid model integrating information-diffusion approximate reasoning and BP neural network.

The procedure includes the following 6 steps:

(1) Changing observation (m_i, g_i) into two fuzzy sets A_i and B_i:

$$A_i : \quad \mu_{m_i}(u) = \exp[-\frac{(u - m_i)^2}{2h_m^2}],$$

$$B_i : \quad \mu_{g_i}(v) = \exp[-\frac{(v - g_i)^2}{2h_g^2}].$$

(2) Changing fuzzy rule $A_i \to B_i$ into a fuzzy relation matrix R_i;

(3) Using sum-product composition and R_i to produce a fuzzy consequent;

(4) Using the max-principle to defuzzify the fuzzy consequent into a crisp value with weight.

(5) Integrating the crisp values from all fuzzy relation matrixes to produce new pattern.

(6) Using the new patterns to train BP network.

The basic advantage of using the information-diffusion method is that we can change observations into a fuzzy subsets which naturally fills up the infor-mation gaps caused by incomplete data. The integration of the relationships thus produced can change the contradictory patterns into more compatible ones which can smoothly and quickly train the BP neural network to get the relationship we want.

It has been demonstrated that the hybrid model proposed in this chapter can effectively estimate the relationship between isoseismal area and earthquake magnitude in which data are usually scanty, incomplete, and contradictory. The model is also applicable to other small-sample problems.

This study is the successor of paper [13], where we use Eq.(7.41) (not Eq.(7.38) as in the chapter) to calculate diffusion coefficient h, and we only change m_i into a fuzzy set to avoid possible error from operator for fuzzy reasoning. Although the ϵ_{HM} in the paper is only 0.1993, its HM curve is not stable as in the chapters. Therefore the new model has more better performance.

References

1. Bollinger, G.A., Chapman, M.C., and Sibol, M.S.(1993), A comparison of earthquake damage areas as a function of magnitude across the United States. Bulletin of the Seismological Society of America Vol.83, pp.1064-80
2. Cavallini, F. and Rebez, A.(1996), Representing earthquake intensity magnitude relationship with a nonlinear function. Bulletin of the Seismological Society of America, Vol. 86, pp.73-78
3. Dickerson, J.A. and Kosko, B.(1993), Fuzzy function learning with covariance ellipsoids. Proceedings of FUZZ-IEEE'93, California, pp. 1162-1167
4. Dickerson, J.A. and Kosko, B.(1993), Fuzzy function approximation with supervised ellipsoidal learning. World Congress on Neural Networks (WCNN), Vol.II, pp. 9-13.
5. Fukushima, Y., Gariel, J.C. and Tanaka, R.(1995), Site-dependent attenuation relations of seismic motion parameters depth using borehole data, Bulletin of the Seismological Society of America, Vol.85, pp.1790-1804
6. Gupta, I.N. and Nuttli,O.W.(1976), Spatial attenuation of intensities fro central U.S. earthquake. Bulletin Seismological Society of America , Vol.66, pp.743-751
7. Hernandez, J.V., Moore, K. and Elphic, R.(1995), Sensor fusion and nonlinear prediction for anomalous event detection. Proceedings of the SPIE - The International Society for Optical Engineering, Vol.2484, pp.102-112
8. Hodgson,J.H.(1964), Earthquakes and Earth Structure. Prentice-Hell, Inc., Englewood Cliffs, New Jersey
9. Howell, B.F. and Schultz, T.R.(1975), Attenuation of Modified Mercalli intensity with distance from the epicenter. Bulletin Seismological Society of America, Vol.65, pp.651-665
10. Huang, C.F.(1995), Fuzziness of incompleteness and information diffusion principle. Proceedings of FUZZ-IEEE/IFES'95, Yokoham, Japan, pp.1605-1612
11. Huang, C.F.(1997), Principle of information diffusion. Fuzzy Sets and Systems, Vol.91, No.1, pp.69-90
12. Huang, C.F. and Jiading, W.(1995), Technology of Fuzzy Information Optimization Processing and Applications, Beijing University of Aeronautics and Astronautics Press, Beijing. (in Chinese)
13. Huang, C.F. and Leung, Y.(1999), Estimating the relationship between isoseismal area and earthquake magnitude by hybrid fuzzy-neural-network method, Fuzzy Sets and Systems, Vol.107, No.2, pp.131-146
14. Huang, C.F. and Liu, Z.R.(1985), Isoseismal area estimation of Yunnan Province by fuzzy mathematical method. Fen Deyi and Liu Xihui (eds): Fuzzy Mathematics in Earthquake Researches. Seismological Press, Beijing, pp. 185-195
15. Kim, H.M. and Kosko, B.(1996), Fuzzy prediction and filtering in impulsive noise. Fuzzy Sets and Systems, Vol.77, pp.15-33
16. Kosko, B.(1992), Neural Networks and Fuzzy Systems, Prentice Hall, Englewood Cliffs, New Jersey
17. Lee, H.M. and Lu, B.H.(1994), FUZZY BP: a neural network model with fuzzy inference. IEEE Internat. Conf. on Neural Networks, Orlando, Florida, pp.1583-1588
18. Lee, H.M., Lu, B.H. and Lin, F.T.(1995), A fuzzy neural network model for revising imperfect fuzzy rules. Fuzzy Sets and Systems, Vol., pp.25-45
19. Liu,Z.R. and Huang, C.F. et al (1987), A fuzzy quantitative study on the effect of active fault distribution on isoseismal area in Yunnan, Journal of seismology, No.1, pp.9-16. (in Chinese)

20. Lomnitz, C. and Rosenblueth, E.(1976), Seismic Risk and Engineering Decisions. Elsevier Scientific Publishing Company, Amsterdam

21. Mamdani, E.H.(1977), Application of fuzzy logic to Approximate reasoning using linguistic synthesis. IEEE Transactions on Computer, Vol.26, pp.1182-1191

22. Monostori, L. and Egresits, C.(1994), Modelling and monitoring of milling through neuro-fuzzy techniques. Proceedings of Intelligent Manufacturing Systems (IMS'94), Vienna, Austria, pp.463-468

23. Pao, Y.H.(1989), Adaptive Pattern Recognition and Neural Networks, Addison-Wesley, Reading, MA

24. Rhee, F.C. -H. and Krishnapuram, R.(1993), Fuzzy rule generation methods for high-level computer vision. Fuzzy Sets and Systems, Vol.60, pp.245-258

25. Rubeis-V, V., Gasparini, C., Maramai, A., Murru, M. and Tertulliani, A.(1992), The uncertainty and ambiguity of isoseismal maps. Earthquake Engineering & Structural Dynamics, Vol.21, Iss: 6, pp.509-

26. Rumelhart, D.E. and McClelland,J.L.(1973), Parallel Distributed Processing: Explorations in the Microstructure of Cognition, Vols. 1 and 2, MIT, Press, Cambridge, MA

27. Stephens, J.E. and VanLuchene, R.D. (1994), Integrated assessment of seismic damage in structures. Microcomputers in Civil Engineering, Vol.9, Iss: 2, pp.119-128

28. Togai, M. and Watanabe, H.(1986), Expert system on a chip: an engine for real-time approximate reasoning. IEEE Expert, Vol.1, No.3, pp.55-62

29. Zadeh, L.A.(1974), On the analysis of large scale system. Gottinger, H. (ed.): Systems Approaches and Environment Problems, Vandenhoeck and Ruprecht, Gottingen, pp.23-37

30. Zadeh, L.A.(1983), Acomputational approach to fuzzy quantifiers in natural languages. Computers and mathematics, Vol.9, pp.149-184

31. Zadeh, L.A.(1995), Fuzzy control, fuzzy graphs, and fuzzy inference. in: Yam, Y. and Leung, K.S. (eds): Future Directions of Fuzzy Theory and Systems, World Scientific, Singapore, pp.1-9.

10. Fuzzy Risk Analysis

Nobody can precisely estimate the future risks — environment, health, and safety — unless he knows all aspects related to the risk system he studies. In practical cases, it is impossible to avoid the gaps with respect to risk assessment, which cause fuzziness (imprecision, vagueness, incompleteness, etc.) Therefore, we have to deal with the fuzziness of a risk system. This chapter reviews some basic concepts of risk assessment and then uses the information diffusion techniques to develop fuzzy risk analysis. Section 10.1 introduces the concept of risk. In section 10.2 we briefly outline the principles of risk recognition and management for environment, health, and safety. Section 10.3 surveys some studies in fuzzy risk analysis. Based on natures of risk, in section 10.4, we give a general definition of fuzzy risk. In section 10.5, we review some classical models to calculate probability-risk. In section 10.6, the principle of information diffusion is employed directly to assess the probability-risk from a given sample. Section 10.7 gives an application in risk assessment of flood disaster. We conclude the chapter with a summary in section 10.8.

10.1 Introduction

Everyone is constantly subjected to many risks, both as individuals and as members of various societal groups. There are voluntary risks which individuals elect to assume, such as those due to personal smoking, and involuntary risks which individuals do not elect to assume, such as those due to being forced to nuclear waste or flood. There are very few situations which create no risk to individuals.

Risks are evaluated qualitatively rather than analysed quantitatively[23]. In spite of that risk had been defined as the probability per unit time of the occurrence of a unit cost burden[26],[27],[28], concept of probability-risk (the definition will be given in Chapter 12) cannot replace the concept of risk. It seems to be no commonly accepted definition of risk in the area of risk and hazard assignment.

Most basic to effective risk management is an understanding of the concept of risk itself. Intuitively, *risk* exists when loss is possible and its financial impact is significant. This linguistic definition captures a property of risk that eludes definition in terms of mathematical formulas. In fact, as Jablonowski stated[17], in the real wold, the possibility and financial significance of loss cannot be defined with precision.

Starr[26] referred to that there are four different types of societal risks:

(T1) *Real risk* — eventually by future circumstances when they fully develop;

(T2) *Statistical risk* — determined by currently available data, typically measured actuarially for insurance premium and other purposes;

(T3) *Predicted risk* — predicted analytically from system models structured from historical studies;

(T4) *Perceived risk* — intuitively seen by individuals.

For example, to the flight insurance company, flying constitutes a known statistical risk. Meanwhile, to the passenger purchasing insurance at the airport, flying constitutes a perceived risk. There are cultural differences in perception of risk issues. Perceived risks might vary from nation to nation and even across industrial groups within a given nation. It is often argued that only the statistical risk involves a measure of the probability and severity of adverse impacts.

The traditional approach used to model risky choicemaking situations is to describe choices involving risk in term of their underlying probability distributions and associated utilities. In the decision theory literature *risk* can be regarded as, at minimum, a two-dimensional concept involving:

(D1) The possibility of an adverse outcome;

(D2) Uncertainty over the occurrence, timing, or magnitude of that adverse outcome.

People talk about risk when there is the chance, but not the certainty, that something they don't want may happen. The main reason why the probability has been used so common is that most researchers consider the probability is the best measure to deal with uncertainty.

In fact, only in situations for which a great amount of data has been collected, such as automobile accidents and fires, probabilistic methods is an extremely effective way of quantifying uncertainty about a risk source (the potential of the source to release a risk agent). For many risk sources, the available data are insufficient to permit estimating reliably the frequencies of release of risk agents or other characteristics of concern. Moreover, probabilistic methods almost can do nothing for *perceived risk* .

In general, uncertain feature of risk is relative to both of randomness and fuzziness. For example, the occurrence of earthquake is a random event, but the earthquake intensity is a fuzzy concept[19]. In the process of risk evaluation, the random is due to a large amount of unknown factors existing, and the fuzziness has concern with the terms of macroscopic grad and incomplete knowledge sample which can be described by fuzzy methods[15].

In many cases, a risk system, including *risk source* (might be flood, earthquake, typhoon, or drought), *exposure process* (must exist by which people or the things they value may be exposed to the released risk agent), and *causal process* (must exist by which exposures produce adverse health or environmental consequences), is too complex to understand from random viewpoint. A natural way to improve risk analysis is to introduce the concept of fuzzy risk to overcome difficulties resulting from fuzzy environments, with macroscopic grad or (and) incomplete knowledge.

10.2 Risk Recognition and Management for Environment, Health, and Safety

Since we celebrated the first Earth Day in 1970, we have made considerable progress in the quality of our environment – air, water, land, and other natural resources. And yet myriad new concerns have emerged: secondhand tobacco smoke, cellular phones, and global warming, to name just a few. Concerns about industrial accidents, chemical and food safety, and natural disasters also have intensified. The heightened public sensitivities to multiple environmental hazards, along with perceptions that the hazards are getting out of government control, spawned a number of public "right to know" and "right to know more" movements. The abundance of information that surfaced, however, proved to be more baffling than enlightening – not really surprising given the degree of scientific uncertainty, the divers interpretations of the same data by experts, and the billions of dollars at stake[18].

Risk recognition provides a systematic framework, grounded in scientific principles, to understand and manage diverse risks. In risk recognition and management, we look at a situation or scenario and ask these types of questions: what can go wrong and why, how likely is it, how bad can it be, and what can we do about it? *Risk*, then, is a function of the natural of the hazard, accessibility or avenue of contact (exposure potential), characteristics of the exposed populations (receptors), the likelihood of occurrence, and the magnitude of exposure and consequences, as well as public values. Table 10-1 presents a classical overview of safety, health, and ecological risk recognitions including major process steps.

Safety risk analysis tend to be much more probabilistic, incorporating the likelihood of initiating events, as well as the likelihood of exposures and the range of consequences. Human health and ecological risk recognitions are typically deterministic and use single-point estimate where valuable information remains hidden from risk managers.

Risk criteria provide an effective frame of reference for prioritizing problems, allocating resources, and reducing risks. However, there is a crucial limitation to realizing the full potential of risk recognition. That is the paucity of professionals who have the broad training and the perspective needed to

Table 10-1 Classical overview of three major types of risk recognitions

Steps	Safety	Human health	Ecological/ environment
1	Hazard identification	Data analysis/ hazard identification	Problem formulation (hazard screening)
2	Probability/ frequency estimation of causes	Exposure assessment	Exposure assessment
3	Consequence analysis	Dose-response or toxicity assessment	Toxicity effects assessment
4	Risk evaluation	Risk characterization	Risk characterization

transcend the "cookbook" approach and focus on the critical issues. Many believe that risk expressions are an attempt to obfuscate responses to a simple question: "Is it safe or unsafe?" In fact, it is very difficult to answer the question due to another limitation of risk recognition that no broad consensus on the purpose, the approach, or the result; inadequate data, speculative and myopic nature of assumptions.

Risk management is the process of evaluating and, if necessary, controlling sources of exposure and risk. Sound environment risk management, whether corporate or regulatory, means weighing many different attributes of a decision and developing alternatives. The scientific information provided by risk recognition is but one input to the process. Other criteria include politics, economics, competing risks, and equity and other social concerns. Although risk recognition is important for scientific analysis, the usage of its results to risk management depends on the questions being structured, conducted and answered. Unfortunately, too many risk recognitions prove to be of little value to risk managers because of inadequate planning.

Risk recognition, risk assessment and risk analysis are often used synonymously, as in this chapter, but *risk assessment* sometimes involves the adjustment of a risk system, and *risk analysis* is sometimes used broadly to include risk management aspects as well.

10.3 A Survey of Fuzzy Risk Analysis

Since the publication of the first paper on fuzzy sets by Zadeh in 1965, there were many researchers [3], [6], [9], [10], [12], [14] who use fuzzy methods to study risk issues.

The initial approach to fuzzy risk analysis is based on the premise that one can provide the input of natural language estimate as probability of failure, severity of loss and reliability of the estimate. A typical technique, suggested by Schmucker[24], is the so-called combining model where the main work is to combine the fuzzy risks of subsystems to form the fuzzy risk of the entire system. Let A be a fuzzy event such as "Johe may loses <u>around</u> 3000 US\$", and B be another fuzzy event such as "Mary may loses <u>around</u> 2000 US\$". Here "around" is a fuzzy concept. Suppose we have known the probabilities of A and B occurring, which are $P(A)$ and $P(B)$. More general, the probabilities can be represented by natural language or fuzzy numbers. Then, using the combining model, we can calculate $P(A \cup B)$.

Strictly speaking, the combining model is as same as to calculate the probabilities of fuzzy events. It is well know that, probability of a fuzzy event is defined by the probabilities of the related crisp events. Let $(\Omega, \mathcal{A}, P)$ be a probability space, A be a fuzzy event of Ω with membership function $\mu_A(x)$, $x \in \Omega$, probability of fuzzy event A is defined by

$$P(A) = \int_\Omega \mu_A(x)dP = E(\mu_A(x)). \tag{10.1}$$

In other words, before we use the combining model, we have to know the probability distribution P. Otherwise, the probabilities of fuzzy events must be given by the so-called risk experts. In general, different person gives different assessment, called experience assessment, which cannot be regarded as the science conclusion we need.

Recent years, the premise is unnecessary. A method has been proposed by Machias and Skikos[21] for the computation of the fuzzy risk index of wind site, where the installation of wind energy conversion systems seems to be very possible in the near future. The proposed method is based on a heuristic rules matrix, which is a relation among linguistic terms of wind speed, frequency and risk value. The matrix which is defined by the authors shows the way that the two sets of independent information for the wind speed and its frequency of occurrence will be combined to yield the combined risk level in a classical risk analysis approach. According to these of the extreme wind events and the matrix, one can calculate membership values and risk index results. The major advantage of the method is that, in the cases where the classical one computes the same risk value, however these sites have completely different risk when fuzzy method is applied. The proposed approach presents a better sensitivity in ranking the wind site according to their risk index.

A fuzzy formalism[17] opens the way for computerization of the process of risk analysis, using a variety of AI techniques. Here, risk is defined, with respect to any probability(p)/loss(l) combination, as:

$< \text{RISK}_{p,l} >$
$=< \text{POSSIBLE}_{p,l} > \cap < \text{SIGNIFICANT FINANCIAL IMPACT}_{p,l} >$

where <> denotes a fuzzy set, and ∩ is the symbol for set intersection. Here, intersection of the fuzzy sets possible and significant financial impact is defined by multiplying the two. The components of risk are combined to produce the definition of risk in probability-loss space and the results appear as a three-dimensional membership function. Based on the definition, contour lines, such as those on a topographic map, can be employed to display membership of risk. Considering the determinations of probabilities of loss themselves being imprecise and fuzzy and combining the fuzziness of the definition of risk with the fuzziness inherent in the measurement of probabilities associated with exposure to accidental loss, the author given something called as a fuzzy risk profile. A neural network model suggested to learn the fuzzy risk profiles to help select insurance policy limits for novel exposures.

For dealing with decision-making problems in which the decision maker has a vague and incomplete information about results and external factors, basic decision rules based on fuzzy risk intervals are developed by Delgado, Verdegay and Vila[7]. Here, from the point of view of semantic, a trapezoidal membership function is to be considered as a fuzzy risk interval.

All of these models can be regarded as fuzzy classifiers in risk space. The fuzzy risk index model of Machias and Skikos divides speed-frequency space of wind into risk and no risk zone in some degree. The fuzzy formalism model of Jablonowski defined a fuzzy set in probability-loss space. And, the fuzzy risk interval model of Delgado and et al. just replace classical intervals by fuzzy numbers. A classical interval is a hard classifier and a fuzzy number is a soft classifier.

The success from these models testify that, fuzzy set theory is an ideal method of incorporation inherent imprecision into risk analysis because our intuitions of risk do not translate well into precise mathematical formulas.

These fuzzy models open ways for computerization of the process of risk analysis, using a variety of classical fuzzy techniques. However, they have not touched the essence of risk, to say nothing of both of risk and fuzzy. Obviously, it is necessary to seek a new way for fuzzy risk analysis. For that, we need to first study concept of fuzzy risk which differs from ones as has been indicated above.

10.4 Risk Essence and Fuzzy Risk

As we well know, risk means many things to many people. The Webster dictionary edited by Anon[1] defines risk as "exposure to the chance of injury or loss." In terms of insurance it defines risk as "the hazard or chance of loss." In terms of natural disaster it may define risk as a probability distribution or similar measures that describe the uncertainty about the magnitude, timing, and other relevant features of a disaster.

Traditionally, the major challenge in risk analysis is considered to find a scientific approach to estimate some probability distribution. It is true, if and only if the risks in a system are statistical risks. However, in many risk systems, randomness is just one of risk natures. Risk essence is based on all risk natures. To study the essence, we review the aim of risk analysis and the situation we usually encounter in a practical system.

It may be argued that, the aim of risk analysis is to describe or understand some states of a system serving for risk management to reduce or control loss. The states are relative to adverse outcomes. Therefore, it is necessary for risk analysis to show the relations among states, time, inputs and so on. A probability distribution is just a relation between events and probabilities of occurrence, where an event and a probability value can be regarded as a state and an input, respectively. For many systems, it is impossible to precisely calculate the relation, and we face the problem of imprecise probability. In other words, the aim of risk analysis is to answer how and why an adverse outcome produces. Based on the view, it may be argued that, **risk essence is dynamics of adverse outcomes**. In fact, a risk system could be studied with some state equations if we could find them.

In many cases, it is impossible to obtain the state equations and all data we need, or it may be unnecessary to study the equations over. Probabilistic methods simplify the procedure. However, it isn't reasonable to replace risk analysis by probability analysis.

When we study a risk system using a probabilistic method, usually, it is difficult to judge if a hypothesis of probability distribution is suitable, and sometimes we may meet the problem of small samples, where the data is too scanty to make a decision. The problem has stimulated empirical Bayes methods[4] and kernel methods[2],[5],[8],[13],[22],[25],[29] for further development. It means that it is difficult to obtain a precise relation between events and probabilities of occurrence.

Going a step further, we know that, if we employ other methods to simplify system analysis, it is also difficult to obtain precise relations we need. In other words, the relations we obtained usually are imprecise. To keep the imprecision, the best way is to employ fuzzy sets to represent the relations. Thus, an event may correspond several probability values in different memberships.

Considering D1 and D2 as stated in section 10.1, a more complete definition for risk can be given as: **risk is a phenomenon** with following natures:

(N1) It is adverse for individuals;

(N2) It is uncertainty over timing, site, or magnitude;

(N3) It is too complex to show precisely by a state equation or a probability distribution.

Clearly, due to N3, we know that, risk is a complex phenomenon. When the complex nature is ignored, concept of risk may be reduced to probability-risk, which means that we can find a probability distribution, obeying a statistical law, to show the risk phenomenon properly. As stated above, it is impossible to obtain a precise probability estimator in many cases.

The concept of fuzzy risk is not strange to us. However, most of us may consider the concept as a copy of simple expansion such as fuzzy group, fuzzy ring, fuzzy topology or fuzzy reliability. By this attitude, we can do nothing to promote risk analysis with fuzzy set theory.

Risk, as a natural or societal phenomenon, is neither precise nor fuzzy. Risk is expressed in terms of the probability-risk only when we know a risk phenomenon can be studied by a probability method.

Fuzzy risk is an engineering concept, which emphasizes a showing. For risk, existence differs from the showing. Similarly, as we well know, mass is a pure physical concept, and weight is one in engineering to show mass. Therefore, roughly speaking, **fuzzy risk can be defined as an approximate representation to show risk with fuzzy theory and techniques.** In general, a fuzzy risk is a fuzzy relation between loss events and concerning factors.

There are three main tasks of fuzzy risk analysis:

(1) Improving probability estimation by fuzzy techniques;
(2) Showing the imprecision of risk assessment by fuzzy expressions;
(3) Managing or controlling risk based on fuzzy-risk-assessment.

In this chapter we use diffusion estimate to perform the first task. In Chapter 11, a model is supposed to assess fuzzy risk represented as fuzzy relation. Chapter 12 gives a formula to calculate possibility-probability risk from a given sample and discuss the way of risk management related to fuzzy risk.

One of difficulties to assess probability-risk in real world, in some cases, is that the data to use for assessment is small in amount. When we are very knowledgeable about the physical process of a risk system, we might use stochastic models to fill the gaps and calculate a predicted risk.

For example, in earthquake engineering, the frequency distribution of earthquake magnitudes, especially in the range $(0 < m < 7)$, is reasonably well approximated by the exponential distribution[20]:

$$f(m) = \beta e^{-\beta m} \qquad (10.2)$$

where $f(m)$ is the probability density function of m in a given volume of the earthquake's crust. Parameter β is a regional constant. Depending on the region, the focal depth, and the stress level, the parameter may fluctuate between 0.3 and 1.5.

Current situation is that, in many cases, the researchers have not any knowledge to fix the function shape for represent the relationship between the probabilities and adverse events. Even we can go around the inside block, we have to collect data in a large region or long time for calculating the statistical parameters as β. The predicted risk is no distinction in place or time. The so-called "predicted risk" calculated by using stochastic models is only to tell us some ambiguous information for making decision. When we limit the area or time, the stochastic models no longer have any effect and we will face small-sample problem again.

In this section, a model would be discussed for studying probability-risk based on a small sample. Before that, let us review some classical models for risk assessment in terms of probability estimation from a given sample.

10.5 Some Classical Models

Let l be an adverse record which might be damage or loss. We would change l into adverse index x that is a scale in which time factor and dimensions are removed. Then, the major task of risk assessment is to estimate the probability distribution about the adverse index based on the observations. Different risk issues connect to different definitions of the adverse index. We will give a definition for flood risk in subsection 10.5.3. In general, $x_i \in [0, 1]$. In the following study, unless stated otherwise, it is assumed that we have n records corresponding, $l_1, l_2, \cdots, l_n$, which can be changed into the adverse index observations $x_1, x_2, \cdots, x_n$.

10.5.1 Histogram

The simplest method to estimate the probability of the index is the histogram.

Let $x_0 = 0$ be the origin and h be the bin width. Then we can construct $m = 1/h$ intervals: $[0, h[, [h, 2h[, \cdots, [(m-1)h, 1]$. According to the definition of the histogram estimate given by Definition 3.5 in Chapter 3, we know that the probability estimation of adverse x from the histogram is

$$\hat{p}(x) = \frac{1}{nh}(\text{number of } x_i \text{ in the same interval as } x).$$

Obviously

$$\lim_{\substack{n \to \infty \\ h \to 0}} \hat{p}(x) = p(x) \tag{10.3}$$

That is, the histogram is an useful tool merely for a sample which size is more large.

10.5.2 Maximum likelihood method

The most common method is parametric density estimation according to maximum likelihood method.

The procedure begins with a so-called likelihood function. For a set of independent and identically distributed (i.i.d.) observations $x_1, x_2, \cdots, x_n$ depending on a parameter θ, the likelihood function is defined as

$$L(x|\theta) = \prod_{i=1}^{n} p(x_i|\theta) \tag{10.4}$$

where $p(\cdot|\theta)$ refers to the density of a single observation.

Fisher[11] introduced the concept of estimating θ from the observations by finding $\hat{\theta}$ such that for $\theta \in \Theta$ (the parameter space)

$$L(x|\hat{\theta}) \geq L(x|\theta) \tag{10.5}$$

The likelihood function has no finite maximum over the class of all densities. It is not possible to use maximum likelihood directly for density estimation without placing restrictions on the class of densities over which the likelihood is to be maximized. Generally, we have to suppose that the distribution model is known. In other words, the parameter cannot be estimated by using maximum likelihood method unless one has specific knowledge of the probability distribution of the estimates as well as the parent population.

In the case that the distribution model is known, differentiating with respect to parameter θ and equating the resulting expressions to zero (see subsection 3.4.2), we can calculate parameter θ and define the distribution function.

10.5.3 Kernel estimation

The most popular method in nonparametric approach is the kernel estimation.

The kernel estimator with kernel K is defined[25] as

$$\hat{p}(x) = \frac{1}{nh} \sum_{i=1}^{n} K\left(\frac{x - x_i}{h}\right), \tag{10.6}$$

where h is the window width, also called the smoothing parameter or bandwidth, and the kernel function K satisfies

$$\int_{-\infty}^{\infty} K(x)dx = 1. \tag{10.7}$$

Usually, K will be a symmetric probability density function.

However, the problem of choosing K and h still is of crucial importance in density estimation. In actual fact, h is invalid when the population density $p(x)$ is unknown.

10.6 Model of Risk Assessment by Diffusion Estimate

For a small sample, the risk assessment from a given sample can be done by using information diffusion method relevant in fuzzy information analysis. The target promoting the model is to improve risk assessment under the conditions : (a) There isn't any knowledge about the population distribution; (b) The given sample is small (size $n < 30$).

Definition 10.1 Let $x_i(i = 1, ..., n)$ be adverse observations drawn from a population with density $p(x)$, $x \in U$, U is the universe of discourse of variable x. Suppose $\mu(y)$ is a diffusion function on U, $\Delta_n > 0$ is a constant,

$$\tilde{p}_n(x) = \frac{1}{n\Delta_n} \sum_{i=1}^{n} \mu(x - x_i) \qquad (10.8)$$

is called a *diffusion estimate* (DE) about probability-risk $p(x)$.

Obviously, the 1-dimensional linear-information-distribution (see formula (4.5)) and the normal diffusion function (see formula (7.39)) are the simplest models to give a diffusion estimate.

In fact, the 1-dimensional linear-information-distribution can only produce a discrete DE because the diffusion function is defined on a discrete universe and we need Δ to serve for calculating. In principle, employing the normal diffusion function, we can produce a continuous DE, with simple coefficient (see formula (7.38)).

It is important to notice that, when we study the risk of natural disasters on the adverse index, the universe of discourse of variable x usually is [0,1]. However, a normal diffusion function is defined on $(-\infty, \infty)$. Now it is time to show that the principle of information diffusion has the advantage of the kernel methods in explaining. Recalling the principle described in Chapter 5, we know that the role of a diffusion function is to change observations into fuzzy sets to partly fill the gap caused by incompleteness and improve non-diffusion estimate.

According to Definition 5.5, we know that any function defined on the universe U of a given sample X is a diffusion function if it is decreasing from an information injecting point to other points. Therefore, when the fuzzy sets must be defined on [0,1] (universe of the adverse index), we can also adjust a diffusion function to be defined on the same universe. Particularly, when we use computer to calculate a DE on a discrete universe, the adjustment is very easy.

Let $U = \{u_j | j = 1, \cdots, m\}$ be the discrete universe. x_i can be diffused on U by using the revised formula of normal information diffusion as the following:

$$\mu_i(u_j) = \exp[-\frac{(x_i - u_j)^2}{2h^2}], \quad u_j \in U \qquad (10.9)$$

Let:

$$\begin{cases} C_i' = \sum_{j=1}^{m} \mu_i(u_j), \\ \mu_i'(u_j) = \frac{\mu_i(u_j)}{C_i'}. \end{cases} \tag{10.10}$$

$\mu_i'(u_j)$ is called the *normalized information distribution*. Let:

$$\begin{cases} q(u_j) = \sum_{i=1}^{n} \mu_i'(u_j), \\ C = \sum_{j=1}^{m} q(u_j). \end{cases} \tag{10.11}$$

$q(u_j)$ is called the *information gain* at u_j, C is called the *total information quantity*. Obviously,

$$\tilde{f}(u_j) = \frac{q(u_j)}{C} \tag{10.12}$$

is the frequency of disaster by level (magnitude, intensity or loss) u_j.

Definition 10.2 Let $x_i (i = 1, ..., n)$ be observations with a discrete universe $U = \{u_j | j = 1, \cdots, m\}$. If $\Delta \equiv u_{j+1} - u_j$, information gains are $q(u_j)$, $j = 1, 2, \cdots, m$, and total information quantity is C, then

$$\tilde{p}_{NNE}(u_j) = \frac{q(u_j)}{C\Delta} \tag{10.13}$$

is called a *normalized normal-diffusion estimate* (NNE) about probability-risk $p(x)$.

10.7 Application in Risk Assessment of Flood Disaster

In China, before NNE model, the agriculture premium is calculated on a unit as large as a province. That is, every county or city in one province has the same premium. But in fact, the risk is different. The same premium is unsuitable to the development of economy more and more, and it cannot bring the function of the insurance company into full play. So an insurance company of China wants to study the risk estimation on the unit of county, at the example of Hunan Province.

However, in the project, it is very difficult to collect enough records to support any classical statistical tool for risk assessment, the database contains statistics only a short time ago. Finally, the company has to employ NNE model to assess risk of agriculture disaster caused by flood[16].

In this section, we use an example to show the calculating by NNE model, meanwhile we also can see benefit from NNE. For that, we need to first introduce the definition of the disaster index for agriculture risk.

Definition 10.3 Let l be the seeded area affected by disasters, S be the seeded area, and let

$$x = \frac{l}{S}. \tag{10.14}$$

x is called the *disaster index* of agriculture.

For example, in Changsha County of Hunan Province, China, in 1984, the peasants seeded the fields in $S = 4958 \times 10^3$ mu (1 hectare $=15$ mu). In the same year, 242×10^3 mu was affected by flood, 142×10^3 mu by drought and 169×10^3 mu by wind. The seeded area affected by disasters can be written as

$$l_f = 242 \times 10^3, \quad l_d = 142 \times 10^3, \quad l_w = 169 \times 10^3.$$

According to Definition 10.3, we obtain disaster indexes of flood, drought and wind, respectively, as the following

$$x_f = \frac{l_f}{S} = \frac{242 \times 10^3}{4958 \times 10^3} = 0.049,$$

$$x_d = \frac{l_d}{S} = \frac{142 \times 10^3}{4958 \times 10^3} = 0.029,$$

$$x_w = \frac{l_w}{S} = \frac{169 \times 10^3}{4958 \times 10^3} = 0.034.$$

Now, from a county of Hunan Province, we collected the records represents disaster index of flood which shows the rate of the loss to the whole of the grain, which are in Table 10-2.

Table 10-2 Disaster index materials

Year	1985	1986	1987	1988	1999
Disaster index	0.363	0.387	0.876	0.907	0.632

The work of estimating the loss risk of grain in one year is to estimate the probability distribution about the disaster index.

Obviously, the given sample is

$$X = \{x_1, x_2, x_3, x_4, x_5\} = \{0.363, 0.387, 0.876, 0.907, 0.632\} \tag{10.15}$$

which is incomplete for risk assessment.

10.7.1 Normalized normal-diffusion estimate

Using formula (7.38), we know that the normal diffusion coefficient is

$$h = 0.6481(b - a) = 0.6841(0.907 - 0.363) = 0.372$$

According to the definition of the disaster index, the universe of discourse of the records must be $U = [0, 1]$, and we would study the risk on the following discrete universe:

$$U = \{u_j | j = 1, \cdots, 11\} = \{0, 0.1, 0.2, 0.3, 0.4, 0.5, 0.6, 0.7, 0.8, 0.9, 1\},$$
$$(10.16)$$

where $\Delta = 0.1$.

$\forall x_i \in X$ can be diffused on U according to formula (10.9). The diffused information $\mu_i(u_j)$ is shown by Table 10-3.

Table 10-3 Diffused information $\mu_i(u_j)$ of the given sample X on U

	u_1	u_2	u_3	u_4	u_5	u_6	u_7	u_8	u_9	u_{10}	u_{11}
x_1	0.621	0.779	0.908	0.986	0.995	0.934	0.816	0.663	0.502	0.353	0.231
x_2	0.582	0.743	0.881	0.973	0.999	0.955	0.849	0.702	0.540	0.386	0.257
x_3	0.062	0.114	0.192	0.302	0.441	0.600	0.759	0.894	0.979	0.998	0.946
x_4	0.051	0.095	0.164	0.264	0.395	0.550	0.711	0.857	0.959	1.000	0.969
x_5	0.236	0.360	0.510	0.671	0.823	0.939	0.996	0.983	0.903	0.771	0.613

For example, for $x_2 = 0.387$ and $u_9 = 0.8$, we obtain

$$\mu_2(u_9) = \exp[-\frac{(x_2 - u_9)^2}{2h^2}]$$
$$= \exp[-\frac{(0.387 - 0.8)^2}{2 \times 0.372^2}]$$
$$= \exp[-\frac{0.171}{0.277}]$$
$$= e^{-0.617}$$
$$= 0.540$$

That is, using formula (10.9), from x_2, we diffuse information, in quantity 0.540, to monitoring point u_9.

Obviously, the total of information quantities from a given observation usually is much greater than 1, i.e.,

$$\forall i, \quad \sum_{j=1}^{m} \mu_i(u_j) \gg 1.$$

In other words, the diffusion by (10.9) is not conservative.

Using formula (10.10) to normalize the diffusion with respect to each observation x_i, we can obtain n conservative distributions $\mu_i'(u_j)$ on U. Table 10-4 gives these normalized information distributions.

Table 10-4 Normalized information distributions $\mu_i'(u_j)$

	u_1	u_2	u_3	u_4	u_5	u_6	u_7	u_8	u_9	u_{10}	u_{11}
x_1	0.080	0.100	0.117	0.127	0.128	0.120	0.105	0.085	0.064	0.045	0.030
x_2	0.074	0.094	0.112	0.124	0.127	0.121	0.108	0.089	0.069	0.049	0.033
x_3	0.010	0.018	0.031	0.048	0.070	0.095	0.121	0.142	0.156	0.159	0.150
x_4	0.008	0.016	0.027	0.044	0.066	0.091	0.118	0.142	0.159	0.166	0.161
x_5	0.030	0.046	0.065	0.086	0.105	0.120	0.128	0.126	0.116	0.099	0.079

For example, for $x_2 = 0.387$, we obtain

$$C_2' = \sum_{j=1}^{11} \mu_2(u_j)$$
$$= 0.582 + 0.743 + \cdots + 0.257$$
$$= 7.867.$$

Therefore, the normalized vector is

$$(\mu_2'(u_1), \mu_2'(u_2), \cdots, \mu_2'(u_{11}))$$
$$= (\frac{0.582}{7.867}, \frac{0.743}{7.867}, \cdots, \frac{0.257}{7.867})$$
$$= (0.074, 0.094, 0.112, 0.124, 0.127, 0.121, 0.108, 0.089, 0.069, 0.049, 0.033).$$

For j, summing all normalized information, we obtain the information gain at u_j, which came from the given sample X. The information-gain vector is

$$(q(u_1), q(u_2), \cdots, q(u_{11}))$$
$$= (0.202, 0.274, 0.352, 0.429, 0.496, 0.547, 0.580, 0.584, 0.564, 0.518, 0.453).r$$

For example, for $j = 3$, we have

$$q(u_3) = \sum_{i=1}^{5} \mu_i'(u_3)$$
$$= 0.117 + 0.112 + 0.031 + 0.027 + 0.065$$
$$= 0.352.$$

Summing all information gains, we have

$$C = \sum_{j=1}^{11} q(u_j)$$
$$= 0.202 + 0.274 + \cdots + 0.453$$
$$= 5$$

Because we have used normalizing formula (10.10), we know that the total information quantity, C, must be equal to the size n of the given sample X.

Using formula (10.13) and considering the step length $\Delta = 0.1$, we obtain NNE:

$$(\tilde{p}_{NNE}(u_1), \tilde{p}_{NNE}(u_2), \cdots, \tilde{p}_{NNE}(u_{11}))$$

$$= (\frac{q(u_1)}{C\Delta}, \frac{q(u_2)}{C\Delta}, \cdots, \frac{q(u_{11})}{C\Delta})$$

$$= (\frac{0.202}{5 \times 0.1}, \frac{0.274}{5 \times 0.1}, \cdots, \frac{0.453}{5 \times 0.1})$$

$$= (0.404, 0.548, 0.704, 0.858, 0.992, 1.094, 1.160, 1.168, 1.128, 1.036, 0.906)$$

Comparing with ones calculated by histogram, information distribution, maximum likelihood, and kernel, we can know that information diffusion, approach is more better.

10.7.2 Histogram estimate

Using the asymptotical optimal formula (5.1) to choose the number m of bins of histogram, we obtain

$$m = 1.87(n-1)^{2/5} = 1.87(5-1)^{2/5} \approx 3.$$

Let

$$a = \max\{x_1, x_2, \cdots, x_5\} = 0.363,$$
$$b = \min\{x_1, x_2, \cdots, x_5\} = 0.907.$$

and

$$h = \frac{b-a}{m} = \frac{0.907 - 0.363}{3} = 0.181.$$

If we use the h as bin width to choose intervals for histogram, We obtain $[0.363, 0.544[, [0.544, 0.725[$ and $[0.725, 0.906[$. Any interval dose not include observation $x_4 = 0.907$. We enlarge h a little. We let

$$\Delta = h + 0.001 = 0.182$$

to be the bin width, and obtain three intervals

$$I_1 = [0.363, 0.545[, I_2 = [0.545, 0.727[, I_3 = [0.727, 0.909[,$$

which will be used to construct the histogram. Because

$$x_1 = 0.363 \text{ and } x_2 = 0.387 \in I_1,$$

$$x_5 = 0.632 \in I_2,$$

$$x_3 = 0.876 \text{ and } x_4 = 0.907 \in I_3,$$

according to formula (3.25), we know that the histogram estimate of $p(x)$ is for $x < 0.363$, $\hat{p}(x) = 0$;

for $x \in I_1$, $\hat{p}(x) = \dfrac{2}{5 \times 0.182} = 2.198$;

for $x \in I_2$, $\hat{p}(x) = \dfrac{1}{5 \times 0.182} = 1.099$;

for $x \in I_3$, $\hat{p}(x) = \dfrac{2}{5 \times 0.182} = 2.198$;

for $x > 0.907$, $\hat{p}(x) = 0$.

10.7.3 Soft histogram estimate

The histogram estimate can be improved by using the 1-dimensional linear-information-distribution.

If we only use the center points of above three intervals,

$$a_1 = 0.454, \quad a_2 = 0.636, \quad a_3 = 0.818,$$

to be the controlling points for constructing the framework space and use the 1-dimensional linear-information-distribution to be the diffusion function, the corresponding diffusion must be non-conserved, i.e.,

$$\exists x_i \in X, \ \sum_{j=1}^{3} \mu_i(a_j) < 1,$$

because $x_1 < a_1$, and $x_4 > a_3$. When the sample size, n, is very small, in order to use all information of the given sample, it is better to add two intervals, located at left and right of the histogram intervals, respectively, and employ all center points of these intervals to be controlling points. In our case, the added left and right intervals, respectively, are $[0.181, 0.363[$ and $[0.909, 1.091]$. Their center points are $a_l = 0.272$, $a_r = 1$. Let

$$u_1 = a_l, \ u_2 = a_1, \ u_3 = a_2, \ u_4 = a_3, \ u_5 = a_r,$$

Therefore, we obtain a framework space (monitoring space)

$$U = \{u_1, u_2, \cdots, u_5\} = \{0.272, 0.454, 0.636, 0.818, 1\}. \tag{10.17}$$

Its step length is $\Delta = 0.182$.

Notice: In NNE model, the step length of the monitoring space can be chosen as short as we want. In SHE model, however, the length is controlled by the sample size n. In general, the larger size n of a given sample, the shorter step Δ for SHE model. Therefore, the U in (10.17) differs from one in (10.16).

Using the linear distribution formula (see (4.5))

$$\mu_i(u_j) = \begin{cases} 1 - \frac{|x_i - u_j|}{0.182}, & \text{for } |x_i - u_j| \leq 0.182, \\ 0, & \text{otherwise.} \end{cases}$$

we obtain all distributed information $\mu_i(u_j)$ shown in Table 10-5.

Table 10-5 Distributed information $\mu_i(u_j)$ on u_j from x_i

	u_1	u_2	u_3	u_4	u_5
x_1	0.5	0.5	0	0	0
x_2	0.368	0.632	0	0	0
x_3	0	0	0	0.681	0.319
x_4	0	0	0	0.511	0.489
x_5	0	0.022	0.978	0	0

Then, the primary information distribution of X on U is

$$Q = (Q_1, Q_2, Q_3, Q_4, Q_5) = (0.868, 1.154, 0.978, 1.192, 0.808),$$

where

$$Q_j = \sum_{i=1}^{5} \mu_i(u_j).$$

According to the definition of soft histogram estimate (SHE, see (4.9)), we have

$$\tilde{p}_{SHE}(u_j) = \frac{Q_j}{n\Delta}.$$

Therefore, using the 1-dimensional linear-information-distribution, we obtain a SHE

$$
\begin{aligned}
&(\tilde{p}_{SHE}(u_1), \tilde{p}_{SHE}(u_2), \tilde{p}_{SHE}(u_3), \tilde{p}_{SHE}(u_4), \tilde{p}_{SHE}(u_5)) \\
&= (\frac{0.868}{5 \times 0.182}, \frac{1.154}{5 \times 0.182}, \frac{0.978}{5 \times 0.182}, \frac{1.192}{5 \times 0.182}, \frac{0.808}{5 \times 0.182}) \\
&= (0.954, 1.268, 1.075, 1.310, 0.888).
\end{aligned}
$$

10.7.4 Maximum likelihood estimate

Generally, we use an exponential distribution function $p(x) = \lambda e^{-\lambda x}$ to estimate disaster risk by maximum likelihood principle. With the material in our case,

$$
\begin{aligned}
\lambda &= n/(x_1 + x_2 + \cdots + x_n) \\
&= 5/(0.363 + 0.387 + 0.876 + 0.907 + 0.632) \\
&= 5/3.165 \\
&= 1.58
\end{aligned}
$$

the estimator of probability density is

$$p_e(x) = 1.58e^{-1.58x}, \quad x > 0.$$

For the discrete universe of disaster index, given by Eq. (10.16), we obtain a maximum likelihood estimate (MLE)

$$(\hat{p}_{MLE}(u_1), \hat{p}_{MLE}(u_2), \cdots, \hat{p}_{MLE}(u_{11}))$$
$$= (1.580, 1.349, 1.152, 0.984, 0.840, 0.717, 0.612, 0.523, 0.446, 0.381, 0.325).$$

For $x = 1.1$, we have

$$\hat{p}_{MLE}(x) = 1.58e^{-1.58 \times 1.1} = 0.278 > 0.$$

10.7.5 Gaussian kernel estimate

With Gaussian kernel and optimal window width $h = 1.06\sigma n^{-1/5}$, see Eq.(3.33), for our case,

$$h = 1.06\sigma n^{-1/5}$$
$$= 1.06n^{-1/5}\sqrt{\frac{1}{n-1}\sum_{i=1}^{n}(x_i - \bar{x})^2}$$
$$= 1.06(5^{-1/5})\sqrt{\frac{1}{5-1}\sum_{i=1}^{5}(x_i - 0)^2}$$
$$= 1.06 \times 0.725 \times 0.259$$
$$= 0.199,$$

Gaussian kernel estimate, see Eq. (3.29) and Eq.(3.32), is

$$\hat{p}_{GKE}(x) = \frac{1}{nh}\sum_{i=1}^{n}K(\frac{x - x_i}{h})$$
$$= \frac{1}{nh\sqrt{2\pi}}\sum_{i=1}^{n}\exp[-\frac{(x - x_i)^2}{2h^2}]$$
$$= 0.401\sum_{i=1}^{5}\exp[-12.626(x - x_i)^2],$$

where $-\infty < x < \infty$, and $x_i \in X$ in Eq.(10.15).

For the discrete universe of disaster index, given by Eq. (10.16), we obtain a Gaussian kernel estimate (GKE)

$$(\hat{p}_{GKE}(u_1), \hat{p}_{GKE}(u_2), \cdots, \hat{p}_{GKE}(u_{11}))$$
$$= (0.139, 0.321, 0.585, 0.855, 1.036, 1.096, 1.095, 1.095, 1.083, 0.986, 0.768).$$

For $x = 1.1$, we have

$$\hat{p}_{GKE}(x) = 0.490 > 0.$$

10.7.6 Comparison

Now, we compare normalized normal-diffusion estimate with others.

It is known that the disaster index always be in interval $[0, 1]$, namely, probability of loss over 1 must be 0. But, $\hat{p}_{MLE}(1.1), \hat{p}_{GKE}(1.1) > 0$. Hence, maximum likelihood estimate $\hat{p}_{MLE}(x)$ and Gaussian kernel estimate $\hat{p}_{GKE}(x)$ cannot be used for the risk assessment in the disaster index.

Obviously, histogram estimation is too rough, and it is not good for our case. Therefore, only information estimates are fit the case. Among them, normalized normal-diffusion estimate $\tilde{p}_{NNE}(u_j)$ is better than soft histogram estimate $\tilde{p}_{SHE}(u_j)$. Fig. 10.1 shows that NNE is the best model to assess risk when the given sample is small and the universe is an interval.

10.8 Conclusion and Discussion

In this chapter, we all-around set forth the concept of fuzzy risk with respect to environment, health, and safety. We also analyze the reasons why we have to make fuzzy risk estimation in many cases.

Risk, as a natural or societal phenomenon, is neither precise nor fuzzy. Risk is expressed in terms of fuzzy risk only when we study it by a fuzzy method.

Fuzzy risk, in probability fashion, is a fuzzy relation between events and probabilities. It differs from tradition fuzzy probability which is defined for

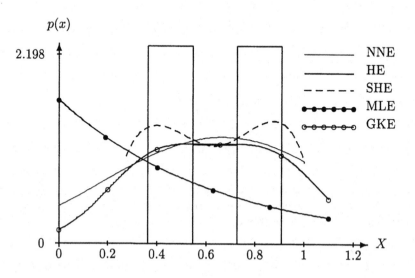

Fig. 10.1 Risk assessment of flood by different models

fuzzy events. The complexity of a risk system leads to that crisp risks agreeing with real situation can not come out of todays knowledge about natural disasters.

We put forward the normalized normal-diffusion to improving probability estimation. From the case calculation, we know that, the proposed estimate $\tilde{p}_{NNE}(u_j)$ is better than maximum likelihood estimate, Gaussian kernel estimate, and histogram estimate.

It is important to note that neither the classical models nor the information diffusion model govern the physical processes in nature. They are introduced as a compensation for their own limitations in the understanding of processes concerned.

Fuzzy risk analysis involves more imprecision, uncertainty and partial truth in a natural and societal phenomenon. The works in fuzzy risks must promote the study of the foundations of fuzzy logic.

References

1. Anon (1989), Webster's Encyclopedic Unabridged Dictionary of the English Language. Gramercy Books, New York
2. Breiman, L., Meisel, W. and Purcell, E.(1977), Variable kernel estimates of multivariate densities. Technometrics **19**, pp.135-144
3. Brown, C.B.(1979), A fuzzy safety measure. J. Engineering Mechanics **105**(5), pp.855-872
4. Carlin B.P. and Louis, T.A.(1986) Bayes and Empirical Bayes Methods for Data Analysis. Chapman & Hall, London
5. Chen, X.R. and et al. (1989), Non-parametric Statistics. Shanghai Science and Technology Press, Shanghai
6. Clement, D.P.(1977), Fuzzy Ratings for Computer Security Evaluation. Ph.D. Dissertation, University of California at Berkeley
7. Delgado, M., Verdegay, J.L. and Vila, M.A.(1994), A model for linguistic partial information in decision-making problems. International Journal of Intelligent Systems **9**, pp.365-378
8. Devroye, Luc and Györfi, László (1985), Nonparametric Density Estimation. John Wiley & Sons, New York
9. Dong, W.M. and et al.(1986), Fuzzy computation in risk and decision analysis. Civil Engineering Systems **2**, pp.201-208
10. Esogbue, A.O., Theologidu, Maria and Guo, Kejiao (1992), On the application of fuzzy sets theory to the optimal flood control problem arising in water resources systems. Fuzzy Sets and Systems **49**(1), pp.155-172
11. Fisher, R.A.(1921), On the mathematical foundations of theoretical statistics. Phil. Trans. Roy. Soc. A **222**, pp.308-368
12. Hadipriono, F.C. and Ross, T.J.(1991), A rule-based fuzzy logic deduction technique for damage assessment of protective structures. Fuzzy Sets and Systems **44**(3), pp.459-468
13. Hand, D.J.(1982), Kernel Discriminate Analysis. Research Studies Press, Chichester, UK
14. Hoffman, L.J., Michelmen, E.H. and Clements, D.P.(1978), SEURAT–Security evaluation and analysis using fuzzy metrics. Proc. of the 1978 National Computer Conference **47** (AFIPS Press, Montvale, New Jersey), pp.531-540
15. Huang, C.F.(1997), Principle of information diffusion. Fuzzy Sets and Systems **91**(1), pp.69-90
16. Huang, C.F. and Liu, X.L.(1998), Risk estimation of natural disaster in Hunan Province by fuzzy mathematical method. in: Da Ruan and et al. Eds., Fuzzy Logic and Intelligent Technologies for nuclear Science and Industry, , Proceedings of 3th International Workshop FLINS'98 (World Scientific, Singapore), pp. 211-217
17. Jablonowski, M.(1994), Fuzzy risk analysis: using AI system. AI Expert **9** (12), pp.34-37
18. Kolluru, R. V., Bartell, S. M., Pitblado, R.M. and Stricoff, R. S. (1996), Risk Assessment and Management Handbook for Environmental, Health, and Safety Professionals. McGraw-Hill, New York
19. Liu, Z.R.(1988), Application of information distribution concept to the estimation of earthquake intensity. in: James C.Bezdek, ed., Analysis of Fuzzy Information, Vol.3 (CRC Press, Boca Raton, Florida), pp.67-73
20. Lomnitz, C. and Rosenblueth, E.(1976), Seismic Risk and Engineering Decisions. Elsevier Scientific Publishing Company, Amsterdam
21. Machias, A.V. and Skikos, G.D.(1992), Fuzzy risk index of wind sites. IEEE Trans. Energy Conversion **7**(4), pp.638-643

22. Parzen, E. (1962), On estimation of a probability density function and mode. Ann. Math. Statist. **33**, pp.1065-1076

23. Sage, A.P. and White, E.B.(1980), Methodologies for risk and hazard assessment: a survey and status report. IEEE Trans. Systems, Man, and Cybernetics **SMC-10**(8), pp.425-446

24. Schmucker, K.J.(1984), Fuzzy Sets, Natural Language Computations, and Risk Analysis. Computer Science Press, Rockvill, Maryland

25. Silverman, B. W. (1986), Density Estimation for Statistics and Data Analysis. Chapman & Hall, London

26. Starr, C.(1977), Risk and Risk Acceptance by Society. Electric Power Res. Inst., Palo Alto, CA

27. Starr, C., Rudman, R., and Whipple, C.(1976), Philosophical basis for risk analysis. Ann. Rev. Energy **1**, pp.629-661

28. Starr, C. and Whipple, C.(1980), Risks of risk decisions. Sci. **208**, pp.1114-1119

29. Wolfgang Wertz (1978), Statistical Density Estimation: a Survey. Vandenhoeck & Ruprecht, Göttingen, Germany

11. System Analytic Model for Natural Disasters

In this chapter, we give a fuzzy system analytic model to assess risk of natural disasters. We suppose that a natural disaster system includes risk source, site, damage and loss. In the model, information distribution technique is used to calculate basic fuzzy relationships showing the historical experience of natural disasters. In section 11.1, we review classical system model for risk assessment of natural disaster and give some definitions to standardize the concepts. In section 11.2, the method of information distribution is employed to calculate the fuzzy relationship between magnitude and probability. Section 11.3 gives fuzzy-system analytic model. In section 11.4, we use the model to study the fuzzy risk of earthquake disaster for a city. The chapter is then summarized with a conclusion in section 11.5.

11.1 Classical System Model for Risk Assessment of Natural Disasters

There are many system models for risk assessment of natural disasters in terms of the classical probability.

In geography, a natural disaster system is regarded as an integration of environment, hazard and object. Particularly, geographers are more interested in what kinds of geographical environment frequently meet natural disasters.

In seismologic, scientists focus on earthquake itself and want to know when, where, how strong, an earthquake will occur. They commonly use probability to measure the uncertainty of the occurrence.

In earthquake engineering, researchers study site intensity of earthquake and develop techniques for buildings to resist earthquakes. Most of them also use probability theory to deal with uncertainty in ground motion attenuation and building damage.

Flood experts pay more attention to hydrographical material and river basin. Some of them study flood frequency of cities with hydrographical material, geography information system and statistics.

When we divide a system of natural disasters into four subsystems: source, site, object and loss, the traditional approaches for risk assessment are almost the same. In this section we review the approach.

11.1.1 Risk assessment of hazard

Let z be a *agent* which may be released by one or several *risk sources*. For example, active faults and rainstorms are risk sources, and constructive earthquake and flood water are agents of active faults and rainstorms, respectively.

Let m be measure of agent z. In many cases, m is a vector including several features. For example, when z is earthquake, m may be the Richter magnitude, or a vector including the Richter magnitude, duration of ground-motion, and so on. If z is flood water, m may be the water level of a dam or a river.

Definition 11.1 If z is a risk agent of a risk source, and m is measure of the agent, then m is called *magnitude* of z.

Example 11.1 The San Andreas Fault is a risk source. Earthquake is a risk agent of the source. The Richter magnitude is a magnitude of the agent. The epicentral intensity, I_0, is also a magnitude of the same agent.

In general, the study of risk source, agent and magnitude is called the *hazard study*.

Traditionally, risk assessment of sources is also called *hazard analysis*, to calculate the probability distribution of exceeding magnitude m, $P_z = \{p(\xi \geq m)|m \in M\}$, where M is the universe (domain) of the magnitude m.

The following five methods are frequently used for the calculation.

(M1) *Random process* — considering disaster systems as dynamic systems driven by some random noise, e.g., seismographers employ Markov process to predict earthquake by probability;

(M2) *Maximum likelihood* — presuming that the type of the population from which observations are taken has been known, e.g. earthquake engineers approximately suppose that the probability distribution of earthquake magnitudes is exponential;

(M3) *Bayesian statistics* — a prior probability distribution may be modified by each observation into a new probability distribution, e.g., because of shortcomings in the input data, in southern Portugal, Bayesian statistics was used for evaluation of spatial correlation to undertake a seismic risk analysis;

(M4) *Frequency estimate* — the relative frequency approach to defining the probability of an event E is to perform an experiment n times. The number of times that E appears is denoted by n_E. Then, the probability of E is given by

$$P[E] = \lim_{n \to \infty} = \frac{n_E}{n}. \tag{11.1}$$

The histogram is also a frequency estimate. The design of highway and railroad stream crossings, delineation of flood plains and flood-prone areas, management of water-control structures, and water supply management are all activities that require frequency distribution of floods;

(M5) *Monte Carlo* — solution from the random variable generated by a computer, e.g., the software "@Risk 4.0" produced by Palisade Corporation permits the user to designate probability distributions and the relevant parameters according to which random draws will be made.

In many cases, more than one method is used. From the earthquake hazard analysis we can see it.

No other natural hazard threatens lives and investments as much as earthquakes. In the magnitude 6.8 January 1994 Los Angeles earthquake alone, the damage may have exceeded $ 25 billion, and about four times more than that again in the magnitude 6.8 January 1995 Kobe earthquake in Japan. Geologists have found that in a given year, the number of earthquakes $n(m)$, that exceed a particular magnitude, m, is approximated by the relation,

$$log_{10}n(m) = a - bm, \tag{11.2}$$

where a is the total number of earthquakes in the region of interest and b is a constant that characterizes the seismicity of a particular region. Eq. (11.2) is called *Gutenberg-Richter relation*. The value of b seems to vary from area to area, but worldwide it seems to be around $b = 1$. The relationship implies that the probability distribution of earthquake magnitudes is[2]

$$F_M(m) = 1 - \exp[-\beta(m - m_0)], \quad m \geq m_0 \tag{11.3}$$

where $\beta = b \ln 10$ and m_0 is some magnitude small enough, say 4, that events of lesser magnitude may be ignored by engineers. This restriction to larger events implies that the probabilities above are conditional on the occurrence of an event of interest, that is, one where $m \geq m_0$. The parameter b is typically such that β is about 1.5 to 2.3. That is, the probability density function of m, $f(m)$, is reasonably well approximated by the exponential distribution (10.2). It is interesting to note that the ranges of β value for Eq.(10.2) and (11.3) are different. They are come from different researchers with respect to different domain of M.

It may be argued that, risk assessment of source in earthquake can be regarded as an integration of Bayesian statistics (Gutenberg-Richter relationship is an prior distribution), frequency estimate (when $n \to \infty$, we obtain probability estimate) and maximum likelihood (exponential assumption is reasonable).

Scientist even want to get more theoretic risk models supported by geophysics.

It has been almost a half of century since Cornell (1968, see [2]) proposed a probabilistic approach for engineering seismic risk. Then Utsu (1972, see [12][13]), Rikitake (1974, see [11]) and Hagiwara (1974, see[4]) proposed a probabilistic approach for forecasting the time of the next earthquake on a specific fault segment. Their proposals were based on a model of earthquake occurrence that assumed that the probability of an earthquake was initially low following a segment-rupturing earthquake, and increased gradually as tectonic processes reloaded the fault.

The conceptual model proposed by Utsu, Rikitake and Hagiwara considers earthquakes as a renewal process, in which the elastic strain energy accumulates over a long period of time after the occurrence of one earthquake before the fault is prepared to release in the next earthquake. To forecast the likelihood that a particular earthquake will occur at some time in the future under these restrictions requires the specification of a probability distribution of failure times. A number of candidate statistical models have been proposed for the computation of conditional probabilities of future earthquakes, for example, Utsu in 1972 (see [13]) suggested to use the Double Exponential. In 1984, see [14], Utsu also suggested to use the Gamma distributions. Other examples are that Rikitake in 1974 (see [11]) suggested to us the Gaussian, Hagiwara in 1974 (see [4]) to Weibull, Nishenko and Buland in 1987 (see [10]) to Log-normal.

The traditional models have been widely used, although none of them has any particular claim as a proper model for earthquake recurrence. They have been used in the past chiefly because they are mathematically well-developed functions with well-known statistical properties; in other words, the usual suspects. Because the predictions obtained from these specific models differ significantly from one another, particularly at times removed from the mean failure time, it is important to consider alternatives to these familiar (and possibly inappropriate) probability models[3].

Obviously, before we discover the physical model of the earthquake cycle that would be accepted commonly, we need a new approach that is based on a simple mathematical model. This model has, at least, a desirable property that make it a suitable candidate for describing the fuzziness of which our knowledge about earthquake recurrence is too incomplete to calculate earthquake risk with the classical probability tools. The fuzzy set theory seems a perfect approach. The method of information distribution might be the simplest model to show the fuzziness due to the incompleteness from a given sample recorded earthquake events. In addition, the new model must be quality to emerge the result from hazard analysis with attenuation, vulnerability and economic loss.

11.1.2 From magnitude to site intensity

To calculate loss risk, it is necessary to study all related objects located a variety of sites.

Definition 11.2 A particular locale is called a *site*. The effect of one or several risk sources on a site is called *site intensity*, denoted as w.

The effects of any one risk source vary greatly from place to place, so there may be many intensity values measured from one risk source.

Example 11.2 The 5th Avenue of New York City is a site and the ozone level of the avenue is the site intensity. Ozone, a major component of smog, is created in the presence of sunlight by reactions of chemicals found in gasoline vapors and emissions from cars and industrial smoke stacks. As ozone levels increase, the severity of the effects increases, so as the number of people affected. Ozone levels in this range can cause a variety of respiratory problems, including coughing, throat irritation, shortness of breath, decreased lung function, increased susceptibility to respiratory infection, aggravation of asthma, and other respiratory ailments.

There are many methods to infer site intensity from risk sources. In general, we would consider the following three factors:

(1) Distance from the sources to the site;

(2) Intermediary, e.g., the rock and soil are the intermediaries of seismic wave;

(3) Environment, e.g., the site intensity of flood is related to geographical environment, the lower the site, the higher the intensity.

The study of the site intensity can be called the *attenuation study*. In earthquake engineering, the simplest attenuation model is to linearly regress the relationship of site intensity with magnitude and distance shown as Eq.(9.1). In environment risk analysis, *exposure assessment* is similar to attenuation study.

In real cases, risk sources are just potential and possible, and their boundaries may be fuzzy. Hence, it is difficult to determine the distance from the risk sources to the site. Aside the boundaries, it is beyond belief to get a real attenuation relationship with a few of records.

Traditionally, if the attenuation relationship as a function is known, as

$$w = w(m, d), \tag{11.4}$$

where w, m and d are site intensity, magnitude and distance, respectively, we can calculate the probability distribution of w from the probability distribution of m by using the theory of *functions of random variables*. In our case, if $w(m)$ is a given function of m with PDF $f_M(m)$, then the probability distribution of w is

$$G(w) = P(W \leq w) = \int_{w(m) \leq w} f_M(m) dm. \tag{11.5}$$

Example 11.3 Consider the linear function

$$w = am + b, \tag{11.6}$$

where m is any random variable, and a and b are any constants with $a \neq 0$. Assume first that $a > 0$. Then

$$G_W(w) = P(W \le w)$$

$$= \int_{am+b \le w} f_M(m)dm$$

$$= \int_{m \le (w-b)/a} f_M(m)dm$$

$$= \int_{-\infty}^{(w-b)/a} f_M(m)dm$$

$$= F_M(\frac{w-b}{a}). \qquad (11.7)$$

Its PDF is obtained by differentiation:

$$g_W(w) = \frac{d}{dw}G_W(w) = \frac{d}{dw}F_M(\frac{w-b}{a}) = \frac{1}{a}f_M(\frac{w-b}{a}). \qquad (11.8)$$

If a has been assumed negative, the division by a in the early stages would have reversed that direction of the inequality; the result would be the same as for $a > 0$, except that the multiplier would be $-1/a$ in replace of $1/a$. The two cases can be summarized in one formula:

$$g_{am+b}(w) = \frac{1}{|a|}f_M(\frac{w-b}{a}). \qquad (11.9)$$

That is, if the attenuation relationship $w = w(m, d)$ and the probability risk $P_z = \{p(\xi \ge m)|m \in M\}$ of the source are known, according to the theory of functions of random variables, we can, by formula (11.5), calculate probability risk $P_W(w) = \{p(\zeta \ge w)|w \in W\}$ of the site intensity.

11.1.3 Damage risk

In probability risk analysis, supported by the theory of functions of random variables, the task of damage risk reduces to calculate damage from site intensity. In the damage subsystem, two terms, *vulnerability* and *dose-response*, are frequently used.

Vulnerability means a susceptibility to physical injury. A dose-response is a mathematical model that describes how a response variable depends on the level of site intensity. The dose-reponse model tells whether the level of response increases or decreases with dose and how rapidly it changes as a function of dose. In fact, the contents of the two terms are broader than damage analysis. For risk analysis of natural disasters, we define the damage as physical injury of objects such as buildings, bridges, highways, even a group of people or others.

Definition 11.3 Let o be an object which will be struck by a destructive force in intensity w. Let $Y = \{y\}$ be the universe of discourse of damage of o. y is called the *damage degree* of o reposing to w.

For convenience sake the damage degree is called *damage*.
In the dose-response model, we denote y as

$$y_o = y(w, c), \tag{11.10}$$

where y, o, w, c are damage, object, site intensity, and characteristic vector of the object, respectively.

Example 11.4 In an earthquake disaster system, when we regard a small city C as an object, then, ground motion acceleration is a site intensity, the characteristic vector may involve the proportion of the buildings to architecture material (reinforced concrete or brick), and the rate of destruction would be considered as damage of the city.

Most dose-response models are derived from statistical data such as that from monitoring or testing. Examples are the linear dose-response models used to estimate human health effects and materials damage of buildings. The dose-response model for a kind of object is a favorite with most engineers. Alternatively, dose-response models may be derived from theoretical considerations with little or no basis in empirical data. In other words, it is very difficult to get mathematical function $y(w, c)$. It is better to present the relationships by fuzzy relations.

If the dose-response relationship $y_o = y(w, c)$ and the probability risk $g_w(w)$ of the site intensity are known, according to the theory of functions of random variables, we can, by formula

$$H(y) = P(Y \leq y) = \int_{y(w) \leq y} g_w(w) dw, \tag{11.11}$$

calculate probability risk $P_Y(y) = \{p(\delta \geq y) | y \in Y\}$ of the damage.

11.1.4 Loss risk

Damage emphasizes the physical injury, but loss relates with cost.

Definition 11.4 Let $L = \{l\}$ be the universe of discourse of loss of o due to damage y, we call l *loss*.

Example 11.5 There is a house o. Before a flood, it costs \$ 250,000, and the property in the house costs \$ 120,000. The flood totally destroyed the house and only 40% property was rescued. The loss is

$$l_o = 250,000 + (120,000 - 120,000 \times 0.4) = 322,000.$$

Even there are a lot of data; it is more difficult to get the relationship between the damage and loss because the changing of the market price of objects and uncertainty of the property in the objects. For example, five years ago, when farmer Mr. Liu built his house near a river, he didn't have any farming machine. Now he has some machines. His property in the house (an object) is increasing. The increasing is uncertainty.

Obviously, the loss does not only depend on the damage of the object and the property but also relates to rescue action during the disaster. To simplify, in general, we don't consider the result from rescue. More simply, we might regard the property as a part of the object.

In the case, the loss is directly a function of the damage, denoted

$$l_o = l(y), \tag{11.12}$$

where l, o and y are loss, object and damage, respectively.

In principle, if we know:
(1) probability of the source;
(2) attenuation relationship;
(3) dose-response relationship;
(4) loss function,

then we can calculate loss risk, denoted as $P_L(l) = \{p(\tau \geq l)|l \in L\}$.

However, almost all subsystems must meet uncertainty, for example, same site intensity often cause different damage to same kind of buildings.

We suggest a fuzzy model to deal with the uncertainties.

11.2 Fuzzy Model for Hazard Analysis

Obviously, the uncertainty with respect to risk is mainly due to the uncertainty of the agent. Nobody can precisely predict when, where, how strong the agent of a disaster will occur. Traditionally, the major challenge in risk analysis of natural disasters is considered to find a scientific approach to estimate the probability distribution of the magnitude.

In Chapter 10, the probability risk is in terms of point probability, where we focus on each point used to measure disaster event. In the system model, we need to introduce a more engineering concept, exceed-probability risk.

Definition 11.5 Let $U = \{u\}$ be the universe of discourse of natural disaster, and $P = \{p(\xi \geq u)|u \in U\}$ be the probability distribution of exceeding magnitude u. P is called *exceed-probability risk*.

For example, suppose that you want to gamble on horse races by $ 1000. If:

the probability that you will lose more than $ 100 is 0.9;
the probability that you will lose more than $ 200 is 0.8;
... ...
the probability that you will lose more than $ 900 is 0.1;
the probability that you will lose more than $ 1000 is 0,

then,

the exceed-probability risk for your gambling on horse races is

$$P = \{p(\xi \geq 100), p(\xi \geq 200), \cdots, p(\xi \geq 900), p(\xi \geq 1000)\}$$
$$= \{0.9, 0.8, \cdots, 0.1, 0\}.$$

Definition 11.6 Let $U = \{u\}$ be the universe of discourse of natural disaster, and $\pi(u, p)$ be the possibility of that disaster magnitude exceeds y in probability p. $\Pi = \{\pi(u, p) | u \in U, p \in [0, 1]\}$ is called *exceed-probability fuzzy-risk*.

For example, suppose that you want to gamble on the stock exchange by $ 1000. It is difficult to calculate the risk.

You may know:

the probability that you will lose more than $ 100 is about 0.9;

the probability that you will lose more than $ 200 is about 0.8;

... ...

the probability that you will lose more than $ 900 is about 0.1;

the probability that you will lose more than $ 1000 is 0.

In your fuzzy term, we suppose that "about p" means fuzzy set as the following:

$$A_{\text{about } p} = \frac{0.8}{p - 0.05} + \frac{1}{p} + \frac{0.8}{a + 0.05},$$

e.g., "about 0.9" means

$$A_{\text{about } 0.9} = \frac{0.8}{0.85} + \frac{1}{9} + \frac{0.8}{0.95}.$$

In this case, the exceed-probability fuzzy-risk for your gambling on the stock exchange is

$$\Pi = \{\pi(100, 0.85), \pi(100, 0.9), \pi(100, 0.95), \pi(200, 0.75), \cdots, \pi(1000, 0)\}$$
$$= \{0.8, 1, 0.8, 0.8, \cdots, 1\}$$

The definition of possibility is given in subsection 5.5.1. For convenience sake, $\pi(u, p)$ is also called *fuzzy risk*, meanwhile, $p(u)$ means $p(\xi \geq u)$, and $p(u)$ is also called probability risk.

Suppose probability risk $p(u)$ is known, it can be turned to fuzzy risk $\pi(u, p)$. In fact,

$$\pi(u, p) = \begin{cases} 1, & \text{when } p = p(u), \\ 0, & \text{others.} \end{cases} \qquad (11.13)$$

That is to say, exceed-probability risk is a special case of the exceed-probability fuzzy-risk.

In general, when we analyse a system of natural disasters, we consider all kinds of natural disasters which may be met in the future. However, the core problem of system analyse for natural disasters is studying a single disaster such as earthquakes, floods, droughts, winds and so on. Combining different disaster risks to a comprehensive disaster risk is relative easy and the paper [6] provides a model to do that.

Definition 11.7 Let $M = \{m\}$ be the magnitude universe of the agent z released by risk sources, and $\pi_z(m,p)$ be the possibility of that the agent exceeds m in probability p. We call

$$\Pi_z = \{\pi_z(m,p)|m \in M, p \in [0,1]\} \tag{11.14}$$

exceed-probability fuzzy-risk of the sources.

Generally, in engineering, $\pi_z(m,p)$ must be presented as a discrete form. We let that the discrete universe of M is

$$M = \{m_j|j = 1, \cdots, J\} \tag{11.15}$$

and the discrete universe of probability p is

$$P = \{p_k|k = 1, 2, \cdots, K\} \tag{11.16}$$

In this case, exceed-probability fuzzy-risk of release assessment is written as $\pi_z(m_j, p_k)$.

When we have a sample from the records in history such as earthquake magnitude, the fuzzy risk can be obtained by using information distribution technique. The model based on subsections 11.2.1-11.2.3 is called the *two-times distribution model* by which we roughly calculate fuzzy risk. In Chapter 12 we will give another model to, in detail, calculate fuzzy risk, called the *interior-outer-set model*.

11.2.1 Calculating primary information distribution

Suppose that the given sample is

$$X = \{x_1, x_2, \cdots, x_n\}, \tag{11.17}$$

then, every x_i can be distributed with information gain q_{ij} into controlling point m_j of Eq.(11.15) by

$$q_{ij} = \begin{cases} 1 - \frac{|x_i - m_j|}{\Delta}, & \text{If } |x_i - m_j| \leq \Delta \\ 0, & \text{otherwise} \end{cases} \tag{11.18}$$

where step length $\Delta \equiv m_{j+1} - m_j$, $j = 1, 2, \cdots, J - 1$.

After all observations have been treated with this simple process and information gains at each controlling point have been summed up, a primary distribution of information gains will turn out. That is

$$Q = \{Q_1, Q_2, \cdots, Q_J\}, \tag{11.19}$$

where $Q_j = \sum_{i=1}^n q_{ij}$.

In fact, Q_j actually means that there are Q_j disasters occurred in the years studied and whose magnitude is around m_j. The concept "around" is described by a triangular shaped fuzzy set

$$\mu_{m_j}(x) = \begin{cases} 0, & \text{if } m < m_{j-1}; \\ \frac{x-m_{j-1}}{m_j-m_{j-1}}, & \text{if } m_{j-1} \leq x \leq m_j; \\ \frac{x-m_{j+1}}{m_j-m_{j+1}}, & \text{if } m_j \leq x \leq m_{j+1}; \\ 0, & \text{if } m_{j+1} < x. \end{cases} \qquad (11.20)$$

11.2.2 Calculating exceeding frequency distribution

If we suppose that any disaster event must be measured by one of $m_1, m_2, \cdots,$ m_J, we know that Q_j can be explain as the number of disasters with magnitude m_j. Using Q, the number of disasters with magnitude greater than or equal to m_j can be obtained as:

$$N_j = \sum_{i=j}^{J} Q_i. \qquad (11.21)$$

They constitute a number distribution of exceeding magnitude as:

$$N = \{N_1, N_2, \cdots, N_J\}. \qquad (11.22)$$

Obviously, the frequency exceeding m_j is $f_j = \frac{N_j}{n}$. We obtain an *exceeding frequency distribution* as

$$\begin{aligned} F &= \{f_1, f_2, \cdots, f_J\} \\ &= \{f_1(M \geq m_1), f_2(M \geq m_2), \cdots, f_J(M \geq m_J)\} \\ &= \{\frac{N_1}{n}, \frac{N_2}{n}, \cdots, \frac{N_J}{n}\}. \end{aligned} \qquad (11.23)$$

When n is small, it must be unreliable to regard f_j as the exceeding probability, the reason is that knowledge provided by the observations is incomplete. In other words, when n is small, F is only a rough relation between magnitude and probability. Using two dimensions information distribution method, F can be represented by a fuzzy relation.

11.2.3 Calculating fuzzy relationship between magnitude and probability

From Eq.(11.23), we obtain J observations which makes a new sample

$$X_J = \{(m_1, f_1), (m_1, f_2), \cdots, (m_J, f_J)\}. \qquad (11.24)$$

We use the simplest formula (11.25) to distribute (m_j, f_j) to discrete point (m_i, p_k), namely,

$$\tilde{f}_j(m_i, p_k)$$
$$= \begin{cases} (1 - \frac{|m_j - m_i|}{\Delta_1})(1 - \frac{|f_j - p_k|}{\Delta_2}), & \text{when } |m_j - m_i| \le \Delta_1 \text{ and } |f_j - p_k| \le \Delta_2 \\ 0, & \text{others} \end{cases}$$

$$\text{(11.25)}$$

where $\Delta_1 \equiv m_{i+1} - m_i$, $i = 1, 2, \cdots, I - 1$, and $\Delta_2 \equiv p_{k+1} - p_k$, $k = 1, 2, \cdots, K - 1$. Let

$$\tilde{f}(m_i, p_k) = \sum_{j=1}^{J} \tilde{f}_j(m_i, p_k), \tag{11.26}$$

and

$$g_i = \max\{\tilde{f}(m_i, p_k)|k = 1, 2, \cdots, K\}. \tag{11.27}$$

If $g_i = 0$, let $g_i = 1$. Then,

$$\pi_z(m_i, p_k) = \frac{\tilde{f}(m_i, p_k)}{g_i} \tag{11.28}$$

is a fuzzy relationship between magnitude, M, and probability, P. The $\pi_z(m_i, p_k)$ is called the fuzzy risk of agent z calculated by *two-times distribution*. We have used two times information distribution formula to get the fuzzy risk from X in (11.17). One time is the 1-dimensional linear-information-distribution, i.e., formula (11.18), another is the 2-dimensional linear-information-distribution, i.e., formula (11.25).

11.3 Fuzzy Systems Analytic Model

In this section, we give a system model to transform the uncertainty in a risk source to site intensity, damage and loss.

11.3.1 Fuzzy attenuation relationship

Definition 11.8 Let $W = \{w\}$ be the universe of discourse of intensity of site s, and $\pi_s(w, p)$ be the possibility of that probability value of exceeding w is p. We call

$$\Pi_s = \{\pi_s(w, p)|w \in W, p \in [0, 1]\} \tag{11.29}$$

fuzzy risk of site s.

Assume that the attenuation relationship, from the source to the site, is as same as Eq.(11.4), i.e., we can get a function $w = w(m, d)$. According to the definition of fuzzy relationship, the function can be changed to be a special fuzzy relationship on $M \times D \times W$, where M, D and W is the universe of discourse of m, d and w, respectively. Let

$$r^{(1)}(m, d, w) = \begin{cases} 1, & \text{when } w = w(m, d), \\ 0, & \text{others.} \end{cases} \tag{11.30}$$

Then,

$$R_1 = R_{(M,D) \to W} = \{r^{(1)}(m,d,w)\} \tag{11.31}$$

is the fuzzy relationship resulted from the given attenuation function. As we stated above, it is difficult to determine the distance from the risk sources to the site. Meanwhile, there is a common phenomenon: two sites with same distance to a source may have different intensity. In other words, in many cases, there does not exist a crisp relation between the site intensity and the distance. Therefore, it is better to, directly from the given sample, calculate R_1 by some fuzzy technique such as information diffusion, or, it can be provided by the experts.

In our case, we suppose that fuzzy attenuation relationship R_1 is given. Only thing we have to do is to calculate the site fuzzy intensity $\underset{\sim}{W}$ by

$$\mu_W(w) = \sup_{m \in M, d \in D} \{r^{(1)}(m,d,w) \wedge \mu_{M,D}(m,d)\} \tag{11.32}$$

where $\mu_{M,D}(m,d)$ is the membership function of fuzzy magnitude and distance. It implies that, as input, one of magnitude and distance, or both of them are fuzzy variables.

The simplest case is as the following:

(a) We know that the nearest distance from the site to the source is d_0 and the farthest distance is d_μ;

(b) We have obtained fuzzy risk of magnitude: $\mu_M(m) = \pi_z(m,p)$.

In the case, it is easy to represent fuzzy input $\mu_{M,D}(m,d)$.

From (a) the fuzzy distance $\underset{\sim}{D}$ can be expressed simply by using a bell function:

$$\mu_D(d) = \exp[-\frac{(\frac{d_\mu+d_0}{2} - d)^2}{\frac{(d_\mu-d_0)^2}{6}}]$$
$$= \exp[-1.5(\frac{d_\mu + d_0 - 2d}{d_\mu - d_0})^2]. \tag{11.33}$$

Hence we have

$$\mu_{M,D}(m,d) = \pi_z(m,p) \wedge \mu_D(d). \tag{11.34}$$

Therefore, we can get the fuzzy risk of the site intensity as the following:

$$\pi_s(w,p) = \sup_{m \in M, d \in D} \{r^{(1)}(m,d,w) \wedge \pi_z(m,p) \wedge \mu_D(d)\} \tag{11.35}$$

11.3.2 Fuzzy dose-response relationship

Definition 11.9 Let $Y = \{y\}$ be the universe of discourse of damage degree of object o, and $\pi_o(y, p)$ be the possibility of that probability value of exceeding y is p. We call

$$\Pi_o = \{\pi_o(y, p)|y \in Y, p \in [0, 1]\} \qquad (11.36)$$

fuzzy risk of damage of object o.

Assume that the relationship between the dose and an object's response can be expressed as:

$$y = y(w) \qquad (11.37)$$

where w is the site intensity, y is the measure of damage. The relationship can be improved by using a fuzzy relationship between W and Y

$$R_2 = R_{W \to Y} = \{r^{(2)}(w, y)\} \qquad (11.38)$$

which can be obtained from Eq.(11.37) or by the experts, where W and Y is the universe of discourse of w and y, respectively.

The fuzzy response $\underset{\sim}{Y}$ can be got by using:

$$\mu_Y(y) = \sup_{w \in W} \{r^{(2)}(w, y) \wedge \mu_W(w)\} \qquad (11.39)$$

where $\mu_W(w)$ is the membership function of site fuzzy intensity which may be in form as Eq.(11.32).

If $\underset{\sim}{W}$ is an estimator of fuzzy risk, calculated by the formula (11.35), namely,

$$\mu_W(w) = \pi_s(w, p) = \sup_{m \in M, d \in D} \{r^{(1)}(m, d, w) \wedge \pi_z(m, p) \wedge \mu_D(d)\},$$

we can get the fuzzy risk of the damage as the following:

$$\pi_o(y, p) = \sup_{w \in W} \left\{ r^{(2)}(w, y) \wedge \pi_s(w, p) \right\}. \qquad (11.40)$$

11.3.3 Fuzzy loss risk

Definition 11.10 Let $L = \{l\}$ be the universe of discourse of loss of city C, and $\pi_C(l, p)$ be the possibility of that probability value of exceeding l is p. We call

$$\Pi_C = \{\pi_C(l, p)|l \in L, p \in [0, 1]\} \qquad (11.41)$$

fuzzy risk of loss of city C.

We strictly discuss the losses in buildings. Suppose that the loss of a building is in direct proportion to its area and damage index. Generally, damage index of building is defined as the damage percentage. The universe of discourse of damage index is [0,1].

There is some fuzzy relationship between damage degree and damage index. For example, in China, for earthquake disaster, damages are defined as

$$
\begin{aligned}
y_1 &= \text{``Good condition''} \\
y_2 &= \text{``Light destruction''} \\
y_3 &= \text{``General destruction''} \\
y_4 &= \text{``Heavy destruction''} \\
y_5 &= \text{``Collapse''}
\end{aligned}
\tag{11.42}
$$

When we take

$$
A = \{a_1, a_2, \cdots a_{11}\} = \{0, 0.1, 0.2, \cdots, 1\}
\tag{11.43}
$$

be the discrete universe of discourse of damage index, using historical data, we can get a fuzzy relationship as the following[7]:

$$
R_{Y \to A} =
\begin{array}{c}
\\
y_1 \\
y_2 \\
y_3 \\
y_4 \\
y_5
\end{array}
\begin{pmatrix}
a_1 & a_2 & a_3 & a_4 & a_5 & a_6 & a_7 & a_8 & a_9 & a_{10} & a_{11} \\
1 & 0.7 & 0.2 & 0 & 0 & 0 & 0 & 0 & 0 & 0 & 0 \\
0.2 & 0.7 & 1 & 0.7 & 0.2 & 0 & 0 & 0 & 0 & 0 & 0 \\
0 & 0 & 0.2 & 0.7 & 1 & 0.7 & 0.2 & 0 & 0 & 0 & 0 \\
0 & 0 & 0 & 0 & 0.2 & 0.7 & 1 & 0.7 & 0.2 & 0 & 0 \\
0 & 0 & 0 & 0 & 0 & 0 & 0.2 & 0.7 & 1 & 0.7 & 0.2
\end{pmatrix}
\tag{11.44}
$$

Moreover, let us presume that every square meter is worth e dollars in city C. If the area of all buildings in city C totalled S square meters, the buildings in city C is worth $E = e \cdot S$ dollars. Corresponding with the universe A of discourse of damage index in (11.43), we can obtain the universe of discourse loss of the city C as:

$$
L_C = \{l_1, l_2, \cdots, l_{11}\} = \{E \cdot a_1, E \cdot a_2, \cdots, E \cdot a_{11}\}
\tag{11.45}
$$

By using (11.44) and (11.45), it is easy to obtain the fuzzy relationship between fuzz damage (degree) and loss as the following:

$$
R_{Y \to L} =
\begin{array}{c}
\\
y_1 \\
y_2 \\
y_3 \\
y_4 \\
y_5
\end{array}
\begin{pmatrix}
l_1 & l_2 & l_3 & l_4 & l_5 & l_6 & l_7 & l_8 & l_9 & l_{10} & l_{11} \\
1 & 0.7 & 0.2 & 0 & 0 & 0 & 0 & 0 & 0 & 0 & 0 \\
0.2 & 0.7 & 1 & 0.7 & 0.2 & 0 & 0 & 0 & 0 & 0 & 0 \\
0 & 0 & 0.2 & 0.7 & 1 & 0.7 & 0.2 & 0 & 0 & 0 & 0 \\
0 & 0 & 0 & 0 & 0.2 & 0.7 & 1 & 0.7 & 0.2 & 0 & 0 \\
0 & 0 & 0 & 0 & 0 & 0 & 0.2 & 0.7 & 1 & 0.7 & 0.2
\end{pmatrix}
\tag{11.46}
$$

Because that, if index is a_i, the loss must be $E.a_i$, i.e., l_i, therefore, the relationship between damage degree and loss is just the relationship between damage degree and damage index.

In general, the relationship between damage and loss can be denoted as:

$$R_3 = R_{Y \to L} = \{r^{(3)}(y,l)\} \tag{11.47}$$

The fuzzy loss $\underset{\sim}{L}$ can be got by using:

$$\mu_L(l) = \sup_{y \in Y}\{r^{(3)}(y,l) \wedge \mu_Y(y)\} \tag{11.48}$$

where $\mu_Y(y)$ is the membership function of fuzzy damage which may be in form as Eq.(11.39).

If $\underset{\sim}{Y}$ is an estimator of fuzzy risk, calculated by the formula (11.40), namely,

$$\mu_Y(y) = \pi_o(y,p) = \sup_{w \in W}\left\{r^{(2)}(w,y) \wedge \pi_s(w,p)\right\},$$

we can get fuzzy risk of loss of city C as the following:

$$\begin{aligned} \pi_C(l,p) &= \sup_{y \in Y}\{r^{(3)}(y,l) \wedge \pi_o(y,p)\} \\ &= \sup_{y \in Y}\left\{r^{(3)}(y,l) \wedge \sup_{w \in W}\{r^{(2)}(w,y) \wedge \pi_s(w,p)\}\right\} \end{aligned} \tag{11.49}$$

11.4 Application in Risk Assessment of Earthquake Disaster

In this section, we give an example of earthquake engineering for showing how to use above fuzzy model. Earthquake engineering is concerned with the design and construction of all kinds of civil and building engineering systems to withstand earthquake shaking. Earthquake engineers, in the course of their work, are faced with many uncertainties and must use sound engineering judgement to develop safe solutions to challenging problems.

The studied city is our imagination according to characteristics of Chinese cities. We suppose that the disaster is earthquake. Let us calculate its exceed-probability fuzzy-risk.

Let there be 50 objects in city C. And suppose all objects are buildings. That is

$$C = \{o_1, o_2, \cdots, o_{50}\} \tag{11.50}$$

11.4.1 Fuzzy relationship between magnitude and probability

The risk source can be regarded as a seismic active belt around or nearby the city. In the belt, 12 magnitude data of historic earthquakes with $x \geq 5.0$ in T years were recorded. The set of these historic earthquakes is:

$$\begin{aligned} X &= \{x_1, x_2, \cdots, x_{12}\} \\ &= \{5.5, 6.8, 5.1, 5.7, 5.0, 6.5, 6.5, 6.0, 6.0, 5.2, 7.4, 5.2\}. \end{aligned} \tag{11.51}$$

Let $u_0 = 4.9$ be the minimum magnitude which used in engineering, and $u_\mu = 7.4$ be the maximum magnitude in the belt. The universe of discourse of earthquake magnitude in the belt is $[u_0, u_\mu] = [4.9, 7.4]$. According to the capacity of the set of these historic earthquakes, take step $\Delta = 0.5$, and let

$$\begin{aligned} U &= \{u_1, u_2, \cdots, u_6\} \\ &= \{4.9, 5.4, 5.9, 6.4, 6.9, 7.4\}. \end{aligned} \qquad (11.52)$$

Then, the universe $[u_0, u_\mu]$ of discourse of earthquake magnitude has been changed into the discrete universe U. Employing information distribution method, we can use U to absorb information from the set of these historic earthquakes and show its information structure.

Step 1: Calculating Primary Information Distribution

Now, the sample is given by Eq. (11.51). Every x_i can be distributed with information gain q_{ij} into controlling point $u_j \in U$ in Eq. (11.52) by

$$q_{ij} = \begin{cases} 1 - \frac{|x_i - u_j|}{0.5}, & \text{If } |x_i - u_j| \le 0.5 \\ 0, & \text{otherwise.} \end{cases} \qquad (11.53)$$

For example, for $x_1 = 5.5$ and $u_2 = 5.4$, we obtain:

$$q_{12} = 1 - |5.5 - 5.4|/0.5 = 1 - 0.2 = 0.8$$

The information gain of u_2 from x_1 is 0.8.

After 12 earthquake data have been treated with this simple process, we obtain all distributed information q_{ij} shown in Table 11-1.

Table 11-1 Distributed information q_{ij} on u_j from x_i

q_{ij}	u_1 4.9	u_2 5.4	u_3 5.9	u_4 6.4	u_5 6.9	u_6 7.4
$x_1 = 5.5$	0	0.8	0.2	0	0	0
$x_2 = 6.8$	0	0	0	0.2	0.8	0
$x_3 = 5.1$	0.6	0.4	0	0	0	0
$x_4 = 5.7$	0	0.4	0.6	0	0	0
$x_5 = 5.0$	0.8	0.2	0	0	0	0
$x_6 = 6.5$	0	0	0	0.8	0.2	0
$x_7 = 6.5$	0	0	0	0.8	0.2	0
$x_8 = 6.0$	0	0	0.8	0.2	0	0
$x_9 = 6.0$	0	0	0.8	0.2	0	0
$x_{10} = 5.2$	0.4	0.6	0	0	0	0
$x_{11} = 7.4$	0	0	0	0	0	1.0
$x_{12} = 5.2$	0.4	0.6	0	0	0	0
$\sum_{i=1}^{12} q_{ij}$	2.2	3.0	2.4	2.2	1.2	1.0

Then, the primary information distribution of X on U is

$$\begin{aligned} Q &= (Q_1, Q_2, Q_3, Q_4, Q_5, Q_6) \\ &= (2.2, 3.0, 2.4, 2.2, 1.2, 1.0). \end{aligned} \qquad (11.54)$$

Step 2: Calculating Exceeding Frequency Distribution

Using Q, we can obtain the number of earthquakes with magnitude greater than or equal to u_j as:

$$N_j = \sum_{i=j}^{6} Q_i$$

They constitute a number distribution of exceeding magnitude as:

$$
\begin{aligned}
N &= \{N_1, N_2, \cdots, N_6\} \\
&= \{12, 9.8, 6.8, 4.4, 2.2, 1\}.
\end{aligned}
\tag{11.55}
$$

Obviously, the frequency value of exceeding u_j is

$$f_j = \frac{N_j}{12}$$

where 12 is the number of the observations in (11.51). We can obtain an exceeding frequency distribution

$$
\begin{aligned}
F &= \{f_1(\xi \geq u_1), f_2(\xi \geq u_2), \cdots, f_1(\xi \geq u_6)\} \\
&= \{1, 0.82, 0.57, 0.37, 0.18, 0.08\}.
\end{aligned}
\tag{11.56}
$$

Step 3: Calculating Fuzzy Relation between Magnitude and Probability

From Eq.(11.56), we obtain 6 observations that makes a new sample

$$
\begin{aligned}
X_6 &= \{(u_1, f_1), (u_1, f_2), \cdots, (u_6, f_6)\} \\
&= \{(4.9, 1), (5.4, 0.82), (5.9, 0.57), (6.4, 0.37), (6.9, 0.18), (7.4, 0.08)\}.
\end{aligned}
\tag{11.57}
$$

We want to use X_6 to get a fuzzy risk of earthquake on $M \times P$, where M is the universe of discourse magnitude and P is the universe of discourse probability. To use the distribution formula, we make the both universes to be discrete. We let discrete universe of discourse magnitude, with step length $\Delta_1 = 0.5$, be

$$
\begin{aligned}
M &= \{m_1, m_2, \cdots, m_{10}\} \\
&= \{4.0, 4.5, 5.0, 5.5, 6.0, 6.5, 7.0, 7.5, 8.0, 8.5\}.
\end{aligned}
\tag{11.58}
$$

meanwhile, we let discrete universe of discourse probability, with $\Delta_2 = 0.2$, be

$$
\begin{aligned}
P &= \{p_1, p_2, \cdots, p_6\} \\
&= \{0, 0.2, 0.4, 0.6, 0.8, 1\}.
\end{aligned}
\tag{11.59}
$$

In our case, 2-dimensional linear-information-distribution formula is

$$
\tilde{f}_j(m_i, p_k) =
\begin{cases}
(1 - \frac{|u_j - m_i|}{0.5})(1 - \frac{|f_j - p_k|}{0.2}), & \text{when } |u_j - m_i| \leq 0.5 \text{ and } |f_j - p_k| \leq 0.2 \\
0, & \text{others}
\end{cases}
\tag{11.60}
$$

For example, for observation $(u_3, f_3) = (5.9, 0.57)$ and controlling point $(m_4, p_3) = (5.5, 0.4)$, we obtain:

$$\tilde{f}_3(m_4, p_3) = (1 - \frac{|u_3 - m_4|}{0.5})(1 - \frac{|f_3 - p_3|}{0.2})$$

$$= (1 - \frac{|5.9 - 5.5|}{0.5})(1 - \frac{|0.57 - 0.4|}{0.2})$$

$$= (1 - 0.8)(1 - 0.85)$$

$$= 0.03$$

Distributing observation (u_3, f_3) to all of controlling points of $M \times P$, we obtain an information matrix based on the observation, which is

$$
\mathcal{Q}_{(u_3, f_3)} =
\begin{array}{c}
m_1 \\ m_2 \\ m_3 \\ m_4 \\ m_5 \\ m_6 \\ m_7 \\ m_8 \\ m_9 \\ m_{10}
\end{array}
\begin{pmatrix}
\begin{array}{cccccc}
p_1 & p_2 & p_3 & p_4 & p_5 & p_6 \\
0 & 0 & 0 & 0 & 0 & 0 \\
0 & 0 & 0 & 0 & 0 & 0 \\
0 & 0 & 0 & 0 & 0 & 0 \\
0 & 0 & 0.03 & 0.17 & 0 & 0 \\
0 & 0 & 0.12 & 0.68 & 0 & 0 \\
0 & 0 & 0 & 0 & 0 & 0 \\
0 & 0 & 0 & 0 & 0 & 0 \\
0 & 0 & 0 & 0 & 0 & 0 \\
0 & 0 & 0 & 0 & 0 & 0 \\
0 & 0 & 0 & 0 & 0 & 0 \\
\end{array}
\end{pmatrix}.
$$

After 6 observations in X_6 have been treated with this simple process, using formula (11.26) and let

$$\mathcal{Q}_{ik} = \tilde{f}(m_i, p_k),$$

we obtain an information matrix based on the given sample, that is shown in Eq.(11.61). Then, employing the formula (11.28), we obtain a fuzzy relationship between magnitude and probability, that is shown in Eq.(11.62).

$$\mathcal{Q}_{X_6} = \{\mathcal{Q}_{ik}\}$$

$$
=
\begin{array}{c}
m_1 \\ m_2 \\ m_3 \\ m_4 \\ m_5 \\ m_6 \\ m_7 \\ m_8 \\ m_9 \\ m_{10}
\end{array}
\begin{pmatrix}
\begin{array}{cccccc}
p_1 & p_2 & p_3 & p_4 & p_5 & p_6 \\
0 & 0 & 0 & 0 & 0 & 0 \\
0 & 0 & 0 & 0 & 0 & 0.20 \\
0 & 0 & 0 & 0 & 0.18 & 0.82 \\
0 & 0 & 0.03 & 0.17 & 0.72 & 0.08 \\
0 & 0.03 & 0.29 & 0.68 & 0 & 0 \\
0.02 & 0.30 & 0.68 & 0 & 0 & 0 \\
0.20 & 0.80 & 0 & 0 & 0 & 0 \\
0.48 & 0.32 & 0 & 0 & 0 & 0 \\
0 & 0 & 0 & 0 & 0 & 0 \\
0 & 0 & 0 & 0 & 0 & 0 \\
\end{array}
\end{pmatrix}. \qquad (11.61)
$$

$$\Pi_z = \begin{array}{c} \\ m_1 \\ m_2 \\ m_3 \\ m_4 \\ m_5 \\ m_6 \\ m_7 \\ m_8 \\ m_9 \\ m_{10} \end{array} \begin{pmatrix} p_1 & p_2 & p_3 & p_4 & p_5 & p_6 \\ 0 & 0 & 0 & 0 & 0 & 0 \\ 0 & 0 & 0 & 0 & 0 & 1.00 \\ 0 & 0 & 0 & 0 & 0.22 & 1.00 \\ 0 & 0 & 0.04 & 0.24 & 1.00 & 0.11 \\ 0 & 0.04 & 0.43 & 1.00 & 0 & 0 \\ 0.02 & 0.44 & 1.00 & 0 & 0 & 0 \\ 0.25 & 1.00 & 0 & 0 & 0 & 0 \\ 1.00 & 0.67 & 0 & 0 & 0 & 0 \\ 0 & 0 & 0 & 0 & 0 & 0 \\ 0 & 0 & 0 & 0 & 0 & 0 \end{pmatrix}. \tag{11.62}$$

11.4.2 Intensity risk

The study of intensity-attenuation relationship for China cases has been somewhat done by Huang(1996, see [5]) who gave a fuzzy relationship between the intensity and distance. To focus on the procedure of calculating intensity risk, we take a more simple intensity-attenuation model, expressed by the formula

$$I = 3.32 + 1.44m - 3.34 \log d, \tag{11.63}$$

where m is local magnitude and d is hypocentral distance in kilometers. The data behind this formula are found in Musson (1994, see [9]) but the workings are currently unpublished[8].

A large number of Modified Mercalli Intensity isoseimal maps for earthquakes that have occurred in the United States have been analyzed by Anderson (1978, see [1]) to find the distribution of distances between the epicenter and isoseismal counters, which is

$$I = I_0 + C_1 + C_2(\gamma d \log_{10} e + \log_{10} d), \tag{11.64}$$

where I_0 is the epicentral intensity of an earthquake and I is the intensity at epicentral distance d (kilometer). $C_1 = 3.2$, $C_2 = -2.7$ and $e = 2.71828$. For the Eastern United States, $\gamma = 0.1$, and for the Western United States, $\gamma = 0.60$.

If we employ formula (11.64), first of all for calculating site-intensity we have to transform earthquake magnitude into epicentral intensity. Although we can use the result in Chapter 8 to do that, it is unnecessary. We directly employ formula (11.63).

For the city we study, let the universe of discourse of distance be

$$D = \{d_1, d_2, \cdots, d_6\} = \{9, 15, 20, 40, 80, 140\}. \tag{11.65}$$

Substituting the points of $M \times D$ into attenuation formula (11.63), we obtain 60 observations, denoted as I_{ij}, shown in Table 11-2. For example, substituting $(m_2, d_3) = (4.5, 20)$, we obtain

$$I_{23} = 3.32 + 1.44m_2 - 3.34\log(d_3)$$
$$= 3.32 + 1.44 \times 4.5 - 3.34\log 20$$
$$= 5.45.$$

Table 11-2 Site intensities I_{ij}

I_{ij}	d_1 9	d_2 15	d_3 20	d_4 40	d_5 80	d_6 140
$m_1 = 4.0$	5.89	5.15	4.73	3.73	2.72	1.91
$m_2 = 4.5$	6.61	5.87	5.45	4.45	3.44	2.63
$m_3 = 5.0$	7.33	6.59	6.17	5.17	4.16	3.35
$m_4 = 5.5$	8.05	7.31	6.89	5.89	4.88	4.07
$m_5 = 6.0$	8.77	8.03	7.61	6.61	5.60	4.79
$m_6 = 6.5$	9.49	8.75	8.33	7.33	6.32	5.51
$m_7 = 7.0$	10.21	9.47	9.05	8.05	7.04	6.23
$m_8 = 7.5$	10.93	10.19	9.77	8.77	7.76	6.95
$m_9 = 8.0$	11.65	10.91	10.49	9.49	8.48	7.67
$m_{10} = 8.5$	12.37	11.63	11.21	10.21	9.20	8.39

It is important to note that, non I_{ij} belongs to set $\{1, 2, \cdots, 12\}$. In Roman numbers, the set is written as $\{I, II, \cdots, XII\}$.

Let $w_0 = V$, and $w_\mu = IX$ be the minimum and maximum intensity, respectively, which used in engineering in the city. We use the method of information distribution to transform the information of the observations to controlling points 5,6,7,8,9. Let the universe of discourse of site intensity be

$$W = \{w_1, w_2, w_3, w_4, w_5\}$$
$$= \{V, VI, VII, VIII, IX\} \qquad (11.66)$$
$$= \{5, 6, 7, 8, 9\}.$$

Then the information gain of w_k from I_{ij} is

$$q_{ijk} = \begin{cases} 1 - |w_k - I_{ij}|, & \text{when } |w_k - I_{ij}| \leq 1 \\ 0, & \text{others} \end{cases} \qquad (11.67)$$

Regarding the gain as the possibility of that the site intensity is just w_k, we know

Poss(when magnitude is m_i and distance is d_j, then site intensity is w_k)=q_{ijk},

The possibility value can also be written as $\pi(m_i, d_j, w_k) = q_{ijk}$, i.e., $\pi(m_1, d_j, w_k) = q_{ijk}$. It is a fuzzy attenuation relationship, i.e., $r^{(1)}(m, d, w) = \pi(m, d, w)$. After all observations of Table 11-2 have been treated with this process, we obtain fuzzy attenuation relationship shown in Eq.(11.68).

$$R_1 \;=\; R_{(M,D)\to W}$$

	V	VI	VII	VIII	IX
$m_1 d_1$	0.11	0.89	0	0	0
$m_1 d_2$	0.85	0.15	0	0	0
$m_1 d_3$	0.73	0	0	0	0
$m_2 d_1$	0	0.39	0.61	0	0
$m_2 d_2$	0.13	0.87	0	0	0
$m_2 d_3$	0.55	0.45	0	0	0
$m_2 d_4$	0.45	0	0	0	0
$m_3 d_1$	0	0	0.67	0.33	0
$m_3 d_2$	0	0.41	0.59	0	0
$m_3 d_3$	0	0.83	0.17	0	0
$m_3 d_4$	0.83	0.17	0	0	0
$m_3 d_5$	0.16	0	0	0	0
$m_4 d_1$	0	0	0	0.95	0.05
$m_4 d_2$	0	0	0.69	0.31	0
$m_4 d_3$	0	0.11	0.89	0	0
$m_4 d_4$	0.11	0.89	0	0	0
$m_4 d_5$	0.88	0	0	0	0
$m_4 d_6$	0.07	0	0	0	0
$m_5 d_1$	0	0	0	0.23	0.77
$m_5 d_2$	0	0	0	0.97	0.03
$m_5 d_3$	0	0	0.39	0.61	0
$m_5 d_4$	0	0.39	0.61	0	0
$m_5 d_5$	0.40	0.60	0	0	0
$m_5 d_6$	0.79	0	0	0	0
$m_6 d_1$	0	0	0	0	0.51
$m_6 d_2$	0	0	0	0.25	0.75
$m_6 d_3$	0	0	0	0.67	0.33
$m_6 d_4$	0	0	0.67	0.33	0
$m_6 d_5$	0	0.68	0.32	0	0
$m_6 d_6$	0.49	0.51	0	0	0
$m_7 d_2$	0	0	0	0	0.53
$m_7 d_3$	0	0	0	0	0.95
$m_7 d_4$	0	0	0	0.95	0.04
$m_7 d_5$	0	0	0.96	0.04	0
$m_7 d_6$	0	0.77	0.23	0	0
$m_8 d_3$	0	0	0	0	0.23
$m_8 d_4$	0	0	0	0.23	0.77
$m_8 d_5$	0	0	0.24	0.76	0
$m_8 d_6$	0	0.04	0.95	0	0
$m_9 d_4$	0	0	0	0	0.51
$m_9 d_5$	0	0	0	0.52	0.48
$m_9 d_6$	0	0	0.33	0.67	0
$m_{10} d_5$	0	0	0	0	0.80
$m_{10} d_6$	0	0	0	0.61	0.39

$$(11.68)$$

For example, from observation $(m_2, d_3, I_{23}) = (4.5, 20, 5.45)$, we have

$$r^{(1)}(m_2, d_3, w_1) = q_{231} = 1 - |w_1 - I_{23}| = 1 - |5 - 5.45| = 0.55,$$

$$r^{(1)}(m_2, d_3, w_2) = q_{232} = 1 - |w_2 - I_{23}| = 1 - |6 - 5.45| = 0.45.$$

Due to limited space for write, the rows whose all elements are zero do not appear in (11.68).

Suppose the nearest and farthest distance from the city to the belt is $d_0 = 0$ km, and $d_\mu = 30$ km, respectively, recalling (11.33), then, the fuzzy distance is

$$\underset{\sim}{D} = 0.79/d_1 + 1/d_2 + 0.85/d_3 + 0.02/d_4 + 0/d_5 + 0/d_6. \qquad (11.69)$$

For example, for d_4, we have

$$\mu_{\underset{\sim}{D}}(d_4) = \exp[-1.5(\frac{d_\mu + d_0 - 2d_4}{d_\mu - d_0})^2]$$

$$= \exp[-1.5(\frac{30 + 0 - 2 \times 40}{30 - 0})^2]$$

$$= \exp[-1.5(\frac{50}{30})^2]$$

$$= \exp(-4.19)$$

$$= 0.02$$

Substituting $r^{(1)}(m, d, w)$, $\pi_z(m, p)$, and $\underset{\sim}{D}$ into the risk formula (11.35), we obtain the fuzzy risk of the site intensity shown in (11.70).

$$
\begin{aligned}
\Pi_s &= \{\pi_s(w, p)\} \\
&= \left\{ \sup_{m \in M, d \in D} \{r^{(1)}(m, d, w) \wedge \pi_z(m, p) \wedge \mu_{\underset{\sim}{D}}(d)\} \right\}
\end{aligned}
$$

$$
= \begin{array}{c} \\ V \\ VI \\ VII \\ VIII \\ IX \end{array}
\begin{array}{c}
\begin{array}{cccccc} p_1 & p_2 & p_3 & p_4 & p_5 & p_6 \end{array} \\
\left(
\begin{array}{cccccc}
0 & 0 & 0.02 & 0.02 & 0.02 & 0.55 \\
0 & 0.02 & 0.04 & 0.11 & 0.22 & 0.87 \\
0.02 & 0.04 & 0.39 & 0.39 & 0.85 & 0.67 \\
0.02 & 0.44 & 0.67 & 0.97 & 0.79 & 0.33 \\
0.25 & 0.85 & 0.75 & 0.77 & 0.05 & 0.05
\end{array}
\right)
\end{array}. \qquad (11.70)
$$

For example, for $w_4 = VIII$ and $p_4 = 0.6$, we obtain $\pi_s(w_4, p_4)$ shown in Eq.(11.71).

$$
\begin{aligned}
\pi_s(w_4, p_4) \quad &= \vee\{r^{(1)}(m_1, d_1, w_4) \wedge \pi_z(m_1, p_4) \wedge \mu_D(d_1), \\
&\quad r^{(1)}(m_1, d_2, w_4) \wedge \pi_z(m_1, p_4) \wedge \mu_D(d_2), \cdots, \\
&\quad r^{(1)}(m_1, d_6, w_4) \wedge \pi_z(m_1, p_4) \wedge \mu_D(d_6), \\
&\quad r^{(1)}(m_2, d_1, w_4) \wedge \pi_z(m_2, p_4) \wedge \mu_D(d_1), \\
&\quad r^{(1)}(m_2, d_2, w_4) \wedge \pi_z(m_2, p_4) \wedge \mu_D(d_2), \cdots, \\
&\quad r^{(1)}(m_2, d_6, w_4) \wedge \pi_z(m_2, p_4) \wedge \mu_D(d_6), \\
&\quad \cdots, \\
&\quad r^{(1)}(m_5, d_1, w_4) \wedge \pi_z(m_5, p_4) \wedge \mu_D(d_1), \\
&\quad r^{(1)}(m_5, d_2, w_4) \wedge \pi_z(m_5, p_4) \wedge \mu_D(d_2), \cdots, \\
&\quad r^{(1)}(m_5, d_6, w_4) \wedge \pi_z(m_5, p_4) \wedge \mu_D(d_6), \\
&\quad \cdots, \\
&\quad r^{(1)}(m_{10}, d_1, w_4) \wedge \pi_z(m_{10}, p_4) \wedge \mu_D(d_1), \\
&\quad r^{(1)}(m_{10}, d_2, w_4) \wedge \pi_z(m_{10}, p_4) \wedge \mu_D(d_2), \cdots, \\
&\quad r^{(1)}(m_{10}, d_6, w_4) \wedge \pi_z(m_{10}, p_4) \wedge \mu_D(d_6)\} \\
&= \vee\{0 \wedge 0 \wedge 0.79, 0 \wedge 0 \wedge 1, \cdots, 0 \wedge 0 \wedge 0, \\
&\quad 0 \wedge 0 \wedge 0.79, 0 \wedge 0 \wedge 1, \cdots, 0 \wedge 0 \wedge 0, \\
&\quad \cdots, \\
&\quad 0.23 \wedge 1 \wedge 0.79, 0.97 \wedge 1 \wedge 1, \cdots, 0 \wedge 1 \wedge 0, \\
&\quad \cdots, \\
&\quad 0 \wedge 0 \wedge 0.79, 0 \wedge 0 \wedge 1, \cdots, 0 \wedge 0 \wedge 0\} \\
&= \vee\{0, 0, \cdots, 0.23, 0.97, \cdots, 0\} \\
&= 0.97.
\end{aligned}
$$

$$\text{(11.71)}$$

In fact, formula (11.35) can also be written as

$$
\Pi_s(w, p) = (R_1 \circ \underset{\sim}{D}) \circ \Pi_z,
$$

where

$$
R_1 \circ \underset{\sim}{D} = \left\{ \sup_{d \in D}\{r^{(1)}(m, d, w) \wedge \mu_D(d)\} \right\}.
$$

Denote

$$
R_{M \to w} = R_1 \circ \underset{\sim}{D} = \{r(m, w)\},
$$

then

$$
\Pi_s = R_{M \to w} \circ \Pi_z = \left\{ \sup_{m \in M}\{r(m, w) \wedge \pi_z(m, p)\} \right\}.
$$

$$\text{(11.72)}$$

In our case, we obtain

$$
R_{M \to W} = \begin{array}{c} \\ m_1 \\ m_2 \\ m_3 \\ m_4 \\ m_5 \\ m_6 \\ m_7 \\ m_8 \\ m_9 \\ m_{10} \end{array} \begin{array}{ccccc} w_1 & w_2 & w_3 & w_4 & w_5 \\ \left(\begin{array}{ccccc} 0.85 & 0.79 & 0 & 0 & 0 \\ 0.55 & 0.87 & 0.61 & 0 & 0 \\ 0.02 & 0.83 & 0.67 & 0.33 & 0 \\ 0.02 & 0.11 & 0.85 & 0.79 & 0.05 \\ 0 & 0.02 & 0.39 & 0.97 & 0.77 \\ 0 & 0 & 0.02 & 0.67 & 0.75 \\ 0 & 0 & 0 & 0.02 & 0.85 \\ 0 & 0 & 0 & 0.02 & 0.23 \\ 0 & 0 & 0 & 0 & 0.02 \\ 0 & 0 & 0 & 0 & 0 \end{array} \right) \end{array}.
$$

11.4.3 Earthquake damage risk

Suppose that every object in city C is a single layer brick pillar factory-building.

Generally, damage index of building is defined as the damage percentage, and set A in (11.43) acts as the discrete universe of discourse of damage index.

From (11.44), we can define fuzzy damage as:

$$
\begin{cases}
y_1 = \text{Good condition} = 1/a_1 + 0.7/a_2 + 0.2/a_3 \\
y_2 = \text{Light destruction} = 0.2/a_1 + 0.7/a_2 + 1/a_3 + 0.7/a_4 + 0.2/a_5 \\
y_3 = \text{General destruction} = 0.2/a_3 + 0.7/a_4 + 1/a_5 + 0.7/a_6 + 0.2/a_7 \\
y_4 = \text{Heavy destruction} = 0.2/a_5 + 0.7/a_6 + 1/a_7 + 0.7/a_8 + 0.2/a_9 \\
y_5 = \text{Collapse} = 0.2/a_7 + 0.7/a_8 + 1/a_9 + 0.7/a_{10} + 0.2/a_{11}
\end{cases}
$$

$$(11.73)$$

In China, the fuzzy relationship[15] between the site intensity and fuzzy damage of a single layer brick pillar factory-building is:

$$
R_{W \to Y} = \begin{array}{c} \\ VI \\ VII \\ VIII \\ IX \end{array} \begin{array}{ccccc} y_1 & y_2 & y_3 & y_4 & y_5 \\ \left(\begin{array}{ccccc} 1.00 & 0.43 & 0.14 & 0.00 & 0.00 \\ 0.21 & 1.00 & 0.36 & 0.00 & 0.00 \\ 0.21 & 0.36 & 1.00 & 0.14 & 0.13 \\ 0.00 & 0.14 & 0.43 & 1.00 & 0.57 \end{array} \right) \end{array} \qquad (11.74)
$$

where $W = \{VI, VII, VIII, IX\}$, and $Y = \{y_1, y_2, y_3, y_4, y_5\}$.

Using formula (11.40) and according to (11.70), we can obtain the fuzzy risk of an object response as the following:

$$
\Pi_o = \begin{array}{c} \\ y_1 \\ y_2 \\ y_3 \\ y_4 \\ y_5 \end{array} \begin{array}{cccccc} p_1 & p_2 & p_3 & p_4 & p_5 & p_6 \\ \left(\begin{array}{cccccc} 0.02 & 0.21 & 0.21 & 0.21 & 0.22 & 0.87 \\ 0.14 & 0.36 & 0.39 & 0.39 & 0.85 & 0.67 \\ 0.25 & 0.44 & 0.67 & 0.97 & 0.79 & 0.36 \\ 0.25 & 0.85 & 0.75 & 0.77 & 0.14 & 0.14 \\ 0.25 & 0.57 & 0.57 & 0.57 & 0.13 & 0.13 \end{array} \right) \end{array} \qquad (11.75)
$$

For example,

$$
\begin{aligned}
\pi_o(y_2, p_3) &= \sup_{w \in W} \left\{ r^{(2)}(w, y_2) \wedge \pi_s(w, p_3) \right\} \\
&= \vee \left\{ r^{(2)}(VI, y_2) \wedge \pi_s(VI, p_3), r^{(2)}(VII, y_2) \wedge \pi_s(VII, p_3), \right. \\
&\qquad \left. r^{(2)}(VIII, y_2) \wedge \pi_s(VIII, p_3), r^{(2)}(IX, y_2) \wedge \pi_s(IX, p_3) \right\} \\
&= \vee \{ 0.43 \wedge 0.04, 1 \wedge 0.39, 0.36 \wedge 0.67, 0.14 \wedge 0.75 \} \\
&= \vee \{ 0.04, 0.39, 0.36, 0.14 \wedge \} \\
&= 0.39.
\end{aligned}
$$

11.4.4 Earthquake loss risk

Suppose that the loss of a building is in direct proportion to its area and damage index. Moreover, let us presume that every square meter is worth $ 490 in city C. If the area of all buildings in city C totalled 50,000 square meters, the buildings in city C is worth 24.5 million dollars. Corresponding with the universe of discourse of damage index in (11.44), we can obtain the universe of discourse of loss

$$
\begin{aligned}
L_C &= \{ l_1, l_2, \cdots l_{11} \} \\
&= \{ 0, 2.5, 4.9, 7.4, 9.8, 12.3, 14.7, 17.1, 19.6, 22.1, 24.5 \}
\end{aligned}
\tag{11.76}
$$

where a unit of loss is million dollars.

According to the definition of fuzzy damage by (11.73), and the relationship between the damage index and universe of discourse of loss, we know that the $R_{Y \to L}$ in (11.46) is just the relationship between damage (degree) and loss (universe is given in (11.76)). It is relationship R_3 in formula (11.48). Substituting the R_3 and Π_0 in (11,73) into formula (11.49), we obtain fuzzy risk of loss of city C as the following.

$$
\Pi_C =
\begin{array}{c}
\begin{array}{ccccccc}
 & p_1 & p_2 & p_3 & p_4 & p_5 & p_6
\end{array} \\
\begin{array}{c}
l_1 \\ l_2 \\ l_3 \\ l_4 \\ l_5 \\ l_6 \\ l_7 \\ l_8 \\ l_9 \\ l_{10} \\ l_{11}
\end{array}
\left(
\begin{array}{cccccc}
0.14 & 0.21 & 0.21 & 0.21 & 0.22 & 0.87 \\
0.14 & 0.36 & 0.39 & 0.39 & 0.70 & 0.70 \\
0.20 & 0.36 & 0.39 & 0.39 & 0.85 & 0.67 \\
0.25 & 0.44 & 0.67 & 0.70 & 0.70 & 0.67 \\
0.25 & 0.44 & 0.67 & 0.97 & 0.79 & 0.36 \\
0.25 & 0.70 & 0.70 & 0.70 & 0.70 & 0.36 \\
0.25 & 0.85 & 0.75 & 0.77 & 0.20 & 0.20 \\
0.25 & 0.70 & 0.70 & 0.70 & 0.14 & 0.14 \\
0.25 & 0.57 & 0.57 & 0.57 & 0.14 & 0.14 \\
0.25 & 0.57 & 0.57 & 0.57 & 0.13 & 0.13 \\
0.20 & 0.20 & 0.20 & 0.20 & 0.13 & 0.13
\end{array}
\right)
\end{array}
\tag{11.77}
$$

According to Π_C, we know that the probability of exceeding loss is not one value but a fuzzy set. For example, when $l = l_5 = 9.8$(million us dollars), the fuzzy probability of loss is:

$$\underset{\sim}{P}(\xi \geq 9.8) = 0.25/0 + 0.44/0.2 + 0.67/0.4 + 0.97/0.6 + 0.79/0.8 + 0.36/1$$

The benefit of this result is that one can easily understand impreciseness of risk estimator of earthquakes due to complexity of a system and insufficient data. It might be useful to set a flexible and more economical strategy, plan and action on disaster reduction.

11.5 Conclusion and Discussion

Until now few researchers consider all stages affecting risks of natural disasters. The traditional method might produce theoretical results being far away real cases, because each stage involves uncertainties.

The model in the chapter gives a simplest approach to deal with all stages of risk assessment. The result is coordinated with engineering practice.

If we say that probabilistic method reduces the degree of blindness and provides more information, it is true that the fuzzy risk method can offer more information which enhances our understanding of probability to avoid acting rashly.

In our model, we only use maximum and minimum operator. In fact, it can be replace by other powerful operator if we want analyse fuzzy risks more carefully.

The two-times distribution model in the chapter is rough to calculate fuzzy risk. In Chapter 12 we discuss a professional algorithm to calculate fuzzy risk more precisely.

References

1. Anderson, J. G.(1978), On the attenuation of Modified Mercalli Intensity with distance in the United States. Bulletin of the Seismological Society of America **68**, pp.1147-1179

2. Cornell,C.A.(1968), Engineering seismic risk analysis. Bulletin of the Seismological Society of America **58**, pp.1583-160

3. Ellsworth, W. L., Matthews, M. V., Nadeau, R. M., Nishenko, S. P., Reasenberg, P.A., and Simpson, R. W. (1998), A physically-based earthquake recurrence model for estimation of long-term earthquake probabilities. Proceedings of the Second Joint Meeting of the UJNR Panel on Earthquake Research, pp. 135-149

4. Hagiwara, Y.(1974) , Probability of earthquake occurrence as obtained from a Weibull distribution analysis of crustal strain. Tectonophysics **23**, pp. 323-318

5. Huang, C.F. (1996), Fuzzy risk assessment of urban natural hazards, Fuzzy Sets and Systems **83**, pp.271-282

6. Huang, C.F., and Ruan, D.(1999), Systems analytic models for fuzzy risk estimation. Zadeh, L.A. and Kacprzyk, J. (eds): Computing with Words in Information/Intelligent Systems 2. Physica-Verlag, Heidelberg, pp.195

7. Huang, C.F. and Wang, J.D. (1995),Technology of Fuzzy Information Optimization Processing and Applications. Beijing University of Aeronautics and Astronautics Press, Beijing (in Chinese)

8. Musson, R. (1998), Seismicity and Earthquake Hazard in the UK. http://www.gsrg.nmh.ac.uk/hazard/hazuk.htm

9. Musson, R.M.W.(1994), A Catalogue of British Earthquakes. British Geological Survey Technical Report, No. WL/94/04

10. Nishenko, S. P., and Buland, R. (1987), A generic recurrence interval distribution for earthquake forecasting. Bulletin of the Seismological Society of America **77**, pp. 1382-1399

11. Rikitake, T. (1974), Probability of an earthquake occurrence as estimated from crustal strain. Tectonophysics **23**, pp. 299-312

12. Utsu, T. (1972a), Large earthquakes near Hokkaido and the expectancy of the occurrence of a large earthquake of Nemuro. Report of the Coordinating Committee for Earthquake Prediction **7**, pp. 7-13

13. Utsu, T.(1972b), Aftershocks and earthquake statistics (IV). Journal of the Faculty of Science, Hokkaido University Series VII (Geophysics) **4**, pp. 1-42

14. Utsu, T. (1984), Estimation of parameters for recurrence models of earthquakes. Bulletin of the Earthquake Research Institute, University of Tokyo **59**, pp. 53-66

15. Xiu, X. W. and Huang, C.F. (1989), Fuzzy identification between dynamic response of structure and structural earthquake damage. Earthquake Engineering and Engineering Vibration, Vol.9, No.2, pp.57, (in Chinese).

12. Fuzzy Risk Calculation

For showing the imprecision of risk assessment in terms of probabilities, in this chapter we introduce the possibility-probability distribution (PPD). We focus on calculation and application of a PPD. Section 12.1 gives the definition of a PPD. Section 12.2 develops the method of information distribution forming an algorithm to calculate a PPD. Section 12.3 introduces the fuzzy expected value of a PPD to rank alternatives. Section 12.4 gives a real example in flood management to show the benefit of the ranking based on a PPD. We summarize this chapter in section 12.5.

12.1 Introduction

Until recently the well-known fuzzy logic almost owes the success to fuzzy control which plays a pivotal role in fuzzy commercial products such as fuzzy washing machines, fuzzy vacuum sweepers, fuzzy microwave ovens and so on. There was no mathematical model to back up a truck-and-trailer in a parking lot to a loading dock if the truck starts from any position. Both human being and fuzzy control can perform this nonlinear control even for a truck with five trailers. However, fuzzy control is just the use of fuzzy logic as the technology underlying the development of human-friendly machines, and it has not involved matters of the foundations of fuzzy logic.

Fuzzy information granulation, discussed recently by Zadeh [13], seems to be a very important issue for a restructuring of the foundations of fuzzy logic. Of particular importance is the need for a better understanding of decision-making in an environment generalized imprecision, uncertainty and partial truth[14].

Risk recognition and management are typical issues with imprecision, uncertainty and partial truth. As we discuss in Chapter 10, a risk system could be studied with some state equations if we could find them, or concept of risk may be reduced to be probability-risk if we can find a probability distribution to show the risk phenomenon properly. However, risk systems are too complex to be shown in state equations or probability distributions.

To show the imprecision of risk assessment, in Chapter 11, we define the exceed-probability fuzzy-risk, meanwhile we give the two-times distribution model to calculate an exceed-probability fuzzy-risk. However, the concept of exceed-probability fuzzy-risk is a technical terminology, and the two-times distribution model is rougher due to that we have to choose controlling points two times. Therefore, it is necessary to give a more general concept and to develop a reasonable algorithm for calculating.

As we know, in Chapter 10, fuzzy risk can be defined as an approximate representation to show risk with fuzzy theory and techniques. A fuzzy risk is a fuzzy relation between loss events and concerning factors. In principle, we can study fuzzy risk without any probability terminologies. However, it is not time to do so. The reason is that both risk science and fuzzy set theory need to be further developed. What we can do is to show the imprecision of probability-risk assessment. We can employ *fuzzy probability* to do it.

12.1.1 Fuzziness and probability

As we well know, probability and fuzziness are related but different concepts. Fuzziness is a type of deterministic uncertainty. Fuzziness measures the degree to which an event occurs, not whether it occurs. Probability arouses from the question whether or not an event occurs. Moreover, it assumes that the event class is crisply defined and that the law of non contradiction holds. That is, $A \cap A^c = \emptyset$.

A fuzzy probability extends the traditional notion of a probability when there are the outcomes that belong to several event classes at the same time but to different degrees. The fuzziness and probability are orthogonal concepts that characterize different aspects of human experience. Hence, it is important to note that neither fuzziness nor probability governs the physical processes in Nature. They are introduced by human being to compensate for their own limitations.

The earliest concept of fuzzy probability is the so-called probability of fuzzy event, given by Zadeh in 1968, which is defined by Eq.(1.17). Even in 1997, the fuzzy probability was still considered[10] as the probability of fuzzy event, meanwhile many papers discuss the fuzzy random variables[7] as a derivation of the fuzzy random numbers.

Fuzzy risks connect to imprecise probability rather than the probability of fuzzy event. Imprecise probability can be also called fuzzy probability.

In statistical applications, imprecise probabilities usually come from subjectively assessed prior probabilities. The fuzzy set theory is applicable to the modeling of imprecise subjective probabilities, suggested by many researchers (for example, Freeling in 1979 [2], Watson et al. in 1980 [11], Dubois and Prade in 1989 [1]). We also can see that the multi-valued probability of a failure has been employed to describe the so-called fuzzy reliability[4]. Traditionally, the term "fuzzy probability" connects to the fuzzy events, fuzzy random numbers, or subjective probabilities.

To avoid any confusion, we restrict ourselves here to express imprecise probability with a fuzzy relation, called the possibility-probability distribution.

12.1.2 Possibility-probability distribution

Definition 12.1 Let m be an adverse event, $p(m)$ be the probability of m occurrence, and M be the space of adverse events. Probability distribution $P = \{p(m)|m \in M\}$ is called the *probability-risk*.

Example 12.1 In May 2000, the experts continue to believe that the stock is worth less if Microsoft Corporation were not to win on appeal with respect to the breakup. In this case, Mr. Buffett would loss some investment in the year. The lost money m is the value of the adverse event and the probability distribution of the m is Buffett's probability-risk in the stock market.

Definition 12.2 Let $M = \{m\}$ be the space of adverse events, $P = \{p\}$ be the universe of discourse of probability, and $\pi_m(p)$ be possibility that probability of m occurring is p.

$$\Pi_{M,P} = \{\pi_m(p)|m \in M, p \in P\} \tag{12.1}$$

is called a *possibility-probability distribution* (PPD).

Example 12.2 Suppose the integration of propositions: "Probability that Buffett will loss 1% investment is very large; Probability that Buffett will loss 2% investment is large; $\cdots$; Probability that Buffett will loss 10% investment is very small" is a fuzzy risk estimate. It can be represented by a PPD. If we define

Small probability $= 1/0 + 0.8/0.1 + 0.2/0.2,$

Large probability $= 0.2/0.8 + 0.8/0.9 + 1/1,$

where, for a/b, a is membership, b is probability, using the concentration operator $\mu_{con(A)}(p) = (\mu_A(p))^2$, we have

Very small probability $= 1/0 + 0.64/0.1 + 0.04/0.2,$

Very large probability $= 0.04/0.8 + 0.64/0.9 + 1/1.$

In this case, Mr. Buffett's PPD with respect to stock risk can be represented by Table 12-1.

Table 12-1 Mr. Buffett's PPD with respect to stock risk

$\pi_m(p)$	p_0	p_1	p_2	$\cdots$	p_8	p_9	p_{10}	Natural language estimate
	0	0.1	0.2	$\cdots$	0.8	0.9	1	
m_1 (loss 1%)	0	0	0	$\cdots$	0.04	0.64	1	Very large probability
m_2(loss 2%)	0	0	0	$\cdots$	0.2	0.8	1	Large probability
$\cdots$	$\cdots$	$\cdots$	$\cdots$	$\cdots$	$\cdots$	$\cdots$	$\cdots$	$\cdots$
m_{10}(loss 10%)	1	0.64	0.04	$\cdots$	0	0	0	Very small probability

Traditionally, the major challenges in fuzzy probability are considered:

(1) Collect or produce expert experience;

(2) Find scientific approaches to express randomness of fuzzy variables;

(3) Process subjectively assessed probabilities.

It is more and more far from engineering, because engineers believe calculated result rather than subjective assessment, and few engineers can understand the so-called super-probability-space such as $\Xi(\Omega, \mathcal{A}, P; U, \mathcal{B}, \check{C})$ in Definition 5.13.

Now we want to find reasonable approaches to calculate fuzzy probability with data.

For fuzzy event $\underset{\sim}{A}$, under the condition that we know the basic probability distribution $p(m)$, it is easy to calculate fuzzy probability by

$$P(\underset{\sim}{A}) = \int_{R^n} \mu_A(m)p(m)dm.$$

In risk analysis, unfortunately, the focus point is just to find $p(m)$. As we indicate in above chapters, it is impossible to get precise estimate of $p(m)$. Therefore, we believe that the first challenge in fuzzy risk analysis is to calculate (not subjectively assess) PPDs.

12.2 Interior-outer-set Model

12.2.1 Model description

Let $X = \{x_1, \cdots, x_n\}$ be a sample, $X \subset \mathbb{R}$, and $U = \{u_1, \cdots, u_m\}$ be the discrete universe of X, $\Delta \equiv u_j - u_{j-1}$, $j = 2, 3, \cdots, m$.

This subsection introduces the interior-outer-set model to calculate a PPD on intervals

$$I_j = [u_j - \Delta/2, u_j + \Delta/2[, \quad u_j \in U,$$

with respect to probability values

$$p_k = \frac{k}{n}, \quad k \in \{0, 1, 2, \cdots, n\}.$$

For interval I_j, we use $\pi_{I_j}(p_k)$, or $\pi_{u_j}(p_k)$, $k = 0, 1, 2, \cdots, n$, to represent its possibility- probabilities.

Definition 12.3 $X_{in-j} = X \cap I_j$ is called the *interior set* of interval I_j. The elements of X_{in-j} are called the *interior points* of I_j.

Obviously, we have

$$\forall i \neq j, \quad I_i \cap I_j = \emptyset,$$

and

$$X = X_{in-1} \cup X_{in-2} \cup \cdots \cup X_{in-m}.$$

Example 12.3 Let

$$X = \{x_i | i = 1, 2, \cdots, 6\}$$
$$= \{7.8, 7.5, 7.25, 7.9, 7.3, 7.2\},$$

and $I_1 = [7.0, 7.3[, I_2 = [7.3, 7.6[, I_3 = [7.6, 7.9[, I_4 = [7.9, 8.2[$. The interior set of interval I_1 is $X_{in-1} = \{x_3, x_6\}$.

Definition 12.4 Let X_{in-j} be the interior set of interval I_j. $X_{out-j} = X \setminus X_{in-j}$ is called *outer set* of the interval I_j. The elements of X_{out-j} are called the *outer points* of I_j.

Example 12.4 For X in Example 12.3, the outer set of interval I_1 is $X_{out-1} = \{x_1, x_2, x_4, x_5\}$.

Obviously, an interval I acts as a classifier to divide a given sample X into two parts. The interior set of I is just $X \cap I$, and the outer set is $X \cap (\mathbb{R} \setminus I)$. It is easy to extend the definitions of the interior and outer sets to serve for a general universe Ω and a sample $X \subset \Omega$.

Let S_j be an index set such that $\forall s \in S_j$ then $x_s \in X_{in-j}$, and $\{x_s | s \in S_j\} = X_{in-j}$.

Let T_j be one for X_{out-j}, that is, $\{x_t | t \in T_j\} = X_{out-j}$.

S_j is called *interior index set*, and T_j *outer index set*.

For example, for X and I_1 in Example 12.3, we obtain $S_1 = \{3, 6\}$ and $T_1 = \{1, 2, 4, 5\}$.

When sample X and universe U are given, we can employ the formula of 1-dimensional linear-information-distribution

$$q_{ij} = \begin{cases} 1 - \frac{|x_i - u_j|}{\Delta}, & \text{if } |x_i - u_j| \leq \Delta; \\ 0, & \text{if } |x_i - u_j| > \Delta. \end{cases} \quad x_i \in X, \ u_j \in U. \quad (12.2)$$

to distribution sample point x_i to controlling point u_j. q_{ij} is called an *information gain*.

$\forall x_i \in X$, if $x_i \in X_{in-j}$ we say that it may lose information, by gain at $1 - q_{ij}$, to other interval, we use $q_{ij}^- = 1 - q_{ij}$ to represent the loss; if $x_i \in X_{out-j}$ we say that it may give information, by gain at q_{ij}, to I_j, we use q_{ij}^+ to represent the addition. q_{ij} means that x_i may leave I_j in possibility q_{ij}^- if $x_i \in X_{in-j}$, or x_i may join I_j in possibility q_{ij}^+ if $x_i \in X_{out-j}$.

q_{ij}^- is called *leaving possibility*, and q_{ij}^+ called *joining possibility*. The leaving possibility of an outer point is defined as 0 (it has gone). The joining possibility of an interior point is defined as 0 (it has been in the interval).

In the case of that the size of X_{in-j} is n_j, the most possible probability that adverse event x occurs in I_j is n_j/n. Hence, we define the possibility of $P\{x \in I_j\} = n_j/n$ as

$$\pi_{I_j}(n_j/n) = 1. \quad (12.3)$$

However, if $x_i \in X_{in-j}$, the loss information q_{ij}^- to u_{j-1} or u_{j+1} implicates that x_i may leave interval I_j when there is a disturbance in the random experiment. The gain, q_{ij}^-, is the possibility that x_i may leave I_j. When one of x_s , $s \in S_j$, has leaven I_j, we know that $P\{x \in I_j\} = (n_j - 1)/n$. Any one of the interior points of I_j can make it true. Therefore, the possibility of $P\{x \in I_j\} = (n_j - 1)/n$ is

$$\pi_{I_j}(\frac{n_j - 1}{n}) = \bigvee_{s \in S_j} q_{sj}^-. \tag{12.4}$$

If two of them leave the interval, we obtain $P\{x \in I_j\} = (n_j - 2)/n$. According to properties of possibility, both of x_{s_1}, x_{s_2} leave I_j, the possibility is $q_{s_1j}^- \wedge q_{s_2j}^-$. Considering all pairing observations in I_j, we obtain the possibility that probability of adverse event x is $(n_j - 2)/n$ as

$$\pi_{I_j}(\frac{n_j - 2}{n}) = \bigvee_{s_1, s_2 \in S_j, s_1 \neq s_2} (q_{s_1j}^- \wedge q_{s_2j}^-). \tag{12.5}$$

In other side, the observations in X_{out-j} may move to I_j when there is a disturbance in the random experiment. $\forall x_t \in X_{out-j}$, it may move to I_j in possibility q_{tj}^+. Therefore, the possibility of $P\{x \in I_j\} = (n_j + 1)/n$ is

$$\pi_{I_j}(\frac{n_j + 1}{n}) = \bigvee_{t \in T_j} q_{tj}^+. \tag{12.6}$$

If two of elements in X_{out-j} go into I_j, we obtain $P\{x \in I_j\} = (n_j + 2)/n$. Its possibility is

$$\bigvee_{t_1, t_2 \in T_j, t_1 \neq t_2} (q_{t_1j}^+ \wedge q_{t_2j}^+). \tag{12.7}$$

Hence, when there are n_j observations $\{x_s | s \in S_j\}$ in interval I_j, we can obtain a formula to calculate a PPD as

$$\pi_{I_j}(p) = \begin{cases} \bigwedge_{s \in S_j} q_{sj}^-, & p = p_0; \\ \cdots & \cdots \\ \bigvee_{s_1, s_2, s_3 \in S_j, s_1 \neq s_2 \neq s_3} (q_{s_1j}^- \wedge q_{s_2j}^- \wedge q_{s_3j}^-), & p = p_{n_j-3}; \\ \bigvee_{s_1, s_2 \in S_j, s_1 \neq s_2} (q_{s_1j}^- \wedge q_{s_2j}^-), & p = p_{n_j-2}; \\ \bigvee_{s \in S_j} q_{sj}^-, & p = p_{n_j-1}; \\ 1, & p = p_{n_j}; \\ \bigvee_{t \in T_j} q_{tj}^+, & p = p_{n_j+1}; \\ \bigvee_{t_1, t_2 \in T_j, t_1 \neq t_2} (q_{t_1j}^+ \wedge q_{t_2j}^+), & p = p_{n_j+2}; \\ \bigvee_{t_1, t_2, t_3 \in T_j, t_1 \neq t_2 \neq t_3} (q_{t_1j}^+ \wedge q_{t_2j}^+ \wedge q_{t_3j}^+), & p = p_{n_j+3}; \\ \cdots & \cdots \\ \bigwedge_{t \in T_j} q_{tj}^+, & p = p_n. \end{cases} \tag{12.8}$$

$j = 1, 2, \cdots, m; \; p_0 = 0, p_1 = 1/n, \cdots, p_{n_j} = n_j/n, \cdots, p_n = 1.$

The formula (12.8) is called the *interior-outer-set model*.

12.2.2 Calculation case

The sample X in Example 12.3 is a batch of natural disaster records. The sample is too small to precisely estimate probability-risk of the disaster. We calculate a PPD with the dada to replace a probability-risk.

We also consider the intervals $I_1 = [7.0, 7.3[, I_2 = [7.3, 7.6[, I_3 = [7.6, 7.9[,$ and $I_4 = [7.9, 8.2[,$ that are used in Example 12.3. Correspondingly, the discrete universe of discourse is

$$U = \{u_j | j = 1, 2, 3, 4\} = \{7.15, 7.45, 7.75, 8.05\}.$$

with step length $\Delta = 0.3$. u_j is the center point of I_j.

Because $n = 6$, universe of discourse of probability is

$$
\begin{aligned}
P = \ & \{p_k | k = 0, 2, \cdots, 6\} \\
= \ & \{0/6, 1/6, 2/6, \cdots, 6/6\} \\
= \ & \{0, 0.17, 0.33, 0.5, 0.67, 0.83, 1\}.
\end{aligned}
$$

Firstly, using the formula (12.2), we distribute all observations in X on U and obtain Table 12-2.

Table 12-2. Information gain from x_i to u_j

q_{ij}	u_1 7.15	u_2 7.45	u_3 7.75	u_4 8.05
$x_1 = 7.8$	0.00	0.00	0.83	0.17
$x_2 = 7.5$	0.00	0.83	0.17	0.00
$x_3 = 7.25$	0.67	0.33	0.00	0.00
$x_4 = 7.9$	0.00	0.00	0.50	0.50
$x_5 = 7.3$	0.50	0.50	0.00	0.00
$x_6 = 7.2$	0.83	0.17	0.00	0.00

Secondly, we calculate all leaving possibilities shown in Table 12-3. For I_1, we know, $X_{in-1} = \{x_3, x_6\}$. Therefore

$$q_{\overline{31}} = 1 - q_{31} = 1 - 0.67 = 0.33, \quad q_{\overline{61}} = 1 - q_{61} = 1 - 0.83 = 0.17.$$

For I_2, $X_{in-2} = \{x_2, x_5\}$, then

$$q_{\overline{22}} = 1 - q_{22} = 1 - 0.83 = 0.17, \quad q_{\overline{52}} = 1 - q_{52} = 1 - 0.5 = 0.5.$$

For I_3, $X_{in-3} = \{x_1\}$, then

$$q_{\overline{13}} = 1 - q_{13} = 1 - 0.83 = 0.17.$$

For I_4, $X_{in-4} = \{x_4\}$, then

$$q_{\overline{44}} = 1 - q_{44} = 1 - 0.5 = 0.5.$$

Table 12-3. Possibilities that x_i may leave I_j.

q_{ij}^{-}	I_1 [7.0,7.3[	I_2 [7.3,7.6[	I_3 [7.6,7.9[	I_4 [7.9,8.2[
$x_1 = 7.8$	0.00	0.00	0.17	0.00
$x_2 = 7.5$	0.00	0.17	0.00	0.00
$x_3 = 7.25$	0.33	0.00	0.00	0.00
$x_4 = 7.9$	0.00	0.00	0.00	0.50
$x_5 = 7.3$	0.00	0.50	0.00	0.00
$x_6 = 7.2$	0.17	0.00	0.00	0.00

Thirdly, we calculate all joining possibilities shown in Table 12-4. For I_1, we know, $X_{out-1} = \{x_1, x_2, x_4, x_5\}$. Therefore,

$$q_{11}^{+} = q_{11} = 0, \quad q_{21}^{+} = q_{21} = 0,$$

$$q_{41}^{+} = q_{41} = 0, \quad q_{51}^{+} = q_{51} = 0.5.$$

Because $X_{out-2} = \{x_1, x_3, x_4, x_6\}$, we obtain

$$q_{12}^{+} = q_{12} = 0, \quad q_{32}^{+} = q_{32} = 0.33,$$

$$q_{42}^{+} = q_{42} = 0, \quad q_{62}^{+} = q_{62} = 0.17.$$

For I_3, $X_{out-3} = \{x_2, x_3, x_4, x_5, x_6\}$, and

$$q_{23}^{+} = q_{23} = 0.17, \quad q_{33}^{+} = q_{33} = 0,$$

$$q_{43}^{+} = q_{43} = 0.5, \quad q_{53}^{+} = q_{53} = 0,$$

$$q_{63}^{+} = q_{63} = 0.$$

For I_4, $X_{out-4} = \{x_1, x_2, x_3, x_5, x_6\}$, and

$$q_{14}^{+} = q_{14} = 0.17, \quad q_{24}^{+} = q_{24} = 0,$$

$$q_{34}^{+} = q_{34} = 0, \quad q_{54}^{+} = q_{54} = 0,$$

$$q_{64}^{+} = q_{64} = 0.$$

Table 12-4. Possibilities that x_i may join I_j.

q_{ij}^{+}	I_1 [7.0,7.3[	I_2 [7.3,7.6[	I_3 [7.6,7.9[	I_4 [7.9,8.2[
$x_1 = 7.8$	0.00	0.00	0.00	0.17
$x_2 = 7.5$	0.00	0.00	0.17	0.00
$x_3 = 7.25$	0.00	0.33	0.00	0.00
$x_4 = 7.9$	0.00	0.00	0.50	0.00
$x_5 = 7.3$	0.50	0.00	0.00	0.00
$x_6 = 7.2$	0.00	0.17	0.00	0.00

Finally, we employ formula (12.8) to calculate all possibilities of p_k on I_j, $k = 0, 1, \cdots, 6$; $j = 1, 2, 3$, shown in Table 12-5.

For example, for I_1, we have

$$\pi_{I_1}(p_0) = \bigwedge_{s \in S_1} q_{s1}^- = q_{31}^- \wedge q_{61}^- = 0.33 \wedge 0.17 = 0.17,$$

$$\pi_{I_1}(p_1) = \bigvee_{s \in S_1} q_{s1}^- = q_{31}^- \vee q_{61}^- = 0.33 \vee 0.17 = 0.33,$$

$$\pi_{I_1}(p_2) = 1,$$

$$\pi_{I_1}(p_3) = \bigvee_{t \in T_1} q_{t1}^+ = q_{11}^+ \vee q_{21}^+ \vee q_{41}^+ \vee q_{51}^+ = 0 \vee 0 \vee 0 \vee 0.5 = 0.5,$$

$$\pi_{I_1}(p_4) = \bigvee_{t_1, t_2 \in T_1, t_1 \neq t_2} (q_{t_1 1}^+ \wedge q_{t_2 1}^+)$$

$$= (q_{11}^+ \wedge q_{21}^+) \vee (q_{11}^+ \wedge q_{41}^+) \vee (q_{11}^+ \wedge q_{51}^+) \vee (q_{21}^+ \wedge q_{41}^+) \vee (q_{21}^+ \wedge q_{51}^+) \vee (q_{41}^+ \wedge q_{51}^+)$$

$$= (0 \wedge 0) \vee (0 \wedge 0) \vee (0 \wedge 0.5) \vee (0 \wedge 0) \vee (0 \wedge 0.5) \vee (0 \wedge 0.5) = 0,$$

$$\pi_{I_1}(p_5) = \bigvee_{t_1, t_2, t_3 \in T_1, t_1 \neq t_2 \neq t_3} (q_{t_1 1}^+ \wedge q_{t_2 1}^+ \wedge q_{t_3 1}^+)$$

$$= (q_{11}^+ \wedge q_{21}^+ \wedge q_{41}^+) \vee (q_{11}^+ \wedge q_{21}^+ \wedge q_{51}^+) \vee (q_{21}^+ \wedge q_{41}^+ \wedge q_{51}^+)$$

$$= (0 \wedge 0 \wedge 0) \vee (0 \wedge 0 \wedge 0.5) \vee (0 \wedge 0 \wedge 0.5) = 0,$$

$$\pi_{I_1}(p_6) = \bigwedge_{t \in T_1} q_{t1}^+ = 0.$$

Table 12-5. The PPD of a natural disaster

$\pi_{I_j}(p_k)$	p_0	p_1	p_2	p_3	p_4	p_5	p_6
I_1	0.17	0.33	1	0.5	0	0	0
I_2	0.17	0.5	1	0.33	0.17	0	0
I_3	0.17	1	0.5	0	0	0	0
I_4	0.5	1	0.17	0	0	0	0

12.2.3 Algorithm and Fortran program

It is not easy to use above model in practice. We transform it into an algorithm and give a Fortran program to run it.

Suppose we have a sample X with universe U. Using the formula of linear-information-distribution, we can get q_{ij}.

Let

$$\mu_Q(x_i, u_j) = q_{ij}$$
$$= \begin{cases} 1 - \frac{|x_i - u_j|}{\Delta}, & \text{if } |x_i - u_j| \leq \Delta; \\ 0, & \text{if } |x_i - u_j| > \Delta. \end{cases} \quad x_i \in X, \ u_j \in U.$$

$$\tag{12.9}$$

where $\Delta \equiv u_j - u_{j-1}, j = 2, 3, \cdots, m.$ Q is a fuzzy set of $X \times U$, called the *fuzzy relation between observations and controlling points*, denoted as $Q = \{q_{ij}\}$.

Let

$$\mu_E(x_i, u_j) = \begin{cases} 1, & \text{if } x_i \text{ is an interior point of } I_j; \\ 0, & \text{otherwise.} \end{cases} \tag{12.10}$$

E is also a fuzzy set of $X \times U$, called the *fuzzy relation between observations and derived intervals from controlling points*, denoted as $E = \{e_{ij}\}$.

Let

$$\mu_{Q^-}(x_i, u_j) = \max\left(0, \mu_E(x_i, u_j) - \mu_Q(x_i, u_j)\right), \tag{12.11}$$

and

$$\mu_{Q^+}(x_i, u_j) = \max\left(0, \mu_Q(x_i, u_j) - \mu_E(x_i, u_j)\right). \tag{12.12}$$

Then,

$$q_{ij}^- = \mu_{Q^-}(x_i, u_j)$$

is just the leaving possibility that x_i may leave interval I_j, and,

$$q_{ij}^+ = \mu_{Q^+}(x_i, u_j)$$

is just the joining possibility that x_i may join interval I_j.

$Q^- = \{q_{ij}^-\}$ and $Q^+ = \{q_{ij}^+\}$ are called the *leaving and joining matrixes*, respectively.

For above calculation case, using formulae (12.9) and (12.10), we have

$$Q = \begin{pmatrix} 0 & 0 & 0.83 & 0.17 \\ 0 & 0.83 & 0.17 & 0 \\ 0.67 & 0.33 & 0 & 0 \\ 0 & 0 & 0.50 & 0.50 \\ 0.50 & 0.50 & 0 & 0 \\ 0.83 & 0.17 & 0 & 0 \end{pmatrix}, \quad E = \begin{pmatrix} 0 & 0 & 1 & 0 \\ 0 & 1 & 0 & 0 \\ 1 & 0 & 0 & 0 \\ 0 & 0 & 0 & 1 \\ 0 & 1 & 0 & 0 \\ 1 & 0 & 0 & 0 \end{pmatrix}.$$

Hence, we obtain

$$E - Q = \begin{pmatrix} 0 & 0 & 0.17 & -0.17 \\ 0 & 0.17 & -0.17 & 0 \\ 0.33 & -0.33 & 0 & 0 \\ 0 & 0 & -0.50 & 0.50 \\ -0.50 & 0.50 & 0 & 0 \\ 0.17 & -0.17 & 0 & 0 \end{pmatrix},$$

$$Q - E = \begin{pmatrix} 0 & 0 & -0.17 & 0.17 \\ 0 & -0.17 & 0.17 & 0 \\ -0.33 & 0.33 & 0 & 0 \\ 0 & 0 & 0.50 & -0.50 \\ 0.50 & -0.50 & 0 & 0 \\ -0.17 & 0.17 & 0 & 0 \end{pmatrix}.$$

The leaving and joining matrixes are

$$Q^- = \begin{pmatrix} 0 & 0 & 0.17 & 0 \\ 0 & 0.17 & 0 & 0 \\ 0.33 & 0 & 0 & 0 \\ 0 & 0 & 0 & 0.50 \\ 0 & 0.50 & 0 & 0 \\ 0.17 & 0 & 0 & 0 \end{pmatrix}, \quad Q^+ = \begin{pmatrix} 0 & 0 & 0 & 0.17 \\ 0 & 0 & 0.17 & 0 \\ 0 & 0.33 & 0 & 0 \\ 0 & 0 & 0.50 & 0 \\ 0.50 & 0 & 0 & 0 \\ 0 & 0.17 & 0 & 0 \end{pmatrix}.$$

Concluded above, we obtain the *MOVING-algorithm*:

Step 1-1. Input X and U;

Step 1-2. Calculate Q by formula (12.9);

Step 1-3. Construct E according to definition of the interior point;

Step 1-4. Calculate $E - Q$ and $Q - E$;

Step 1-5. Change the negative values in $E - Q$ and $Q - E$ into 0, and then, respectively, we obtain Q^- and Q^+.

Employing the fuzzy relation $\underset{\sim}{E}$, we can construct index sets S and T.

$$S_j = \{i | i \in N, \mu_E(x_i, u_j) = 1\}, \quad T_j = \{i | i \in N, \mu_E(x_i, u_j) = 0\}, \quad (12.13)$$

where, $N = \{1, 2, \cdots, n\}$ and $j = 1, 2, \cdots, m$. Then, we obtain

$$n_j = |S_j|, \quad j = 1, 2, \cdots, m. \quad (12.14)$$

According to formula (12.8), we know that, for $k = n_j$, possibility-probability $\pi_{I_j}(p_k) = 1$.

We use S_j and Q^- to calculate the possibility-probability $\pi_{I_j}(p_k)$ for $k < n_j$, and T_j and Q^+ for $k > n_j$. We need a *combination algorithm* to do that.

As we well know, the number of all combinations of n different elements taken l at a time is

$$C_n^l = \frac{n(n-1)\cdots(n-l+1)}{1 \cdot 2 \cdot \cdots \cdot l}.$$

$\forall l \in \{1, 2, \cdots, n_j\}$, we let $k = n_j - l$. According to formula (12.8), we have

$$\pi_{I_j}(p_k) = \bigvee_{s_1, s_2, \cdots, s_l \in S_j, s_1 \neq s_2 \neq \cdots \neq s_l} (q_{s_1 j}^- \wedge q_{s_2 j}^- \wedge \cdots \wedge q_{s_l j}^-), \quad k = n_j - l. \quad (12.15)$$

It involves all combinations of n_j different elements (of S_j) taken l at a time. We need an algorithm to get all combinations of S_j.

$\forall S$, no loss in generality, we suppose

$$S = \{a_1, a_2, \cdots, a_n\}.$$

Our task is to get an index set to indicate which of them are in a combination. For example, $A = \{a_1, a_3, a_n\}$ is a combination of n different elements taken 3 at a time. The combination can be identified by the element $\{1, 3, n\}$ of the index set.

Let A_t be t-th combination of different elements taken l. In most common order, we have

$$A_1 = \{a_1, a_2, \cdots, a_{l-1}, a_l\}, \quad A_2 = \{a_1, a_2, \cdots, a_{l-1}, a_{l+1}\}, \quad \cdots,$$
$$A_{n-l+1} = \{a_1, a_2, \cdots, a_{l-1}, a_n\}, \quad \cdots, \quad A_{C_n^l} = \{a_{n-l+1}, a_{n-l+2}, \cdots, a_{n-1}, a_n\}.$$

The corresponding index sets are

$$\{1, 2, \cdots, l-1, l\}, \{1, 2, \cdots, l-1, l+1\}), \cdots, \{1, 2, \cdots, l-1, n\},$$
$$\cdots, \{n-l+1, n-l+2, \cdots, n-1, n\}.$$

We employ a matrix, called INDEX, to store the index sets.

$$\text{INDEX} = \begin{pmatrix} 1 & 2 & \cdots & l-1 & l \\ 1 & 2 & \cdots & l-1 & l+1 \\ \cdots & \cdots & \cdots & \cdots & \cdots \\ 1 & 2 & \cdots & l-1 & n \\ \cdots & \cdots & \cdots & \cdots & \cdots \\ n-l+1 & n-l+2 & \cdots & n-1 & n \end{pmatrix}.$$

An element of the matrix is denoted as $\text{INDEX}(t, i)$, $t = 1, 2, \cdots, C_n^l; i = 1, 2, \cdots, l$.

Analyzing the INDEX, we obtain the *INDEX-algorithm*:

Step 2-1. Let $t = 1$, $C(i) = i, i = 1, 2, \cdots, l$;

Step 2-2. Let $INDEX(t, i) = C(i), i = 1, 2, \cdots, l$.

Step 2-3. Let $t = t + 1$. If $C(1) = n - l + 1$, stop;

Step 2-4. Let $i = 1$;

Step 2-5. If $C(i) \geq n - l + i$, go to step 2-8;

Step 2-6. If $i = l$, go to step 2-9

Step 2-7. Let $i = i + 1$ and go to step 2-5;

Step 2-8. Let $i = i - 1$;

Step 2-9. Let $C(i) = C(i) + 1$;

Step 2-10. Let $C(j) = C(j - 1) + 1$, $j = i + 1, i + 2, \cdots, l$, and go to step 2-2.

Employing the MOVING-algorithm, formula (12.15) and INDEX-algorithm, we can calculate $\pi_{I_j}(p_k)$ for $k < n_j$. Considering the symmetry of formula (12.8), also by above two algorithms and

$$\pi_{I_j}(p_k) = \bigvee_{t_1, t_2, \cdots, t_l \in T_j, t_1 \neq t_2 \neq \cdots \neq t_l} (q_{t_1 j}^+ \wedge q_{t_2 j}^+ \wedge \cdots \wedge q_{t_l j}^+), \quad k = n_j + l. \quad (12.16)$$

we can calculate $\pi_{I_j}(p_k)$ for $k > n_j$.

Naturally, we obtain the algorithm for the interior-outer-set model:

Step 1. Use the MOVING-algorithm to calculate Q^- and Q^+;

Step 2. Construct S_j and T_j according to formula (12.13);

Step 3. Calculate n_j according to formula (12.14);

Step 4. Use the INDEX-algorithm to construct INDEX matrixes of S_j, $j = 1, 2, \cdots, m$, for $l = 1, 2, \cdots, n_j$;

Step 5. Calculate $\pi_{I_j}(p_k)$, $k = n_j - l$, with the INDEX matrix of S_j by using formula (12.15);

Step 6. Let $\pi_{I_j}(p_{n_j}) = 1$;

Step 7. Use the INDEX-algorithm to construct INDEX matrixes of T_j, $j = 1, 2, \cdots, m$, for $l = 1, 2, \cdots, n - n_j$;

Step 8. Calculate $\pi_{I_j}(p_k)$, $k = n_j + l$, with the INDEX matrix of T_j by using formula (12.16);

Appendix 12.A gives a Fortran program to perform this algorithm to calculate the case in subsection 12.2.2.

12.3 Ranking Alternatives Based on a PPD

One of the main tasks of risk management is to rank alternatives so that we would avoid or control risks.

12.3.1 Classical model of ranking alternatives

Let O be a risk system facing n possible adverse outcomes, and let g be an alternative to reduce the loss from the adverse outcomes. For example, a building in reinforced concrete frame can be considered as an earthquake-risk system. Suppose that the building will be used 50 years, and in the future 50 years the building may meet 4 possible earthquake measured by magnitude, $m_1 = 5, m_2 = 6, m_3 = 7, m_4 = 8$. When we are going to construct the building, one of the alternatives is to increase the ductility of the joints and strengthen the frame in rank to meet the requirements of earthquake resistance. To execute the alternative increasing the ductility, we have to spend more money. The addition is called the alternative cost, denoted by c. If we do not take any measure, suppose the loss is l_0. Taking the alternative, we lose l when an adverse outcome, in m, occurs. Then, the benefit output from the alternative is $b = (l_0 - l) - c$. The larger the benefit output, the better the alternative. In general, the benefit output of an alternative is determined by the adverse outcome that really occurs in the future, and the relation between the benefit output and the adverse outcome is fuzzy. To be easy, we suppose the relation is a mathematical crisp mapping f from the space of adverse events to the benefit-output space.

Definition 12.5 Let M be the space of adverse events, B be the benefit-output space. mapping

$$f: \quad M \to B$$
$$m \mapsto f(m)$$

is called *benefit-output function*.

In general, an alternative g defines a mapping f, and a set of alternatives $G = \{g_1, g_2, \cdots, g_n\}$ defines a set of mappings $F = \{f_1, f_2, \cdots, f_n\}$.

If we can precisely calculate the probability distribution of the adverse outcomes, it is easy to choice the best alternative. Let $p(m)$ be a probability distribution we obtained, and $f(m)$ be the benefit-output function of an alternative. Then the expected value of the benefit output is

$$E(f(m)) = \int_M f(m)p(m)dm. \qquad (12.17)$$

Obviously, the best alternative is the one whose expected value is the largest among the alternatives.

Example 12.5 Suppose that, in a seismic active belt near the city we study, the probability distribution of earthquake magnitudes is

$$p(m) = 0.5e^{-0.5m}, \quad m \in M = [0, 7].$$

For a project, we have two alternatives, g_1 and g_2, to resist the earthquake force. For magnitude $m \in [0, 7]$, suppose the benefit-output functions of the alternatives are

$$g_1 : f_1(m) = -20 + 11m;$$
$$g_2 : f_2(m) = \begin{cases} -30 + 20m, & \text{for } 0 \le m \le 5; \\ 245 - 35m, & \text{for } 5 \le m \le 7. \end{cases}$$

as shown in Fig. 12.1.

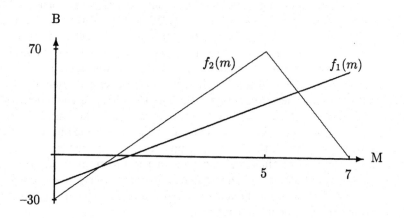

Fig. 12.1 Benefit-output functions corresponding to two alternatives

According to formula (12.17) we have

$$E(f_1(m)) = \int_M f_1(m)p(m)dm$$
$$= \int_0^7 (-20 + 11m)0.5e^{-0.5m}dm$$
$$= -19.4 + 20.7$$
$$= 1.3$$

and

$$E(f_2(m)) = \int_M f_2(m)p(m)dm$$
$$= \int_0^5 (-30 + 20m)0.5e^{-0.5m}dm + \int_5^7 (245 - 35m)0.5e^{-0.5m}dm$$
$$= -27.6 + 28.8 + 12.3 - 10.5 \qquad\qquad = 3$$

Above, we have employed formulae

$$\int_a^b e^x dx = e^b - e^a,$$

and

$$\int_a^b xe^{-x}dx = -[xe^{-x}]_a^b + \int_a^b e^{-x}dx.$$

Because $E(f_2(m)) > E(f_1(m))$, alternative g_2 is better than g_1.

12.3.2 Fuzzy expected value

Now, we turn to the case of a PPD; i.e., suppose that we cannot obtain a probability-risk $p(m)$, but obtain a fuzzy risk $\Pi_{M,P}$, where M and P are the universes of the adverse outcome and concerning probability, respectively.

The oldest definition of the *fuzzy expected value* (FEV) was given for any $\mathcal{B}$-measurable function μ_A such that $\mu_A \in [0,1]$ (i.e., the membership function of a fuzzy set $\underset{\sim}{A}$) with a fuzzy measure $\chi(\cdot)$. That is[6]

$$\text{FEV} = \sup_{T \in [0,1]} \{\min[T, \chi(\xi_T)]\},$$

where

$$\xi_T = \{x | \mu_A(x) \geq T\}.$$

The fuzzy measure $\chi(\cdot)$, in fact, is a set function defined on $\mathcal{B}$, a Borel field of subsets of $\Omega \subset \mathbb{R}$, and satisfies
   (i) $\chi(\emptyset) = 0$, $\emptyset$ is the empty set and $\chi(\Omega) = 1$;

(ii) if A, $B \in \mathcal{B}$ with $A \subset B$, then $\chi(A) \leq \chi(B)$;

(iii) if $\{A_n\}$ is a monotone sequence in $\mathcal{B}$ then

$$\lim_{n \to \infty} \chi(A_n) = \chi(\lim_{n \to \infty}(A_n)).$$

By the FEV definition for measuring some sort of central tendency, Zadeh [12] and Kandel [5] attempted to generalize to fuzzy sets certain notions of the axiomatic probability theory. In their framework, fuzzy probability is the probability of fuzzy event rather than the possibility-probability.

In 1993, with the α-cut, Heilpern [3] given a definition of the expected value of a variable induced by fuzzy evidence

$$G \equiv \Pr(V \text{ is } A) \text{ is } Q,$$

where A is a fuzzy subset of the space S and Q is a fuzzy probability. V takes values in the space S. The evidence G induces on the space $\mathcal{M}(S)$ a fuzzy subset M_A. The FEV of V is defined by

$$(E(V))_\alpha = [e_1(\alpha), e_2(\alpha)],$$

where

$$e_1(\alpha) = \min\{E(F)|F \in (M_A)_\alpha\} \text{ and } e_2(\alpha) = \max\{E(F)|F \in (M_A)_\alpha\}.$$

Römer and Kandel (1995, see [9]) suggested

$$[\overline{X}_n(\delta)]_\alpha = [[\underline{X_n(\delta)}]_\alpha, \overline{[X_n(\delta)]}_\alpha]$$
$$= [\min_{x \in [\delta]_\alpha} \overline{X}_n(x), \max_{x \in [\delta]_\alpha} \overline{X}_n(x)]$$

to calculate FEV of fuzzy data, where $\delta = (\delta_1, \delta_2, \cdots, \delta_m)$ is a fuzzy sample vector and $x = (x_1, x_2, \cdots, x_m)$ is a point in the Cartesian space $(\delta_1)_\alpha \times (\delta_2)_\alpha \times \cdots \times (\delta_m)_\alpha$, $\alpha \in [0, 1]$. They assumed the following forms to calculate the FEV.

$$[\underline{X_n(\delta)}]_\alpha = \frac{1}{n}\sum_{i=1}^{n}[\underline{\delta_i}]_\alpha, \quad \overline{[X_n(\delta)]}_\alpha = \frac{1}{n}\sum_{i=1}^{n}\overline{[\delta_i]}_\alpha,$$

and

$$[\underline{\delta_i}]_\alpha = \min[\delta_i]_\alpha = \min\{x | x \in \mathbb{R}, \mu_{\delta_i}(x) \geq \alpha\},$$

$$\overline{[\delta_i]}_\alpha = \max[\delta_i]_\alpha = \max\{x | x \in \mathbb{R}, \mu_{\delta_i}(x) \geq \alpha\}.$$

Qiao, Zhang and Wang (1994, see [8]) also given a formula to calculate the expectation of a fuzzy random variable $\tilde{a}$ by

$$E(\tilde{a}) = \bigcup_{\alpha \in]0,1]} \alpha[E(a_\alpha^-), E(a_\alpha^+)],$$

where

$$a_\alpha^- = \min\{x | x \in \mathbb{R}, \mu_a(x) \ge \alpha\},$$

and

$$a_\alpha^+ = \max\{x | x \in \mathbb{R}, \mu_a(x) \ge \alpha\}.$$

Based on all existent models of FEV, we suggest the following definitions.

Definition 12.6 Let M be the space of adverse events, P be universe of discourse of probability, and

$$\underset{\sim}{p}(m) \overset{\Delta}{=} \{\pi_m(p) | m \in M, p \in P\} \tag{12.18}$$

be a PPD. $\forall \alpha \in [0, 1]$, let

$$\underline{p}_\alpha(m) = \min\{p | p \in P, \pi_m(p) \ge \alpha\}, \tag{12.19a}$$

$$\overline{p}_\alpha(m) = \max\{p | p \in P, \pi_m(p) \ge \alpha\}. \tag{12.19b}$$

$\underline{p}_\alpha(m)$ is called the *minimum probability* in α-cut with respect to m, and $\overline{p}_\alpha(m)$ is called the *maximum probability* in α-cut with respect to m.
The finite closed interval

$$p_\alpha(m) \overset{\Delta}{=} [\underline{p}_\alpha(m), \overline{p}_\alpha(m)] \tag{12.20}$$

is called α-*cut* of $\underset{\sim}{p}(m)$ with respect to m.

Example 12.6 Let

$$M = \{7.15, 7.45, 7.75, 8.05\}, \quad P = \{0, 0.17, 0.33, 0.5, 0.67, 0.83, 1\}$$

be the space of adverse events and the universe of discourse of probability, respectively, and

$$\begin{aligned}
\underset{\sim}{p}(m) = \ & \{\pi_{m_i}(p_j) | m_i \in M, p_j \in P\} \\
= \ & \{\pi_{m_1}(p_1), \pi_{m_1}(p_2), \cdots, \pi_{m_4}(p_6), \pi_{m_4}(p_7)\} \\
= \ & \{0.17, 0.33, 1, 0.5, 0, 0, 0, 0.17, 0.5, 1, 0.33, 0.17, 0, 0, \\
& 0.17, 1, 0.5, 0, 0, 0, 0, 0.5, 1, 0.17, 0, 0, 0, 0\}
\end{aligned}$$

be a PPD that can also be represented by a matrix

$$\underset{\sim}{p}(m) = \begin{array}{c} \\ \\ m_1 = 7.15 \\ m_2 = 7.45 \\ m_3 = 7.75 \\ m_4 = 8.05 \end{array} \begin{array}{ccccccc} p_0 & p_1 & p_2 & p_3 & p_4 & p_5 & p_6 \\ 0 & 0.17 & 0.33 & 0.5 & 0.67 & 0.83 & 1 \\ \left(\begin{array}{ccccccc} 0.17 & 0.33 & 1 & 0.5 & 0 & 0 & 0 \\ 0.17 & 0.5 & 1 & 0.33 & 0.17 & 0 & 0 \\ 0.17 & 1 & 0.5 & 0 & 0 & 0 & 0 \\ 0.5 & 1 & 0.17 & 0 & 0 & 0 & 0 \end{array} \right) \end{array}.$$

For $m = 7.45$ and $\alpha = 0.25$, we have

$$\begin{aligned}
\underline{p}_{0.25}(7.45) = \ & \min\{p | p \in P, \pi_{7.45}(p) \ge 0.25\} \\
= \ & \min\{0.17, 0.33, 0.5\} \\
= \ & 0.17,
\end{aligned}$$

$$\overline{p}_{0.25}(7.45) = \max\{p|p \in P, \pi_{7.45}(p) \geq 0.25\}$$
$$= \max\{0.17, 0.33, 0.5\}$$
$$= 0.5,$$

so, the 0.25-cut of $\underset{\sim}{p}(m)$ with respect to $m = 7.45$ is

$$[\underline{p}_{0.25}(7.45), \overline{p}_{0.25}(7.45)] = [0.17, 0.5].$$

Definition 12.7 Let

$$\underline{E}_\alpha(m) = \int_M m\underline{p}_\alpha(m)dm, \quad \overline{E}_\alpha(m) = \int_M m\overline{p}_\alpha(m)dm. \qquad (12.21)$$

We call

$$E_\alpha(m) \overset{\triangle}{=} [\underline{E}_\alpha(m), \overline{E}_\alpha(m)] \qquad (12.22)$$

the *expected interval* of α-cut of $\underset{\sim}{p}(m)$ with respect to m.

$$\underset{\sim}{E}(m) \overset{\triangle}{=} \int_M m\underset{\sim}{p}(m)dm \overset{\triangle}{=} \bigcup_{\alpha \in]0,1]} \alpha E_\alpha(m) \qquad (12.23)$$

is called the *fuzzy expected value* of $\underset{\sim}{p}(m)$.

Example 12.7 For $\underset{\sim}{p}(m)$ in Example 12.6, we have
(1) With respect to m_1,

$\forall \alpha \in]0, 0.17]$, $p_\alpha(m_1) = [0, 0.5]$; $\forall \alpha \in]0.17, 0.33]$, $p_\alpha(m_1) = [0.17, 0.5]$;

$\forall \alpha \in]0.33, 0.5]$, $p_\alpha(m_1) = [0.33, 0.5]$; $\forall \alpha \in]0.5, 1]$, $p_\alpha(m_1) = [0.33, 0.33]$.

(2) With respect to m_2,

$\forall \alpha \in]0, 0.17]$, $p_\alpha(m_2) = [0, 0.67]$; $\forall \alpha \in]0.17, 0.33]$ $p_\alpha(m_2) = [0.17, 0.5]$;

$\forall \alpha \in]0.33, 0.5]$, $p_\alpha(m_2) = [0.17, 0.33]$; $\forall \alpha \in]0.5, 1]$, $p_\alpha(m_2) = [0.33, 0.33]$.

(3) With respect to m_3,

$\forall \alpha \in]0, 0.17]$, $p_\alpha(m_3) = [0, 0.33]$; $\forall \alpha \in]0.17, 0.33]$, $p_\alpha(m_3) = [0.17, 0.33]$;

$\forall \alpha \in]0.33, 0.5]$, $p_\alpha(m_3) = [0.17, 0.33]$; $\forall \alpha \in]0.5, 1]$, $p_\alpha(m_3) = [0.17, 0.17]$.

(4) With respect to m_4,

$\forall \alpha \in]0, 0.17]$, $p_\alpha(m_4) = [0, 0.33]$; $\forall \alpha \in]0.17, 0.33]$, $p_\alpha(m_4) = [0, 0.17]$;

$\forall \alpha \in]0.33, 0.5]$, $p_\alpha(m_4) = [0, 0.17]$; $\forall \alpha \in]0.5, 1]$, $p_\alpha(m_4) = [0.17, 0.17]$.

For the discrete random variable, according to formula (3.5), we have

$$\underline{E}_\alpha(m) = \int_M m\underline{p}_\alpha(m)dm = \sum_{i=1}^n m_i\underline{p}_\alpha(m_i), \qquad (12.24a)$$

$$\overline{E}_\alpha(m) = \int_M m\overline{p}_\alpha(m)dm = \sum_{i=1}^n m_i\overline{p}_\alpha(m_i). \qquad (12.24b)$$

In our case, we obtain

$$\forall \alpha \in]0, 0.17], \quad \underline{E}_\alpha(m) = 0,$$

$$\begin{aligned}\overline{E}_\alpha(m) \quad &= 7.15 \times 0.5 + 7.45 \times 0.67 + 7.75 \times 0.33 + 8.05 \times 0.33 \\ &= 3.58 + 4.99 + 2.56 + 2.66 \\ &= 13.79.\end{aligned}$$

That is, $\forall \alpha \in]0, 0.17]$, the expected interval of α-cut of $\underline{p}(m)$ is $E_\alpha(m) = [0, 13.79]$. Similarly, we obtain

$$E_\alpha(m) = [3.81, 11.23], \quad \forall \alpha \in]0.17, 0.33],$$

$$E_\alpha(m) = [4.95, 9.97], \quad \forall \alpha \in]0.33, 0.5],$$

$$E_\alpha(m) = [7.51, 7.51], \quad \forall \alpha \in]0.5, 1].$$

12.3.3 Center of gravity of a fuzzy expected value

In many cases, we need a crisp value for decision-making. The simplest approach to defuzzify a FEV $\underline{E}(m)$ into a crisp value is the so-called center-of-gravity method:

$$CG(\underline{E}(m)) = \int_{-\infty}^{+\infty} m\mu_E(m)dm \Big/ \int_{-\infty}^{+\infty} \mu_E(m)dm. \qquad (12.25)$$

For discrete case, the expected intervals of α-cuts can be written as

$$E_\alpha(m) = [a_1, b_1], \quad \forall \alpha \in]\gamma_1, \beta_1],$$

$$E_\alpha(m) = [a_2, b_2], \quad \forall \alpha \in]\gamma_2, \beta_2],$$

$$\cdots, \cdots$$

$$E_\alpha(m) = [a_s, b_s], \quad \forall \alpha \in]\gamma_s, \beta_s],$$

satisfying

$$a_s = b_s, \ \gamma_1 = 0, \ \gamma_2 = \beta_1, \ \cdots, \ \gamma_s = \beta_{s-1}, \ \beta_s = 1.$$

Set

$$\Lambda = \{a_1, b_1, a_2, b_2, \cdots, a_s\}$$

is called the *feature set of the universe* of the discrete FEV. Sets

$$\underline{\Theta} = \{\gamma_1, \gamma_2, \cdots, \gamma_s\}, \quad \overline{\Theta} = \{\beta_1, \beta_2, \cdots, \beta_s\}$$

are called the *min-α and max-α feature subsets*, respectively. Mappings

$$\underline{\mu}_E : \begin{array}{ccc} \Lambda & \to & \underline{\Theta} \\ a_1 & \mapsto & \gamma_1 \\ b_1 & \mapsto & \gamma_1 \\ a_2 & \mapsto & \gamma_2 \\ b_2 & \mapsto & \gamma_2 \\ \cdots & \cdots & \\ a_s & \mapsto & \gamma_s \end{array} \qquad (12.26a),$$

and

$$\overline{\mu}_E : \begin{array}{ccc} \Lambda & \to & \overline{\Theta} \\ a_1 & \mapsto & \beta_1 \\ b_1 & \mapsto & \beta_1 \\ a_2 & \mapsto & \beta_2 \\ b_2 & \mapsto & \beta_2 \\ \cdots & \cdots & \\ a_s & \mapsto & \beta_s \end{array} \qquad (12.26b)$$

are called the *min-α and max-α membership functions*, respectively.

Definition 12.8 Let $\underline{\mu}_E(m)$ and $\overline{\mu}_E(m)$ are the min-α and max-α membership functions of a discrete FEV, respectively.

$$CG(\underline{E}(m))_D$$
$$= \frac{1}{2}[(\sum_{m\in\Lambda} m\underline{\mu}_E(m) / \sum_{m\in\Lambda} \underline{\mu}_E(m)) + (\sum_{m\in\Lambda} m\overline{\mu}_E(m) / \sum_{m\in\Lambda} \overline{\mu}_E(m))] \quad (12.27)$$

is called the *center of gravity* of the discrete FEV.

Example 12.8 For the discrete FEV given by Example 12.7, we have the feature set of the universe

$$\begin{aligned} \Lambda &= \{a_1, b_1, a_2, b_2, a_3, b_3, a_4\} \\ &= \{0, 13.79, 3.81, 11.23, 4.95, 9.97, 7.51\}, \end{aligned}$$

and the min-α and max-α feature subsets are

$$\underline{\Theta} = \{\gamma_1, \gamma_2, \gamma_3, \gamma_4\} = \{0, 0.17, 0.33, 0.5\},$$

$$\overline{\Theta} = \{\beta_1, \beta_2, \beta_3, \beta_4\} = \{0.17, 0.33, 0.5, 1\}.$$

Then, we obtain the min-α and max-α membership functions shown in Table 12-6.

Table 12-6 Min-α and max-α membership functions

m	0	3.81	4.95	7.51	9.97	11.23	13.79
$\underline{\mu}_E(m)$	0	0.17	0.33	0.5	0.33	0.17	0
$\overline{\mu}_E(m)$	0.17	0.33	0.5	1	0.5	0.33	0.17

Therefore,

$$\sum_{m\in\Lambda} m\underline{\mu}_{E}(m)$$
$$= 3.81 \times 0.17 + 4.95 \times 0.33 + 7.51 \times 0.5 + 9.97 \times 0.33 + 11.23 \times 0.17$$
$$= 11.24,$$

$$\sum_{m\in\Lambda} \underline{\mu}_{E}(m)$$
$$= 0.17 + 0.33 + 0.5 + 0.33 + 0.17$$
$$= 1.5,$$

$$\sum_{m\in\Lambda} m\overline{\mu}_{E}(m)$$
$$= 3.81 \times 0.33 + 4.95 \times 0.5 + 7.51 \times 1 + 9.97 \times 0.5 + 11.23 \times 0.33 + 13.79 \times 0.17$$
$$= 22.28,$$

$$\sum_{m\in\Lambda} \overline{\mu}_{E}(m)$$
$$= 0.17 + 0.33 + 0.5 + 1 + 0.5 + 0.33 + 0.17$$
$$= 3.$$

Hence, we obtain

$$CG(\underset{\sim}{E}(m))_D = \frac{1}{2}(\frac{11.24}{1.5} + \frac{22.28}{3}) = 7.46$$

We notice that the operator changing $\underset{\sim}{p}(m)$ into $\underset{\sim}{E}(m)$ is linear. Therefore, when we rank the alternatives according to the center of gravity, the procedure can be simplified to avoid α-cuts computation.

In the case, firstly, we calculate the center of gravity of $\underset{\sim}{p}(m)$ with respect to m, which is

$$c(m) = \int_{P} p\pi_m(p)dp \Big/ \int_{P} \pi_m(p)dp, \qquad (12.28)$$

It indicates the average probability that m occurring. Employing

$$p(m) = c(m) \Big/ \int_{M} c(m)dm, \qquad (12.29)$$

to normalize the average probability distribution, we obtain a *simple center-of-gravity formula*

$$E(m) = \int_{M} mp(m)dm. \qquad (12.30)$$

to calculate the expected value of $\underset{\sim}{p}(m)$.

Example 12.9 We use the simple formula to calculate the center of gravity of the discrete FEV given by Example 12.7. As we know, the FEV is produced from the $\underset{\sim}{p}(m)$ in Example 12.6. In the case, the center of gravity of $\underset{\sim}{p}(m_i)$ with respect to m_i is

$$c(m_i) = \left(\sum_{j=0}^{6} p_j \pi_{m_i}(p_j)\right) \Big/ \sum_{j=0}^{6} \pi_{m_i}(p_j), \quad i = 1, 2, 3, 4.$$

We have

$$c(m_1) = \frac{0.17 \times 0.33 + 0.33 \times 1 + 0.50 \times 0.50}{0.17 + 0.33 + 1 + 0.50} = 0.32,$$

$$c(m_2) = \frac{0.17 \times 0.50 + 0.33 \times 1 + 0.50 \times 0.33 + 0.67 \times 0.17}{0.17 + 0.50 + 1 + 0.33 + 0.17} = 0.32,$$

$$c(m_3) = \frac{0.17 \times 1 + 0.33 \times 0.50}{0.17 + 1 + 0.50} = 0.20,$$

$$c(m_4) = \frac{0.17 \times 1 + 0.33 \times 0.17}{0.50 + 1 + 0.17} = 0.14.$$

Normalizing them, we obtain

$$\{p(m_1), p(m_2), p(m_3), p(m_4)\} = \{0.33, 0.33, 0.20, 0.14\}.$$

Therefore, by above simple formula, we obtain the expected value of $\underset{\sim}{p}(m)$

$$\begin{aligned}
E(m) &= \int_M m p(m) dm \\
&= 7.15 \times 0.33 + 7.45 \times 0.33 + 7.75 \times 0.20 + 8.05 \times 0.14 \\
&= 7.50.
\end{aligned}$$

12.3.4 Ranking alternatives by FEV

Definition 12.9 Let $f(m)$ be a benefit-output function with a PPD $\underset{\sim}{p}(m)$. We call

$$\underset{\sim}{E}(f(m)) = \int_M f(m) \underset{\sim}{p}(m) dm \tag{12.31}$$

the fuzzy expected value of $f(m)$ with $\underset{\sim}{p}(m)$.

Mathematically speaking, we can also employ so-called α-cuts tool to run the formula. That is,

$$\underset{\sim}{E}(f(m)) = \bigcup_{\alpha \in]0,1]} \alpha E_\alpha(f(m)). \tag{12.32}$$

The simplest criterion for the ranking is the center of gravity of $\underset{\sim}{E}(f(m))$, calculated by

$$CG(\underline{E}(f(m))) = \int_{-\infty}^{+\infty} f(m)\mu_{E_f}(m)dm \Big/ \int_{-\infty}^{+\infty} \mu_{E_f}(m)dm, \qquad (12.33)$$

where $\mu_{E_f}(m)$ is the membership function of FEV $\underline{E}f(m)$. Corresponding to formula (12.30), we have a simple formula to calculate the center-of-gravity of FEV of $f(m)$ with $\underline{p}(m)$, which is

$$E(f(m)) = \int_M f(m)p(m)dm, \qquad (12.34)$$

where $p(m)$ is calculated by formula (12.28) and (12.29). The form of the formula (12.34) is just the same as the formula (12.17), however their $p(m)$ is different. For the former $p(m)$ is given by using a classical statistic method; for the latter $p(m)$ is given by using a fuzzy statistic method such as the interior-outer-set model.

If $E(f_i(m)) > E(f_j(m))$, we say that the alternative g_i defined by $f_i(m)$ is better than the alternative g_j defined by $f_j(m)$.

Example 12.10 Let discrete universe of disaster index

$$U = \{u_1, u_2, \cdots, u_{10}\} = \{0.05, 0.15, 0.25, \cdots, 0.95\}$$

be the space of adverse events, and

$$\begin{aligned} P &= \{p_0, p_1, p_2, p_3, p_4, p_5, p_6, p_7, p_8, p_9\} \\ &= \{0, 0.11, 0.22, 0.33, 0.44, 0.56, 0.67, 0.78, 0.89, 1\}. \end{aligned}$$

be universe of discourse of probability. Give a PPD

$$\underline{p}(u) = \begin{matrix} & \begin{matrix} p_0 & p_1 & p_2 & p_3 & p_4 & p_5 & p_6 & p_7 \\ 0 & 0.11 & 0.22 & 0.33 & 0.44 & 0.56 & 0.67 & 0.78 \end{matrix} \\ \begin{matrix} u_1 = 0.05 \\ u_2 = 0.15 \\ u_3 = 0.25 \end{matrix} & \begin{pmatrix} 0.01 & 0.10 & 0.10 & 0.12 & 0.28 & 0.40 & 1 & 0.20 \\ 0.20 & 0.33 & 1 & 0 & 0 & 0 & 0 & 0 \\ 0 & 1 & 0.33 & 0 & 0 & 0 & 0 & 0 \end{pmatrix} \end{matrix}.$$

Suppose that we have two benefit-output functions

$$g_1 : f_1(u) = -1 + 11u;$$
$$g_2 : f_2(u) = \begin{cases} -0.5 + 21u, & \text{for } 0 \le u \le 0.5; \\ 20 - 20u, & \text{for } 0.5 \le u \le 1. \end{cases}$$

Let us calculate the excepted values of $f_1(u)$ and $f(u_2)$ with the given $\underline{p}(u)$, respectively. Using formula (12.28), we have

$$c(u_1) = 0.56, \ c(u_2) = 0.17, \ c(u_3) = 0.14.$$

Normalizing the distribution we have

$$p(u_1) = 0.64, \ p(u_2) = 0.20, \ p(u_3) = 0.16.$$

Because
$$f_1(u_1) = -0.45, \ f_1(u_2) = 0.65, f_1(u_3) = 1.75,$$
$$f_2(u_1) = 0.55, \ f_2(u_2) = 2.65, f_2(u_3) = 4.75,$$

we obtain

$$E(f_1(u)) = -0.45 \times 0.64 + 0.65 \times 0.20 + 1.75 \times 0.16 = 0.13,$$

$$E(f_2(u)) = 0.55 \times 0.64 + 2.65 \times 0.20 + 4.75 \times 0.16 = 1.64.$$

Because $E(f_2(u)) > E(f_1(u))$, we know that the alternative g_1 defined by $f_2(u)$ is better than the alternative g_1 defined by $f_1(u)$.

12.4 Application in Risk Management of Flood Disaster

In this section we discuss a real case in risk management of flood disaster to show the application of a PPD. The studied region is Huarong county, located in the north of Hunan Province, China, where flood frequently occurs.

12.4.1 Outline of Huarong county

Huarong county, belongs to Dongting Lake plain belt, is an agriculture district and high flood risk. The population is 0.7 million and the total area is 1612.57 km^2. It has fertile soil and adequate riverhead. The seeded area is 157,390 hectare. The main crop is rice and rape. The crop rotation system of rice field is rape – early rice – late rice.

The hazard of crop flood in Huarong is the local precipitation and the discharge of river and lake which pass through the county. The annual precipitation of Huarong county is 1,206 millimeter. Due to the influence of subtropical monsoon climate, the average precipitation from May to July (657 millimeter) amounts to 55 percentage of the annual. The peaks of natural precipitation of and the discharge of river and lake, which pass through the county both, are in the month from May to July. The period from May to July is both the rainy season and the high peak of discharge of river and lake. So, crop flood occurs easily in this period.

12.4.2 PPD of flood in Huarong county

Although Huarong county frequently met flood, there are only 8 observations recorded flood events, shown in Table 12-7, are available. Any risk assessment from the data must be imprecise. Certainly, we can also employ the NNE model given in chapter 10, as we have done, to improve the assessment, however, we never eliminate the imprecision. We must be clearly aware that the imprecise is so marked that we cannot ignore it when we manage the risk system. Particularly, for the real case, nobody can give expert experience to form subjective probability. In the case, it is better to use the interior-outer-set model to calculate a PPD representing the flood risk in the county.

Table 12-7 Flood affecting agriculture in Huarong county

Year	1980	1985	1987	1988	1989	1990	1991	1993
S	233.80	237.50	228.40	233.20	245.55	247.79	249.06	228.51
l	199738	40000	195000	673000	88005	181995	525000	612000
x	0.0854	0.0168	0.0854	0.2886	0.0358	0.0734	0.2108	0.2678

S: Total seeded area (10^4mu, 1 hectare=15 mu);
l: Seeded area affected by flood (mu);
x: Disaster index, i.e., $x = l/S$.

From Table 12-7, we obtain a sample

$$X = \{x_i | i = 1, 2, \cdots, 8\}$$
$$= \{0.0854, 0.0168, 0.0854, 0.2886, 0.0358, 0.0734, 0.2108, 0.2678\}. \tag{12.35}$$

Now, we use the interior-outer-set model to calculate a PPD.

The first of all, we construct some intervals. We divide the range approximately covered by X, [0, 0.30] into three intervals $I_1 = [0, 0.10[$, $I_2 = [0.10, 0.20[$, $I_3 = [0.20, 0.30[$. The midpoints of the intervals form the discrete universe of discourse of the disaster index.

$$U = \{u_j | j = 1, 2, 3\} = \{0.05, 0.15, 0.25\}, \tag{12.36}$$

with step length $\Delta = 0.10$. Let the universe of discourse of probability be

$$P = \{p_k | k = 0, 1, 2, \cdots, 8\}$$
$$= \{0/8, 1/8, 2/8, \cdots, 8/8\} \tag{12.37}$$
$$= \{0, 0.125, 0.25, 0.375, 0.50, 0.625, 0.75, 0.875, 1\}.$$

Then, employing the algorithm given in subsection 12.2.3 and substituting X and U into Program 12.A, we obtain a PPD

$$\underset{\sim}{p}(u) = \begin{array}{c} \\ 0.05 \\ 0.15 \\ 0.25 \end{array} \begin{matrix} 0 & 0.125 & 0.25 & 0.375 & 0.5 & 0.625 & 0.75 & 0.875 & 1 \\ \left(\begin{matrix} 0.142 & 0.234 & 0.332 & 0.354 & 0.354 & 1 & 0 & 0 & 0 \\ 1 & 0.392 & 0.354 & 0.354 & 0.234 & 0 & 0 & 0 & 0 \\ 0.178 & 0.386 & 0.392 & 1 & 0 & 0 & 0 & 0 & 0 \end{matrix} \right) \end{matrix}. \tag{12.38}$$

In other words, by the sample X in (12.35), we cannot certainly assign a probability value to a possible adverse event u, but several probability values. For example,

$$\underset{\sim}{p}(0.05) = 0.142/0 + 0.234/0.125 + 0.332/0.25 + 0.354/0.375 + 0.354/0.5 + 1/0.625$$

means that although the possibility that flood measured as $u = 0.05$ occurs in probability $p = 0.625$ is the largest, we cannot ignore the possibilities for other values. Even for $p = 0$ (never occurs), the possibility is 0.142.

12.4.3 Benefit-output functions of farming alternatives

For the sake of computing and comparing easily, we designed only three alternatives in terms of ecological to reduce disaster. We design them based on the current farming mode: conventional rape – conventional early rice – hybrid late rice, called classical mode, denoted as g_0. The designed alternatives are denoted as g_1, g_2, g_3.

Our mitigation means is to replace conventional breeds with hybrid ones based on the fuzzy risk assessment. Alternative g_1 and g_2 are the same mode: hybrid rape – hybrid early rice – hybrid late rice, but the breed of hybrid early rice is different. The alternative g_3 is two seasonal mode, that is, only planting hybrid rape and hybrid late rice, and letting the field fallow from June to July, a period of highest risk.

The crop breeds of different alternatives and their characteristic are showed in table 12-8.

Table 12-8 Outputs from difference breeds and fields

	Mode	Crop	Breed	Submergence resistance	Field grade	Output (kg/mu)
g_0	Three crops a year	Rape	XY #13	Medium	High	100
		Early rice	XZC #7	Medium	Medium	420
		Late rice	SY #63	High	Low	530
g_1	Three crops a year	Rape	HZ #3	High	High	120
		Early rice	WY #1126	High	High	480
		Late rice	SY #63	High	Low	530
g_2	Three crops a year	Rape	HZ #3	High	High	120
		Early rice	WY #402	Medium	Medium	520
		Late rice	SY #63	High	Low	530
g_3	Two crops a year	Rape	HZ #3	High	High	120
		Late rice	SY #63	Higher	Medium	580

Abbr. of breed name: XY—Xiangyou; XZC—Xiangzaocan;

SY—Shanyou; HZ—Huaza

WY—Weiyou;

Among these breeds, only Xiangzaocan #7, Weiyou # 1126 and Weiyou # 402 may meet floods because they are early rice growing during flood season.

According to our experience in agriculture disaster with respect to flood, we infer the following reduction-output functions

Xiangzaocan #7:

$$L_1(u) = \begin{cases} 3u, & \text{for } 0 \le u \le 0.3; \\ 0.9, & \text{for } 0.3 < u \le 1. \end{cases} \tag{12.39}$$

Weiyou #1126:

$$L_2(u) = \begin{cases} 2u, & \text{for } 0 \le u \le 0.45; \\ 0.9, & \text{for } 0.45 < u \le 1. \end{cases} \tag{12.40}$$

Weiyou #402:

$$L_3(u) = \begin{cases} 3u, & \text{for } 0 \le u \le 0.3; \\ 0.9, & \text{for } 0.3 < u \le 1. \end{cases} \tag{12.41}$$

where u is the disaster index. The functions mean that we will lose more as the disaster index increases before some degree. The loss is restricted, because when the loss attains a certain degree, there is no anything to be lost.

Based on a review of available information sources related to grain market, we suppose that the price of rape, early rice and late rice is 0.90, 0.50 and 0.60 RMB Yuan per kilogram, respectively. Meanwhile, we suppose that the prices do not change whether a flood occurs or not.

Then, we can calculate output value for each alternative with respect to u.

From the PPD in (12.38) we know that, it is enough to consider $u = 0.05, 0.15, 0.25$ for comparing the alternatives. The output value associated with alternative g_i with respect to u_j is denoted as V_{ij}.

Considering that only early rice grows in flood season, i.e., we suffer losses from only early rice, we have

$$V_{ij} = \sum (\text{output of non-flood-season breed})_i \times (\text{price of non-flood-season breed})_i +$$
$$(\text{output of early-rice breed})_i \times (\text{price of early-rice breed})_i \times (1 - L_{\sigma(i)}(u_j)). \tag{12.42}$$

where σ is a permutation determined by the chosen alternative and its early-rice breed. For example, in alternative g_1, early-rice breed is Weiyou #1126 with reduction-output function $L_2(u)$ shown in formula (12.40), therefore $\sigma(1) = 2$, i.e., $L_{\sigma(1)}(u) = L_2(u)$. For alternatives in Table 4, we have:

$$\sigma(0) = 1, \quad \sigma(1) = 2, \quad \sigma(2) = 3.$$

$\sigma(3)$ is not defined because there is not early-rice breed in alternative g_3. A breed growing as the rape or the late rice is called a non-flood-season breed. For example, Huaza # 3 is a non-flood-season breed.

Using the formula, we obtain all output values of the alternatives shown in Table 12-9.

For example, for g_1 and $u_2 = 0.15$, we have

$$V_{12} = \sum (\text{output of non-flood-season breed})_1 \times (\text{price of non-flood-season breed})_1 +$$
$$(\text{output of early-rice breed})_1 \times (\text{price of early-rice breed})_1 \times (1 - L_{\sigma(1)}(u_2)).$$
$$= ((\text{output of Huaza #3}) \times (\text{price of rape}) +$$
$$(\text{output of Shanyou #63}) \times (\text{price of late rice})) +$$
$$(\text{output of Weiyou #1126}) \times (\text{price of Weiyou #1126}) \times (1 - L_2(u_2)).$$
$$= (120 \times 0.9 + 530 \times 0.6) + 480 \times 0.5 \times (1 - 2 \times 0.15)$$
$$= 594.$$

Table 12-9 Output values (Yuan/mu) of alternatives
with respect to flood u_i

Alternative	$u_1=0.05$	$u_2=0.15$	$u_3=0.25$
g_0	586.5	523.5	460.5
g_1	642	594	546
g_2	647	569	491
g_3	456	456	456

The cost for executing different alternative is not the same. The planting cost of unit area, denoted as c, includes matter cost and manpower cost. The matter cost embodies the level of investment in the process of crop growing, including the cost of seeds, fertilizer and so on. The manpower cost reflects the labor investment, which including productive manpower and managing manpower. Through investigation, we obtain the planting cost of different alternatives as shown in table 12-10.

Table 12-10 Planting cost (Yuan/mu) of the alternatives

Alternative	Matter cost	Manpower cost	Total cost
g_0	214.51	131.78	346.29
g_1	214.51	131.78	346.29
g_2	214.51	131.78	346.29
g_3	128.01	85.24	213.25

Subtract the total cost as shown in Table 12-10 from the output value as shown in Table 12-9, we obtain the net income, denoted as NI, of alternatives shown in Table 10-11 with respect to different flood level u. The higher flood u, the lower net income NI.

For example, a farmer still executes traditional alternative g_0 and in the year he meets a flood measured as $u_2 = 0.15$, then he can have net income $NI_{02} = 177.21$ RMB Yuan per mu. Another farmer changes to executes alternative g_1, with respect to same flood. The reformer can get net income $NI_{12} = 247.71$ RMB Yuan per mu.

Table 12-11 Net income (Yuan/mu) of alternative
with respect to flood u_i

Alternative	$u_1=0.05$	$u_2=0.15$	$u_3=0.25$
g_0	240.21	177.21	114.21
g_1	295.71	247.71	199.71
g_2	300.71	222.71	144.71
g_3	242.75	242.75	242.75

We employ the benefit-output functions $f_1(u), f_2(u), f_3(u), u \in \{0.05, 0.15, 0.25\}$ to measure the new farming alternatives. The functions are defined as

alternative g_i has function: $f_i(u_j) = NI_{ij} - NI_{0j}, \quad i, j = 1, 2, 3.$ (12.43)

Using formula (12.43) and Table 12-11, we obtain Table 12-12 showing all $f_i(u_j), i, j = 1, 2, 3$.

Table 12-12 Benefit-output functions in discrete form

Alternative	$u_1=0.05$	$u_2=0.15$	$u_3=0.25$
f_1	55.50	70.50	85.50
f_2	60.50	45.50	30.50
f_3	2.54	65.54	128.54

The table indicates that, in the case of $u=0.05$ (or equally say, when flood is light), the best alternative is g_2 because $f_2(u_1) > f_1(u_1), f_3(u_1)$; in the case of $u=0.15$ (i.e., when flood is general) the best alternative is g_1 because $f_1(u_2) > f_2(u_2), f_3(u_2)$; when $u=0.25$ (i.e., flood heavy), the best one is g_3. Obviously, it needs to consider the flood risk of the county for ranking the alternatives.

12.4.4 Ranking farming alternative based on the PPD

For i-th row of the matrix given by (12.38) we use formula (12.28) to calculate the center of gravity with respect to the row and obtain $c(u_i)$. For u_1, we have,

$$c(u_1) = \left(\sum_{j=0}^{8} p_j \pi_{u_1}(p_j) \right) \Big/ \left(\sum_{j=0}^{8} \pi_{u_1}(p_j) \right)$$

$$= \frac{0.125 \times 0.234 + 0.25 \times 0.332 + 0.375 \times 0.35 + 0.5 \times 0.354 + 0.625 \times 1}{0.142 + 0.234 + 0.332 + 0.354 + 0.354 + 1}$$

$$= 0.43.$$

Similarly, we obtain $c(u_2) = 0.17$, $c(u_3) = 0.27$. Normalizing the distribution by using formula (12.29), we obtain $p(u_1) = 0.49$, $p(u_2) = 0.20$, $p(u_3) = 0.31$.

Corresponding to the center-of-gravity of FEV defined by formula (12.34), in our case, the center-of-gravity of FEV is calculated by

$$E(f_i(u)) = \sum_{j=1}^{3} f_i(u_j)p(u_j), \quad i = 1, 2, 3,$$

where $f_i(u_j), i, j = 1, 2, 3$, are give by Table 12-12. Therefore

$$E(f_1(u)) = 55.50 \times 0.49 + 70.50 \times 0.20 + 85.50 \times 0.31 = 67.8,$$

$$E(f_2(u)) = 60.50 \times 0.49 + 45.50 \times 0.20 + 30.50 \times 0.31 = 48.2,$$

$$E(f_3(u)) = 2.54 \times 0.49 + 65.54 \times 0.20 + 128.54 \times 0.31 = 54.2.$$

Because $E(f_1(u)) > E(f_3(u)) > E(f_2(u))$, we know that for Huarong county, under the condition that flood risk is given by (12.38), the best farming alternative is g_1, i.e. three crops a year with Huaza #3 for rape, Weiyou #1126 for early rice and Shanyou #63 for late rice is the best farming.

12.4.5 Comparing with the traditional probability method

Until now nobody knows which shape of probability distribution, as normal or exponential, is fit to describe a flood risk with respect to agriculture. In the case, if we have to choose one of the traditional probability methods to estimate probability-risk, only the histogram method can guarantee explicability of the estimation.

As we know, in our model, discrete points $u_1 = 0.05$, $u_2 = 0.15$, $u_3 = 0.25$ correspond to intervals $I_1 = [0, 0.10[$, $I_2 = [0.10, 0.20[$, $I_3 = [0.20, 0.30[$.

According to the definition of the histogram given by (3.25), in our case, we have

$$\hat{p}_j = \hat{p}(x \in I_j) = \frac{1}{n}(\text{the number of } x_i \text{ in } I_j),$$

$j = 1, 2, 3; i = 1, 2, \cdots, 8, x_i \in X$ in (12.35). We obtain a probability distribution in the intervals

$$\hat{p}(x \in I_1) = 0.625, \quad \hat{p}(x \in I_2) = 0, \quad \hat{p}(x \in I_3) = 0.375.$$

Obviously, for histogram estimate $\hat{p}_j, j = 1, 2, \cdots, m$, the expected value of an alternative with the benefit-output functions $f(u)$ can be calculated by

$$\hat{E}(f(u)) = \sum_{j=1}^{m} f(u)\hat{p}_j. \tag{12.44}$$

For Huarong county, we obtain

$$\hat{E}(f_1(u)) = 55.50 \times 0.625 + 85.50 \times 0.375 = 66.75.$$

$$\hat{E}(f_2(u)) = 60.50 \times 0.625 + 30.50 \times 0.375 = 49.25,$$

$$\hat{E}(f_3(u)) = 2.54 \times 0.625 + 128.54 \times 0.375 = 49.74.$$

The ranking is also that g_1 is the best, g_3 is better than g_2. However, in the decision-making which based on traditional probability, the difference between g_2 and g_3 is insignificant (only 0.54). A little error of probability estimation may change their order. In this condition, it is difficult to determine which alternative, g_3 or g_2, is much better.

Contrast to that, the expected values from the PPD have significant difference $E(f_3(u)) - E(f_2(u)) = 54.2 - 48.2 = 6$, and a small error of probability estimation cannot change their order.

Recalling Table 12-8, we know that g_2 and g_3 are so different. Clearly, there must be a significant difference between their expected values. Obviously, the ranking from the PPD is more near real case. In other words, the ranking based on fuzzy risk is more reliable and stable.

12.5 Conclusion and Discussion

Traditional, fuzzy probability connects to the fuzzy events, fuzzy random numbers and subjective probabilities. It is easy to be used for subjective risk assessment. In our point of view, calculated fuzzy risk may be more useful than subjective-assessed fuzzy risk. The PPD as a kind of fuzzy risk can be used to show the imprecision of risk assessment in terms of probabilities. The main results of this chapter are

(1) A PPD can be calculated by interior-outer-set model.

(2) The ranking from a PPD is better than one from a histogram being a risk assessment.

The reason why the method of information distribution can be developed as an algorithm to calculate a PPD is that the method can show the fuzziness of an incomplete sample. Therefore it can show the fuzziness of probability estimation.

When a sample is strongly incomplete, it is impossible to get a precise probability estimate. If we insist to regard the sample as one from which we also can rank alternatives by some classical statistics tool, we may obtain a wrong result. In fact, a histogram is equal to an information matrix on crisp intervals that cannot show the position difference between the observations in the same interval. It means, by the histogram method to estimate probability-risk, we must lose some information in the given sample. The risk management depending on the estimate must be unreliable and unstable. If we blindly use some theory probability distribution to estimate a flood risk, the result may be much worse than one from the histogram method.

The study of calculated fuzzy risks is relatively primary. However, the engineering value of calculated fuzzy risks is so obvious that any advance in the field can prompt both risk science and fuzzy set theory.

References

1. Dubois, D., Prade, H.(1989), Fuzzy sets, probability, and measurement. European J. Operns. Res. **40**, pp.135-154
2. Freeling, A.N.S.(1980), Fuzzy sets and decision analysis. IEEE Trans. Sys. Man & Cyb. **SMC-10**(7), pp.341-354
3. Heilpern, S.(1993), Fuzzy subsets of the space of probability measures and expected value of fuzzy variable. Fuzzy Sets and Systems **54**(3), pp.301-309
4. Kai-Yuan, C. (1996), System failure engineering and fuzzy methodology an introductory overview. Fuzzy Sets and Systems **83**(2), pp.113-133
5. Kandel, A. and Byatt, W.J.(1978), Fuzzy sets, fuzzy algebra, and fuzzy statistics. Proc. IEEE **66**(12), pp.1619-1639
6. Pal, S.K.(1986), Fuzzy Mathematical Approach to Pattern Recognition. John Wiley, New York.
7. Korner, R.(1997),On the variance of fuzzy random variables. Fuzzy Sets and Systems **92**(1), pp.83-93
8. Qiao, Z., Zhang, Y. and Wang, G.Y.(1994), On fuzzy random linear programming. Fuzzy Sets and Systems **65**(1), pp.41-49
9. Römer, C. and Kadel, A.(1995), Statistical tests for fuzzy data. Fuzzy Sets and Systems **72**(1), pp.1-26
10. Song, Q., Leland, R.P. and Chissom, B.S.(1997), Fuzzy stochastic fuzzy time series and its models. Fuzzy Sets and Systems **88**(3), pp.333-341
11. Watson, S.R., Weiss, J.J., Donnell, M.L.(1979), Fuzzy decision analysis. IEEE Trans. Sys. Man & Cyb. **SMC-9**(1), pp.1-9
12. Zadeh, L.A. (1968), Probability measures of fuzzy events. Journal of Mathematics Analysis and Applications **23** (1), pp.421-427
13. Zadeh, L.A.(1997), Toward a theory of fuzzy information granulation and its centrality in human reasoning and fuzzy logic, *Fuzzy Sets and Systems* **90**(2), pp.111-127.
14. Zadeh, L.A.(1998), Toward a restructuring of the foundations of fuzzy logic (FL), *Proceedings of FUZZ-IEEE'98*, Anchorage, USA, pp.1676-1677

Appendix 12.A: Algorithm Program for Interior-outer-set Model

Fortran program to do calculation case given in subsection 12.2.2
(PPD.DAT restores sample and the discrete universe of discourse)

```
      PROGRAM MAIN
      INTEGER N,M,SJ(100),NJ,L,W,R(1000,100),TJ(100),NT
      REAL X(100),U(50),QL(100,50),QJ(100,50),E(100,50)
      REAL Y(100),PS(50,100),PB(100),P
C   N: sample size, M: number of controlling points
C   X: sample, U: discrete universe of X
C  QL: leaving matrix, QJ: joining matrix
C   E: fuzzy relation between observations and intervals
C  SJ: the j-th interior index set, TJ: the j-th outer index set
C  NJ: number of elememts of the j-th interior index set
C  NT: number of elememts of the j-th outer index set
C   L: taken L elements to combination
C   W: number of all combinations of NJ different elements taken L at a time
C   R: INDEX matrix, R(W,N)
C   Y: the j-th column of QL or QJ
C  PS: possibility distributions,   PB: probability values
C   P: a possibility, it gives value to PS
      CALL SAMPLEU(N,M,X,U)
      CALL MOVING(N,M,X,U,QL,QJ,E)
      DO 50 J=1,M
      CALL GETSJ(N,J,E,SJ,NJ)
      DO 10 I=1,N
      Y(I)=QL(I,J)
10    CONTINUE
      DO 20 L=1,NJ
      CALL INDEXMATRIX(NJ,L,R,W)
      CALL LEAVINGPI(SJ,Y,L,R,W,P)
      PS(J,NJ-L)=P
20    CONTINUE
      PS(J,NJ)=1
      DO 30 I=1,N
      Y(I)=QJ(I,J)
30    CONTINUE
      CALL GETTJ(N,J,E,TJ,NT)
      DO 40 L=1,NT
      CALL INDEXMATRIX(NT,L,R,W)
      CALL LEAVINGPI(TJ,Y,L,R,W,P)
```

```
        PS(J,NJ+L)=P
40      CONTINUE
50      CONTINUE
        AN=N
        DO 60 K=0,N
        AK=K
        PB(K)=AK/AN
60      CONTINUE
        WRITE(*,70)(PB(K),K=0,N)
70      FORMAT(1X,'probability values:',10F5.2)
        DO 80 J=1,M
        WRITE(*,90)J,(PS(J,K),K=0,N)
80      CONTINUE
90      FORMAT(1X,'The',I2,'-th   interval:',10F5.2)
        STOP
        END


        SUBROUTINE SAMPLEU(N,M,X,U)
        INTEGER N,M
        REAL X(100),U(50)
        CHARACTER*26 DAT,A
        DAT='PPD.DAT'
        OPEN(1,FILE=DAT,STATUS='OLD')
        READ(1,*)A,N,(X(I),I=1,N)
        WRITE(*,*)A,' N=',N
        DO 10 I=1,N
        WRITE(*,*)I,X(I)
10      CONTINUE
        READ(1,*)A,M,(U(J),J=1,M)
        WRITE(*,*)A,' M=',M
        DO 20 J=1,M
        WRITE(*,*)J,U(J)
20      CONTINUE
        CLOSE(1)
        RETURN
        END


        SUBROUTINE MOVING(N,M,X,U,QL,QJ,E)
        INTEGER N,M
        REAL X(100),U(50),QL(100,50),QJ(100,50),H,Q(100,50),E(100,50)
C Q: fuzzy relation between observations and controlling points
C E: fuzzy relation between observations and intervals
        CALL DISTR(N,M,X,U,Q)
        CALL GETE(N,M,X,U,E)
```

```
      CALL LEAVING(N,M,Q,E,QL)
      CALL JOINING(N,M,Q,E,QJ)
      RETURN
      END

      SUBROUTINE GETSJ(N,J,E,SJ,NJ)
      INTEGER N,J,NJ,SJ(100)
      REAL E(100,50)
      NJ=0
      DO 10 I=1,N
      IF(E(I,J).LT.1) GOTO 10
      NJ=NJ+1
      SJ(NJ)=I
10    CONTINUE
      RETURN
      END

      SUBROUTINE GETTJ(N,J,E,TJ,NJ)
      INTEGER N,J,NJ,TJ(100)
      REAL E(100,50)
      NJ=0
      DO 10 I=1,N
      IF(E(I,J).GT.0) GOTO 10
      NJ=NJ+1
      TJ(NJ)=I
10    CONTINUE
      RETURN
      END

      SUBROUTINE INDEXMATRIX(N,L,R,W)
      INTEGER N,L,R(1000,100),W,C(100)
C C: medium variable to adjust combinations
      W=1
      DO 10 I=1,L
      C(I)=I
10    CONTINUE
20    CONTINUE
      DO 30 I=1,L
      R(W,I)=C(I)
30    CONTINUE
      W=W+1
      IF(C(1).EQ.N-L+1) GOTO 40
      CALL ADJUSTC(N,L,C)
```

```
        GOTO 20
40      CONTINUE
        W=W-1
        RETURN
        END


        SUBROUTINE ADJUSTC(N,L,C)
        INTEGER N,L,C(100)
        I=1
10      IF(C(I).GE.N-L+I) GOTO 20
        IF(I.EQ.L) GOTO 30
        I=I+1
        GOTO 10
20      I=I-1
30      C(I)=C(I)+1
        DO 40 J=I+1,L
        C(J)=C(J-1)+1
40      CONTINUE
        RETURN
        END


        SUBROUTINE LEAVINGPI(SJ,Y,L,R,W,P)
        INTEGER SJ(100),L,R(1000,100),W
        REAL Y(100),P
        P=0
        DO 20 I=1,W
        A=Y(SJ(R(I,1)))
        DO 10 J=1,L
        IF(A.GT.Y(SJ(R(I,J)))) A=Y(SJ(R(I,J)))
10      CONTINUE
        IF(P.LT.A) P=A
20      CONTINUE
        RETURN
        END


        SUBROUTINE DISTR(N,M,X,U,Q)
        INTEGER N,M
        REAL X(100),U(50),Q(100,50)
        H=U(2)-U(1)
        DO 20 J=1,M
        DO 10 I=1,N
        Q(I,J)=0
        B=ABS(X(I)-U(J))
        IF(B.GT.H) GOTO 10
```

```
      Q(I,J)=1-B/H
10    CONTINUE
20    CONTINUE
      WRITE(*,*)'Q is:'
      DO 30 I=1,N
      WRITE(*,40)(Q(I,J),J=1,M)
30    CONTINUE
40    FORMAT(1X,10F6.2)
      RETURN
      END

      SUBROUTINE GETE(N,M,X,U,E)
      INTEGER N,M,N1
      REAL X(100),U(50),E(100,50)
C N1: medium variable to avoid expression error in computer
C     to guarantee boundary observations just fall the bounds

      H=U(2)-U(1)
      DO 10 I=1,N
      N1=INT(X(I)*100.+0.5)
      X(I)=N1/100.
10    CONTINUE
      DO 40 J=1,M
      V1=U(J)-H/2.
      V2=U(J)+H/2.
      DO 30 I=1,N
      N1=INT(V1*100.+0.5)
      V1=N1/100.
      N1=INT(V2*100.+0.5)
      V2=N1/100.
      E(I,J)=0
      IF(X(I).LT.V1) GOTO 30
      IF(X(I).GT.V2) GOTO 30
      IF(X(I).EQ.V2) GOTO 20
      E(I,J)=1
      GOTO 30
20    IF(J.LT.M) GOTO 30
      E(I,J)=1
30    CONTINUE
40    CONTINUE
      WRITE(*,*)'E is:'
      DO 50 I=1,N
```

```
        WRITE(*,60)(E(I,J),J=1,M)
50      CONTINUE
60      FORMAT(1X,10F6.2)
        RETURN
        END


        SUBROUTINE LEAVING(N,M,Q,E,QL)
        INTEGER N,M
        REAL QL(100,50),Q(100,50),E(100,50)
        DO 20 J=1,M
        DO 10 I=1,N
        QL(I,J)=E(I,J)-Q(I,J)
10      CONTINUE
20      CONTINUE
        WRITE(*,*)'E-Q=:'
        DO 30 I=1,N
        WRITE(*,40)(QL(I,J),J=1,M)
30      CONTINUE
40      FORMAT(1X,4F6.2)
        DO 50 J=1,M
        DO 50 I=1,N
        IF(QL(I,J).LT.0) QL(I,J)=0
50      CONTINUE
        WRITE(*,*)'QL is:'
        DO 60 I=1,N
        WRITE(*,40)(QL(I,J),J=1,M)
60      CONTINUE
        RETURN
        END


        SUBROUTINE JOINING(N,M,Q,E,QJ)
        INTEGER N,M
        REAL QJ(100,50),Q(100,50),E(100,50)
        DO 20 J=1,M
        DO 10 I=1,N
        QJ(I,J)=Q(I,J)-E(I,J)
10      CONTINUE
20      CONTINUE
        WRITE(*,*)'Q-E=:'
        DO 30 I=1,N
        WRITE(*,40)(QJ(I,J),J=1,M)
30      CONTINUE
40      FORMAT(1X,4F6.2)
        DO 50 J=1,M
```

```
        DO 50 I=1,N
        IF(QJ(I,J).LT.0) QJ(I,J)=0
50      CONTINUE
        WRITE(*,*)'QJ is:'
        DO 60 I=1,N
        WRITE(*,40)(QJ(I,J),J=1,M)
60      CONTINUE
        RETURN
        END
```

PPD.DAT
'earthquake magnitude (n=6):',6,7.8,7.5,7.25,7.9,7.3,7.2
'controlling points (m=4):',4,7.15, 7.45,7.75,8.05

Subject Index

List of Special Symbols

$[a, b]$	interval, left closed and right closed
$[a, b[$	interval, left closed and right open
$]a, b]$	interval, left open and right closed
(a, b)	point of a space
$\{a, b, \cdots\}$	set
$\triangleq$	is defined as
A^c	complement of A
$\emptyset$	empty set
A_α	α-cut of A
$|A|$	cardinality of A (number of elements in A)
$|a - b|$	absolute value
$\|a - b\|$	norm
X	sample
U	universe of discourse
R	relationship
$\mathbf{R}$	set of real numbers
R^n	Euclidian n-space (i.e., $\mathbf{R}^n$)
$\prod_{k=1}^m A_k$	$A_1 \times A_2 \times \cdots \times A_m$
Ω	space of elementary events
$\mathcal{A}$	sigma-field of events
P	probability measure
$\mathcal{P}$	power set
$\Pi(A)$	possibility measure
	(i.e., a function $\Pi \colon \mathcal{P}(U) \to [0, 1]$)
$\pi(u)$	possibility distribution
$\Pi_{M,P}$	possibility-probability distribution
	(i.e., $\Pi_{M,P} = \{\pi_m(p)|m \in M, p \in P\}$)
$\mathcal{F}(X)$	fuzzy power set of X
$\mathcal{F}(X \times U)$	class of all information distributions of X on U
$\mathcal{D}(X)$	sample of fuzzy sets derived from X on U by information diffusion
$\mathcal{N}$	set $\{1, 2, \cdots, n\}$
r.v.	random variable
i.i.d.	independent and identical distribution
$E(x)$	expected value for the random variable x
$Var(x)$	variance for the random variable x